Planetary Cinema

Planetary Cinema

Film, Media and the Earth

Tiago de Luca

Amsterdam University Press

The publication of this book is made possible by a grant from the British Academy.

Cover illustration: Cover of the Urbanora catalogue (1909), held at the National Science and Media Museum, Bradford, UK.

Cover design: Coördesign, Leiden
Lay-out: Crius Group, Hulshout

ISBN	978 94 6372 962 8
e-ISBN	978 90 4855 096 8
DOI	10.5117/9789463729628
NUR	670

Printed and bound by CPI Group (UK) Ltd, Croydon, CR0 4YY

Table of Contents

List of Figures

Acknowledgements

The beginnings of this project coincided with my appointment in the Department of Film and Television Studies at the University of Warwick. I could not have asked for a more supportive department and a more supportive institution. I would like to pay special thanks to my friends and colleagues Helen Wheatley, Karl Schoonover, Jon Burrows, Rachel Moseley, Catherine Constable and Alastair Phillips, who, at various stages, read grant applications, draft proposals and specific chapters. Their comments and suggestions, but most importantly, their enthusiasm for this project, have proved invaluable. I taught materials contained in this book on two modules in the department: 'Envisioning the World' (2017) and 'Global Visions' (2021). Many of the ideas featured here were modified and refined thanks to enlightening and stimulating discussions with students. My heartfelt thanks to all the students who took these two modules.

This book is coming out as an immediate open-access publication thanks to generous funding from the Faculty of Arts and the Habitability GRP at the University of Warwick, for which I am grateful. This book is also the main output of a British Academy Mid-Career Fellowship, which I was awarded in the academic year 2018–2019 and which allowed me to dedicate myself entirely to this project for the period of a year. I am thankful for the British Academy for granting me time to write and research: a precious thing these days. The funding from this fellowship also enabled me to put together public-engagement activities. One of these was held at the Bill Douglas Cinema Museum, and I cannot thank Phil Wickham enough for his support in helping me select the materials for the illustrated lecture 'All Mountains of the Earth: Visual Culture and Global Travel in Victorian Britain', which I gave at the museum in September 2019. Another event was the three-day film season 'Envisioning the World: How Film Shapes the Earth', held at the Close-up Film Centre in London in June 2019, which also received support from the Faculty of Arts at Warwick in the form of an Impact grant. I thank Martin Roberts, Elena Gorfinkel, Michael Cowan, Joanna Zylinska and Tim Boon, who all kindly agreed to participate at this event as guest speakers and discussants.

A large portion of the research undertaken for this book was carried out at archives and I must thank some institutions and the people who work for them. In 2017 I was awarded a research stipend to conduct research at the Bill Douglas Cinema Museum, and once again, I thank Phil Wickham, as well as Helen Hanson, for their warm hospitality in Exeter. Research for this project was also conducted at the National Science and Media Museum, in Bradford, and I thank Sarah Coggrave for her generous assistance. Thanks are also due to Delphine Allannic and Anne Sigaud at the Musée Albert-Kahn, where I spent some memorable weeks in the summer of 2019. Many of the artefacts explored here are housed at the British Library. My thanks to all staff members at the British Library for their help and assistance.

I have presented the findings contained in this book as research papers and talks throughout the past years and I must thank the friends and colleagues who have given me this opportunity: Lukas Brasiskis and the team from the amazing Cinema Camp group, in Lithuania; Laura McMahon, from the University of Cambridge; Felicity Gee, from the University of Exeter; Denilson Lopes, from the Federal University of Rio de Janeiro; my friend and research partner Cecília Mello, from the University of São Paulo; Catherine Constable, from my department at Warwick; and all the PGR students who invited me to speak at the University of Birmingham. Thanks also to Janine Marchessault, Zoë Druick and Chris Russill, who immediately and enthusiastically accepted my invitation to be part of a panel on 'Cinema and Planetary Consciousness' at SCMS 2019, in Seattle.

I am thankful to Martin Roberts, who has always been so generous with his suggestions and whose foundational work on the global in the cinema has been a source of inspiration. My friend Cecilia Sayad kindly read some drafts and was always available when I needed to vent. I am moved by the continued support of my former MA supervisor, C. Claire Thomson, without whose encouragement I would never have embarked on an academic career. And I am equally moved by the support of my former PhD supervisor and dear friend Lúcia Nagib, whose dizzying knowledge of and passion for the cinema will forever dazzle me.

When this book was contracted with Amsterdam University Press, its Film Culture in Transition series was under the editorship of the late Thomas Elsaesser, who embraced this project from the start. It is an honour to be part of the series he originally conceived and this book, I hope, is a testament to the indelible impact of his scholarship. Thanks are also due to Maryse Elliott, who has been an enthusiastic and encouraging commissioning editor, and the new Film Culture in Transition series editors, Malte Hagener, Patricia Pisters and Wanda Strauven, for their support and patience. Many thanks also for the two reviewers for their attentive reading and helpful suggestions and recommendations.

I must also thank the people, companies and institutions who have granted me permission to publish pictures and illustrations in this book: the British Library, the Bill Douglas Cinema Museum, the National Science and Media Museum, Natalie Bookchin, Perry Bard, Nikolaus Geyrhalter Filmproduktion, GoodPlanet Foundation and Magidson Films. Thanks also to Jon Burrows and Luke McKernan for their helpful advice relating to image reproduction rights. My talented friend Victor Antonio Leão has helped me with many issues concerning the quality of the images reproduced in this book and helped design the book cover. He also designed the most amazing flyers for the 'Envisioning the World' film season in 2019, which I now keep as mementos. My heartfelt thanks to him for capturing my ideas so beautifully.

As with any long-term project, I have published excerpts of this book in journal articles and chapters, although these reappear here in significantly modified form. Materials contained in chapter 1 were previously published in 'Global Visions: Around-the-world Travel and Visual Culture in Early Modernity' in Louis Bayman and Natália Pinazza (eds.) *Journeys on Screen: Theory, Ethics and Aesthetics* (Edinburgh: Edinburgh University Press, 2018). Portions of chapter 5 were published in 'Figuring a Global Humanity: Cinematic Universalism and the Multinarrative Film', *Screen*, 58:1 (2017) and 'Earth Networks: *The Human Surge* and Cognitive Mapping', *NECSUS: European Journal of Media Studies*, 7:2 (2018).

Thanks are finally due to my family: my parents, Roney and Angela, and my sister, Bianca. Their kindness, integrity and generosity furnish me with reasons to not give up on this world.

Introduction

The world picture would be a painting, so to speak, of what is as a whole.
Martin Heidegger

When a photograph is cropped, the rest of the world is cut out.
Stanley Cavell

The present environmental crisis is the gravest challenge of our time. And it is so because it is planetary. To say that it is planetary is not to say that 'we are all in this together' or that anthropogenic disturbances act as a welcome equaliser: inequities are still the currency of the day in a vastly unjust world. Rather, it is to say that it is no longer possible to delimit or suppress the consequences arising from events pertaining to one particular geographical area or to even predict the areas that will be affected by an uncontrollable and extreme climate that sometimes does not spare even the rich. It also means confronting the fact that the environmental transformations currently on course are unprecedented in human history in terms of both the rapidity at which they are occurring and their global scope. Whether we like it or not, our world is an interconnected sphere made up of delicately interdependent ecosystems and lifeworlds. It urgently demands solutions on the planetary scale. This book hopes to contribute to this task by exploring how film and related media have both shaped and responded to the history of our planetary consciousness.

The story is now familiar. In the late 1960s humans were finally able to see photographic evidence of the Earth in space for the first time. Taken during the Apollo missions, two images in particular have lodged in the public consciousness: one of a half-shadowed Earth in the distance seen from a lunar landscape, taken in 1968 during the Apollo 8 mission and subsequently referred to as Earthrise; and another showing the entire planet as an enframed disk floating in space, taken in 1972 during the Apollo 17 mission and known as the Blue Marble. These two photographs are among the most disseminated in human history, ubiquitous in their visibility and

de Luca, T. *Planetary Cinema. Film, Media and the Earth.* Amsterdam: Amsterdam University Press, 2022
DOI: 10.5117/9789463729628_INTRO

adorning screensavers all over the world. As the story goes, the impact of such images in cementing a planetary awareness is yet to be matched.[1] Not only did they represent the apex of post-war globalism, they quickly became the emblem of the new environmentalism.[2] According to this well-trodden narrative, if globalisation and the global environmental crisis are the talk of today, their imaginaries must be traced back to that historical moment for a deeper understanding of their technologies, ideologies and mythologies.

This book tells a different story. It contends that this narrative has failed to account for the vertiginous global imagination undergirding late nineteenth-century media culture. Advancements in the sciences, technology, transportation and communication, boosted and supported by integrated economic networks in the context of imperial global expansion, had a dramatic impact on the conception and representation of the Earth in Western metropolitan culture. Panoramas, giant globes, world exhibitions, photography and stereography: all promoted and hinged on the idea of a world made whole and newly visible. When it emerged, cinema did not simply contribute to this effervescent globalism so much as become its most significant and enduring manifestation. One of the main arguments of this book is that an exploration of this media culture can help us understand contemporary planetary imaginaries and the way we see the world.

The (Whole) World in Motion

As a visual object, the world has never been and will never be seen in its totality. As Kelly Oliver reminds us: 'Whether we are looking at a table and chairs a few feet away or the Earth from space, we see only one side, one perspective, and cannot, and never will, see the whole in its entirety'.[3]

1 Whether directly or indirectly, a number of recent publications, many important for the present study, have reinforced this narrative, including: Kelly Oliver, *Earth and World: Philosophy after the Apollo Missions* (New York: Columbia University Press, 2015); Ursula K. Heise, *Sense of Place and Sense of Planet: The Environmental Imagination of the Global* (New York: Oxford University Press, 2008); Christopher Potter, *The Earth Gazers: On Seeing Ourselves* (New York: Pegasus Books, 2017); Robert Poole, *Earthrise: How Man First Saw the Earth* (New Haven, CT, and London: Yale University Press, 2008). The recent fiftieth anniversary of the Apollo 11 mission has cemented this narrative further, with a number of films released in the last years, including *First Man* (2018) and the documentaries *Mission Control: The Unsung Heroes of Apollo* (2017) and *Apollo 11* (2019), the last entirely comprising original footage.

2 For an excellent study of post-war media globalism, see Janine Marchessault, *Ecstatic Worlds: Media, Utopia, Ecologies* (Cambridge, MA, and London: MIT Press, 2017).

3 Oliver, *Earth and World*, 23.

Oliver's observation arises from her argument that 'global thinking only emerges after [the] Apollo pictures', which gave humans the first sight of the planet from afar.[4] And yet, as Denis Cosgrove has shown in his monumental *Apollo's Eye: A Cartographic Genealogy of the Earth in the Western Imagination* (2001), even if 'humans are unable to embrace more than a tiny part of their planetary surface', and for all the 'radical newness' of the Apollo photographs, 'actually witnessing the globe culminates a long genealogy of imagining the possibility of doing so' that goes back in time at least as far as ancient Rome and Greece.[5]

In fact, as discussed in chapter 1 of this book, it was in order to resolve the very conundrum that one can never see the entirety of the Earth that a giant public attraction such as the georama was first created in France in 1825, to then become the model for the Great Globe in the UK in 1851, both of which existed within a panoramic lineage that aimed to incite sublime experiences of the world's vastness. A gigantic sphere containing a concave world map, the aim of the georama, as Jean-Marc Besse tells us, was 'to make possible a type of perception that neither the flat map nor the convex globe allow: a global, and so to speak immediate, view of the totality of the surface of the Earth'.[6] Not coincidentally, georamas were erected in Paris and London, nineteenth-century capitals of global modernity and empire: betraying expansionist and ordering urges, these cartographic globes crystallised the longstanding alignment between whole-world figuration and a geographical imagination.

In her illuminating *La pensée cartographique des images: Cinéma et culture visuelle* (2011), Teresa Castro argues that this alignment gained a new intensity with the advent of cinema, which became characterised by a 'mapping impulse' undergirding a number of descriptive techniques and structuring devices.[7]

4 Ibid., 25.

5 Denis Cosgrove, *Apollo's Eye: A Cartographic Genealogy of the Earth in the Western Imagination* (Baltimore, MD, and London: John Hopkins University Press, 2001), xi.

6 Jean-Marc Besse, *Face au monde: Atlas, jardins, géoramas* (Paris: Desclée de Brouwer, 2003), 14.

7 Teresa Castro, *La pensée cartographique des images: Cinéma et culture visuelle* (Lyon: Aléas Éditeur, 2011). On mapping and cinema, see also Teresa Castro, 'Cinema's Mapping Impulse: Questioning Visual Culture', *The Cartographic Journal* 46:1 (2013): 9–15; Giorgio Avezzù, Teresa Castro and Giuseppe Fidota, 'The Exact Shape of the World? Media and Mapping', *NECSUS* 7:2 (2018): 85–95, https://necsus-ejms.org/the-exact-shape-of-the-world-media-and-mapping/; Alberto Toscano and Jeff Kinkle, *Cartographies of the Absolute* (Alresford, UK: Zero Books, 2015). Many of these works are indebted to Fredric Jameson's influential concept of 'cognitive mapping'; see his *The Geopolitical Aesthetic: Cinema and Space in the World System* (London: BFI, 1992), 1–35.

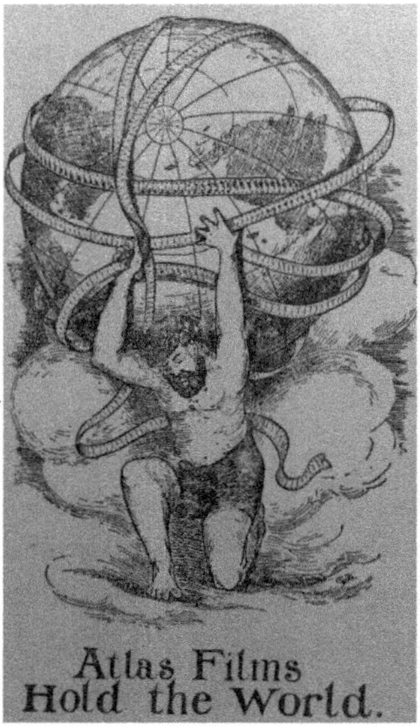

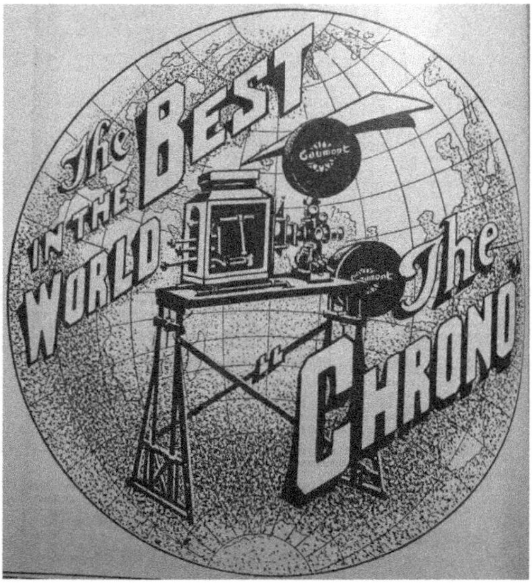

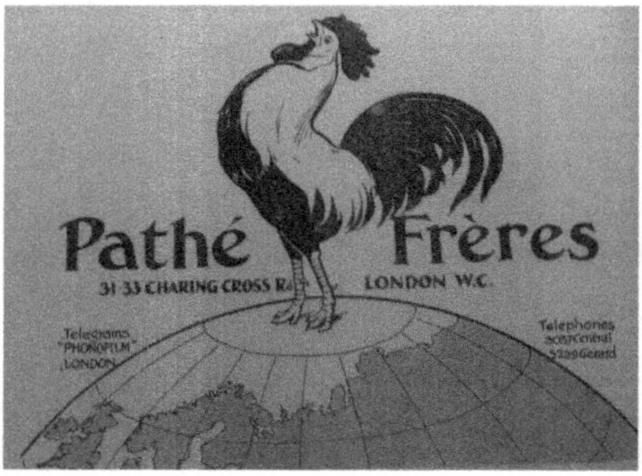

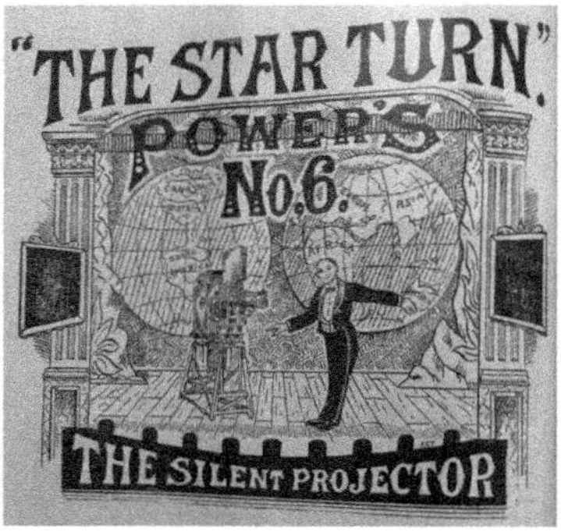

Figures 0.1, 0.2, 0.3 & 0.4 Advertisements from film companies and hirers published in *The Bioscope* (1910–1913) reveal the definitionally constitutive globalism of early cinema. Source held at the British Library.

These included the surveying panoramic shot and its aerial counterpart, the view from above, as well as film's ability to capture, inventory and organise visual chunks of the world in the guise of an atlas, which helps explain the ubiquitous figure of the globe in early cinema's promotional machinery.

To say that cinema acquired global contours as it developed in its early days does not quite do justice to its definitionally constitutive globalism, one that starts with the Lumière brothers' widely advertised travels around the world, finds verbal expression in taglines such as 'The Whole World Within Reach' (Méliès's Star Film, 1896) and 'We Put the World Before You' (Charles Urban Trading Company, 1903), and becomes materialised in the omnipresent global figures adopted by the nascent companies and hirers, many of which duly

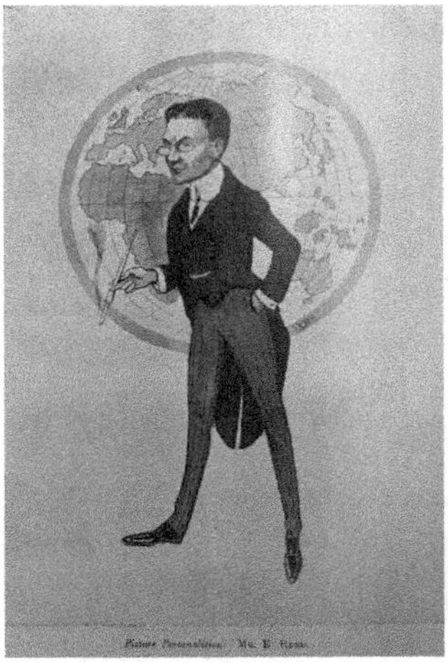
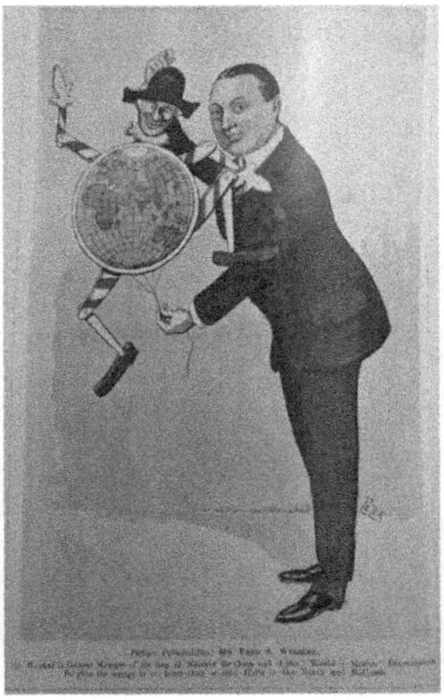
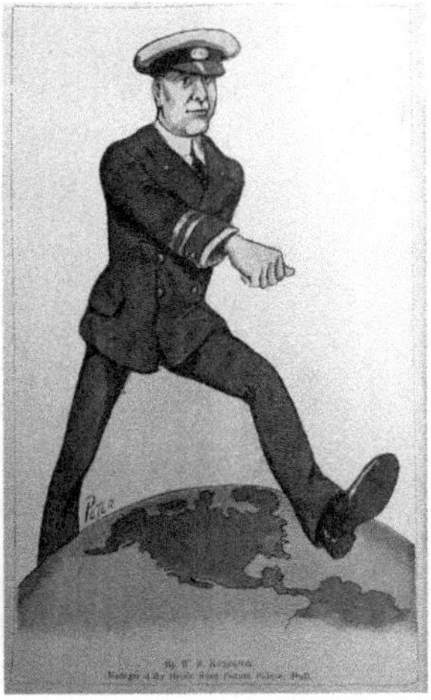

Figures 0.5, 0.6 & 0.7 Conquering the world: eminent businessmen and personalities in the UK were regularly depicted in the trade magazine *The Bioscope* (1910–1913) alongside globes and world maps so as to highlight the global expansionism of the nascent film industry. Source held at the British Library.

attending by the names of Globe, Cosmopolitan, Atlas Films and The World on Wings, to cite but a few in the UK. The globes varied in terms of composition, layout and design but on leafing through any film trade magazine or periodical of the time, there was no way you could miss them. They appear on Atlas's hunched back 'holding the world', here enveloped by a tangled web of film reels (figure 0.1); as a spherical container within which Gaumount's Chrono camera is proudly lauded as 'the best in the world' (figure 0.2); as the curved surface on which Pathé's famous rooster perches (figure 0.3); and as the flattened-out background against which Walturdaw's silent projector Powers No. 6 sits at the center of a stage (figure 0.4).[8] Globes and world maps also regularly featured in cartoons and illustrations depicting eminent personalities and businessmen, leaving no doubt as to the expansionist aspirations of the embryonic film industry (figures 0.5, 0.6 and 0.7).[9]

It is tempting to dismiss such imagery as the hyperbolic promotional rhetoric accompanying the emergence of a wildly popular new medium. Yet other evidence suggests that globalism and its variations were foundational structuring concepts underpinning early articulations of medium specificity

8 These advertisements respectively appeared in the following issues of *The Bioscope*: 258 (21 September 1911), 598; 361 (11 September 1913), 864; 186 (5 May 1910), 12; 351 (3 July 1913), viii.

9 These illustrations respectively appeared in the following issues of *The Bioscope* (no page numbers): 216 (01 December 1910); 240 (18 May 1911); 361 (11 September 1913).

in a variety of discourses and practices, including the prevalent idea of film as a universal language explored in chapter 5. Above all, it was cinema's ability to travel that thrilled metropolitan commentators and viewers alike, so much so that one writer could claim: 'I can learn more of what the world is like from the armchair of a picture than I can from travelling because I have neither the means nor the time to travel to all parts of the world'.[10] As Jennifer Lynn Peterson and Alison Griffiths have shown, the trope of global armchair travel became especially attached to the travelogue genre, which, with its ethnographic affiliations, held the promise of a mediated encounter with 'exotic' peoples and faraway locales across the world.[11] Titles such as 'Round the World in Two Hours', 'The World in Motion', 'The World on Wings' and 'Globe Express Excursions', to cite those of a few UK programmes, accordingly doubled down on the kinetic properties of the medium through the suggestion of a world that itself moved.[12]

To be sure, cinema's self-ascribed mission to reveal the planet anew was preceded by other media. Its planetary aspirations must be viewed, as Tom Gunning has emphasised, as the last in an illustrious lineage of 'feverish production of views of the world' that gained traction in the second half of the nineteenth century, when new travel routes opened up by colonial networks and intercontinental trading businesses enacted a 'geographical extension of the field of the visible', to quote Jean-Louis Comolli.[13] As I explore in chapter 4, before cinema claimed to show the faces of peoples of the world, photography had, since the early 1860s, partially fulfilled this task in carte de visite collections, anthropological compendia and photographic galleries of a racialised global humanity. Other earlier media and attractions, such as panoramas, around-the-world stereocard boxes,

10 The Man About Town, 'Popular Travel Pictures', *Pathé Cinema Journal* 6 (8 November 1913), 23.

11 Jennifer Lynn Peterson, *Education in the School of Dreams: Travelogues and Early Nonfiction Film* (Durham, NC, and London: Duke University Press, 2013), esp. 137–174; Alison Griffiths, *Wondrous Difference: Cinema, Anthropology, & Turn-of-the-Century Visual Culture* (New York: Columbia University Press, 2002), esp. 171–252.

12 These titles appear in the following articles and programmes: 'A "Travel Feature", *The Bioscope* 347 (5 June 1913); '"The World in Motion" at Wigan', *The Bioscope* 169 (18 November 1909), 39; 'The World in Motion at Loughborough', *The Bioscope* 181 (3 March 1910), 37; 'The World on Wings', *The Bioscope* 217 (8 December 1910), 15; 'Globe Express Excursions by the Bigograph', programme n. 18 for the Scenorama and Globe Choir Federation, n/d, held at the Bill Douglas Cinema Museum (BDCM), University of Exeter, UK. Item number: 18536.

13 Tom Gunning, '"The Whole World Within Reach": Travel Images without Borders' in Jeffrey Ruoff (ed.) *Virtual Voyages: Cinema and Travel* (Durham and London: Duke University Press, 2006), 25–41; 33; Jean-Louis Comolli, 'Machines of the Visible' in Stephen Heath and Teresa de Lauretis (eds.) *The Cinematic Apparatus* (London: Palgrave, 1980), 121–142; 122–123.

travel picture collections and world exhibitions, had equally testified, as Brooke Belisle has shown, 'to a broader interest in picturing the world as a coherent whole'.[14] Thus, already in 1852, a poster for a moving panorama described the medium as a 'new method [...] which brings to the very doors of those who cannot go to the mountains, all the mountains of the earth'.[15] Yet, no doubt, cinema added a new frisson to this idea. This was not only because it mechanically registered landscapes and peoples of the Earth *in motion*, it similarly broadened and diversified what constituted a subject worthy of attention beyond places and humans. One of the aims of this book is to show that formulations of worlding in early cinema and related media already harboured connections with what we would today call the nonhuman and even the posthuman.

Early cinema's voracious and promiscuous appetite to show anything and everything, which Mary Ann Doane has theorised as a consequence of its indexical rapport with real-world contingency, was repeatedly asserted in critical and promotional discourses.[16] This is how one commentator described the cinema experience in 1910: as a visual concatenation of 'perhaps Macbeth, motor skating, the Victoria Falls, glass blowing, a Passion play, the latest aviation meeting in France, a Texas melodrama, and King Edward opening a bazaar'.[17] Cinematic capaciousness, as I examine in chapter 3, was then conflated with globalism as part of a semiotic process whereby the whole world was meant to signify 'all things in the world'; or to put it differently, that any thing was potentially *a thing* worth recording and inventorying. A case in point is the suggestion for a poster design, published in the trade journal *The Bioscope* (1910), where miscellaneous figures and captions are crowded inside a globe to drive the point home that the variety of film subjects enumerated – travel, sport, humour, drama, pathos – are both part and parcel of cinema's ability to bridge terrestrial distances and encompass the globe (figure 0.8).[18]

Among cinema's favoured things were nonhuman life forms and their environs. To cite another dominant perception, the idea that film has

14 Brooke Belisle, 'Nature at a Glance: Immersive Maps from Panoramic to Digital', *Early Popular Visual Culture* 13:4 (2015): 313–335; 317. See also Brooke Belisle, 'Picturing Networks: Railroads and Photographs', *Amodern* 2 (October 2013), https://amodern.net/article/picturing-nineteenth-century-networks/ (last accessed 2 September 2020).

15 'The Great European Diorama' (1852), promotional leaflet, BDCM. Item number: 70431.

16 Mary Ann Doane, *The Emergence of Cinematic Time: Modernity, Contingency, the Archive* (Cambridge, MA, and London: University of Harvard Press, 2002).

17 'The Bioscope a Necessity', *The Bioscope* 179 (17 March 1910), 43.

18 *The Bioscope* 192 (16 June 1910), 86.

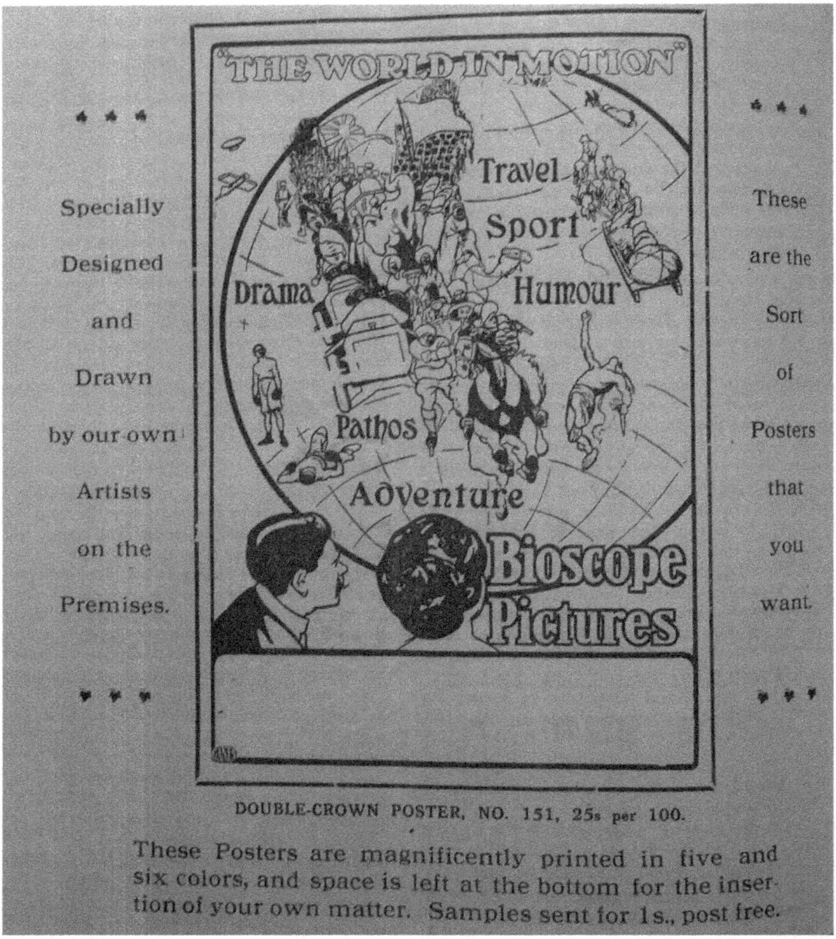

Figure 0.8 All things in the world: a suggested poster in *The Bioscope* makes visible cinema's perceived ability to show anything and everything on the planet. Source held at the British Library.

'literally brought the world – sea and land, animal and vegetable, insect, reptile, and so on – before me' was tied to its wondrous revelation of an entirely unsuspected living world, whether via film-specific devices, such as magnification and speed manipulation processes, or by showing animals in their natural habitat in distant places.[19] Put differently, the unseen world unveiled by film was mapped on to its capacity to travel and record wildlife across the world as part of a levelling impulse to equate infinitesimal organic phenomena and wild animal life, which, as I explore in chapter 2, revitalised

19 The Man About Town, 'Impressions Here and There', *Pathé Cinema Journal* 1:6 (8 November 1913): 23.

the static representational systems of natural history as an added experience of wonder by way of motion.

Cinema thus promised to reveal and organise the previously unseen: the places, faces, things, humans, animals and other life forms making up and crowding the world. Its appeal derived from its ability to make the world visible, in motion, for the first time, and its kinetic and evidentiary properties legitimised its veridical claims. But film was similarly understood, already at its dawn and like photography before it, as a medium whose duty was to record phenomenal realities on the verge of disappearance. Its appeal was therefore also inseparable from the idea that it could preserve vanishing worlds as they presented themselves in the world, for a camera, for the last time. These endangered realities included certain animals, believed to be on the brink of extinction, and nature's wilderness, threatened by an unrelenting urbanisation (chapter 2); certain humans, from presumably inferior races, who had their days counted on Earth thanks to the onslaught of 'progress' (chapter 4); and certain traditions and lifestyles about to be swept away by a wholesale and homogenising global modernity, a belief that provided the conceptual raison-d'être for Albert Kahn's Les Archives de la Planète (1909–1931), or Archives of the Planet, discussed in chapter 6.

These two poles, disclosure and disappearance, or the never-before-seen and the never-to-be-seen-again, structure many chapters in this book, most of which are correspondingly concerned with nonfiction genres and practices. In this sense, *Planetary Cinema* insists on the continuing relevance of the concept of indexicality, that is, cinema's automatic ability to record the world, for an understanding of historical (and contemporary) screen cultures.[20] But there is a significant exception in chapter 5, where my focus falls instead on a single fiction film, D. W. Griffith's *Intolerance* (1916), which I explore in relation to the contemporaneous discourse of film as a universal language. Even here, however, the notion that film could show the world

20 As is well known, the term 'indexicality' derives from Charles S. Peirce's theory of signs, the index being the sign that attests to the existence of its referent through a physical connection. In the 1960s, Peter Wollen borrowed the term to account for André Bazin's concept of the 'ontology' of the photographic image. See Peter Wollen, *Signs and Meanings in the Cinema* (Bloomington: Indiana University Press, 1972), 116–154. As many theorists have further noted, indexicality, as a property that attests to an automatic transference from reality to its reproduction, remains in digital capture. See Doane, *The Emergence of Cinematic Time*, 208. For studies that have questioned the usefulness of indexicality, or the idea of cinema as an indexical medium, see: Lev Manovich, *The Language of New Media* (Cambridge, MA: MIT Press, 2001), esp. 286–334; Tom Gunning, 'Moving Away from the Index: Cinema and the Impression of Reality', *differences* 18:1 (2007): 29–52; Jordan Schonig, 'Contingent Motion: Rethinking the "Wind in the Trees" in Early Cinema and CGI', *Discourse* 40:1 (2018): 30–61.

anew is also paramount, though not connected with its indexical properties but, rather, with its ability to reveal the links and connections between otherwise disparate places and times, that is, with its editing properties understood as the enabler of a networked figuration of the planet.

As can be glimpsed from the scholars mentioned above, a solid literature has critically engaged, whether directly or indirectly, with the global dimension and imagination of nineteenth-century and early twentieth-century film and media culture. To this scholarship I would add the names of Tanya Agathocleous, Alison Byerly, Erkki Huhtamo and Paula Amad, and further underline the unparalleled importance of Tom Gunning.[21] This book hopes to extend this important work in three ways.

First, *Planetary Cinema* proposes a theoretical, comprehensive (though certainly not exhaustive) and critical exploration of how 'the world', as a totalising concept and figure, was variously constituted and constructed across film and related media as they commingled, interacted and overlapped in the second half of the nineteenth century and into the twentieth century. Put simply, how was the planet imaged and imagined at the time? More specifically, what were the conceptual coordinates, aesthetic strategies and structural parameters delineating the globality of giant globes, the worldhood of unseen and thingly worlds, and the planetarity of planetary archives? Do these world visions replicate the same discourses and beliefs, and what explains their conjoined appearance? For reasons that will hopefully become clear by the end of this introduction, my historical case studies are mostly drawn from Britain and France. The 'world' envisioned in the historical projects and artefacts discussed here is therefore rooted in a Western, often imperial, global imagination.

Second, this book proposes a novel methodological approach to account for world-building processes in film and media. To be sure, whenever appropriate I will make recourse to the usual tools from geography, anthropology and natural history, understood as disciplines and fields of knowledge that

21 See, for example, Alison Byerly, *Are We There Yet? Virtual Travel and Victorian Realism* (Ann Arbor: University of Michigan Press, 2012); Erkki Huhtamo, *Illusions in Motion: Media Archaeology of the Moving Panorama and Related Spectacles* (Cambridge, MA, and London: MIT Press, 2013); Paula Amad, *Counter-Archive: Film, the Everyday, and Albert Kahn's Archives de la Planète* (New York: Columbia University Press, 2010); Tom Gunning, 'The World as Object Lesson: Cinema Audiences, Visual Culture and the St. Louis World's Fair', *Film History* 6:4 (1994): 422–444; Tom Gunning, 'Early Cinema as Global Cinema: the Encyclopedic Ambition' in Richard Abel, Giorgio Bertellini and Rob King (eds.) *Early Cinema and the 'National'* (Bloomington: Indiana University Press, 2008), 11–16; Tanya Agathocleous, *Urban Realism and the Cosmopolitan Imagination in the Nineteenth Century: Visible City, Invisible World* (Cambridge: Cambridge University Press, 2013).

necessitate the global as ontological and epistemological categories; or, to put it differently, as fields that both produce and reproduce the global, especially in the nineteenth century. Yet it is also part of the argument of this book that such tools are insufficient in themselves if we want to grasp the *aesthetic* foundations of media worlding. To that end, individual chapters are organised around a specific philosophical concept, each in turn related to a specific formal technique or category. These are: the sublime and the overview (chapter 1); wonder and the unseen view (chapter 2); things and the catalogue (chapter 3); faciality and the posed portrait (chapter 4); universalism and the network narrative (chapter 5); disappearance and ruinous imagery (chapter 6). Whereas these couplings are not mutually exclusive, it is my hope that the adoption of these ordering descriptors will provide a simultaneously rigorous and capacious conceptual approach where a number of mediums, discourses and ideas can be productively brought together in terms of their world-making affiliations and proclivities.

Third, these pairings are motivated by a quest to enlarge the timeframe and accordingly provide methodological bridges through which contemporary planetary imaginaries can be illuminated in terms of meaningful divergences and continuities with earlier imaginaries. Some issues related to terminology are, however, in need of clarification before I explore the rationale for this comparative methodology and issues of corpus selection and historical periodisation in more detail.

Earth · World · Globe · Planet

Earth, world, globe, planet: we can use a number of words to refer to our terrestrial home, each of which carries a distinct semiotic baggage.[22] The most capacious and flexible signifier of these is *world*: one can be or make a world, worlds can be many and everywhere, one can world a world. When preceded by the definite article, *the world* can still disturb and elude the imagination in its impossibly simultaneous embrace of all things, peoples, connections and relations crisscrossing the planet. World is formless. By contrast, *globe* has an irreducibly spherical shape that denotes spatiality: doubtless the most maligned of these descriptors, it generates associations with maritime routes, digital networks and electronic signals spreading

22 My understanding of these terms is informed by the following essay and books: W. J. T. Mitchell, 'World Pictures: Globalization and Visual Culture', *Neohelicon* 34:2 (2007): 49–59; Oliver, *Earth and World*; Cosgrove, *Apollo's Eye*; Heise, *Sense of Place and Sense of Planet.*

over the world's surface, from global imperial expansionism all the way to global finance capital. *Earth* and *planet*, for their part, partly resist abstract and anthropocentric connotations by evoking materialities, temporalities and processes above and beyond the human. Whereas Earth shares its name with the ground and soil, thus producing an adherence to ideas of land nourishment, rootedness and organic life, to speak of the world as a planet is also to picture it as a physical entity, but as a rounded, solid object floating in outer space alongside other celestial bodies.

As a result of the physicality inherent in Earth and planet, these two descriptors have latterly gained a renewed significance thanks to another increasingly widespread but also increasingly maligned concept: the Anthropocene. Coined in 2000 by Paul Crutzen and Eugene Stoermer, the Anthropocene describes a new geological phase the planet has entered because of momentous human-induced changes in its biogeophysical constitution, including ocean acidification, deforestation and biodiversity loss.[23] According to this idea, humanity has remodelled the planet to such an extent that it now comprises a geological force in its own right, paradoxically jeopardising the conditions that make possible the existence and sustenance of human life on Earth. In this sense, the Anthropocene functions as a bifurcating discourse of human supremacism and insignificance: elevated to the status of a telluric force on the one hand (the Anthropos), humanity is reduced on the other to a brief interlude in the planet's temporal trajectory due to the geological periodicity of the concept (the '-ocene').

One could argue that the Anthropocene, or at least the age of the Anthropocene, has precipitated a crisis of the 'age of the world picture'. As we know, this was an expression famously coined by Martin Heidegger in relation to what he saw as the technological instrumentalisation of the world – a world that, rather than lived and felt in its unknowable worldliness, was instead 'conceived and grasped as a picture', that is, as an enclosed totality that stood rationally 'at man's disposal as conquered'.[24] For Hannah Arendt, the Archimedean sight of the Earth in space made available in the 1960s crystallised this techno-imperialist world view whilst contravening our phenomenological and cognitive limitations as Earth-bound humans.[25] There is no doubt that a utilitarian and objectifying perspective on the

23 Paul J. Crutzen, 'The "Anthropocene"', *Journal de Physique IV*, 12:10 (2002): 1–6.

24 Martin Heidegger, *The Question Concerning Technology and Other Essays*, trans. William Lovitt (New York and London: Garland, 1977), 127, 133.

25 Hannah Arendt, *The Human Condition* (Chicago and London: The University of Chicago Press, 1998 [1958]); Hannah Arendt, 'Man's Conquest of Space', *The American Scholar* 32:4 (Autumn 1963): 527–540. For an illuminating study of Heidegger's and Arendt's Earth and world thinking,

planet still prevails in our time: one needs only to turn to military optics, satellite surveillance and discourses of geoengineering for actualisations of this perspective.[26] Yet, in a philosophical sense, there is also the growing perception that the idea of the world as an enframed picture one can measure and control from a distance is becoming untenable.[27] Or, to cite Jean-Luc Nancy and Aurélien Barrau, that 'we can no longer be certain of a distinction between "the world" and "us", between some thing that is in front of or around us and ourselves as "subjects" of this object', a process that is as much the result of 'the complexity of our interactions with the given (matter, life, space, and time)' as of 'the upheavals that affect all forms of civilization (knowledge, power, and values)'.[28]

The curious career of the Anthropocene during its 20-year span is that it has managed to escape the confines of the biological and Earth sciences and successfully migrate to the humanities as a concept that can illuminate the scalar transmutations of human-world relationships. Yet, as with any concept that proves too fashionable, the Anthropocene has generated important counter-arguments rejecting its universalising bent according to which 'we' – that is to say, the whole of humanity – are responsible for the current state of the world. Underlining the intertwined histories of colonialism, genocide, racism, slavery, industrial capitalism, extractivism and current ecological calamities, these arguments have ranged from outright rejections of the concept through to terminological adjustments intended to more adequately reflect on the real causes, agents and factors of the global environmental crisis.[29]

see Oliver, *Earth and World*, esp. 71–161. For an equally insightful discussion, see Brooke Belisle, 'Whole World Within Reach: Google VR', *Journal of Visual Culture* 19:1 (2020): 112–136; 125–131.

26 On geoengineering, see Christophe Bonneuil and Jean-Baptiste Fressoz, *The Shock of the Anthropocene* (London: Verso, 2017), 79–96. On global military surveillance, see Nicholas Mirzoeff, *The Right to Look: A Counterhistory of Visuality* (Durham and London: Duke University Press, 2011), 18–22.

27 See Anna Lowenhaupt Tsing, *The Mushroom at the End of the World: On the Possibility of Life under Capitalist Ruins* (Princeton and Oxford: Princeton University Press, 2015); Timothy Morton, *Hyperobjects: Philosophy and Ecology after the End of the World* (Minneapolis: University of Minnesota Press, 2013); Déborah Danowski and Eduardo Viveiros de Castro, *The Ends of the World*, trans. Rodrigo Nunes (Cambridge and Malden, MA: Polity Press, 2017).

28 Jean-Luc Nancy and Aurélien Barrau, *What's These Worlds Coming to?*, trans. Travis Holloway and Flor Méchain (New York: Fordham University Press, 2015), 5.

29 See, for example, Bonneuil and Fressoz, *The Shock of the Anthropocene*; Donna Haraway, *Staying with the Trouble: Making Kin in the Chthulucene* (Durham and London: Duke University Press, 2016); T. J. Demos, *Against the Anthropocene: Visual Culture and Environment Today* (Berlin: Stenberg Press, 2017); Kathryn Yusoff, *A Billion Black Anthropocenes or None* (Minneapolis: University of Minnesota Press, 2018); Nicholas Mirzoeff, 'It's Not the Anthropocene, It's the

Although this book shares many of the concerns voiced by these critiques and rejects the all-levelling rhetoric that the Anthropocene can assume, to provide nomenclatural or conceptual replacements for it is not one of my goals. This book instead takes *the occasion* of the Anthropocene – or however one may wish to name it – as a valuable opportunity to rethink how we understand and conceptualise the world away from 'world', and towards the Earth both in terms of its life-sustaining and life-generating processes and as a planet with its own past and future. This is in no way a suggestion to turn a blind eye to the violent histories of exclusion, domination and exploitation of which the global ecological breakdown is but a consequence, but it is a suggestion that, in addition to confronting these histories, there is the urgent need to confront the planet Earth as *a planet* and as *the Earth*.[30]

If world and the global may be deemed inadequate terms to account for the irruption of planetarity ushered in by the Anthropocene, when it comes to the cinema such concepts have never enjoyed as much prominence as they do today. Now established sub-disciplines within film studies, 'world cinema' and 'global cinema' feature across university curricula worldwide and stamp an ever-growing number of anthologies.[31] Whereas world cinema has gained currency since the early 2000s in studies that, broadly speaking, have attempted to reclaim or refigure its positive valence away from the market-driven exoticism with which it was identified (as with 'world music' and 'world literature'), global cinema has emerged as an alternative but related concept to account for the networks of transnationalism upon which film cultures and industries have indelibly relied. Yet, regardless of their differences, both terms have been recruited so as to underline the circulation, distribution and flows of films and directors as they break national

White Supremacy Scene; or, The Geological Color Line' in Richard Grusin (ed.) *After Extinction* (Minneapolis and London: University of Minnesota Press, 2018), 123–149.

30 This is moreover where my approach differs from Jameson's 'cognitive mapping', which is concerned with mapping out the invisibility of global capitalism rather than the physical planet as such. I have expanded on this in 'Earth Networks: *The Human Surge* and Cognitive Mapping', *NECSUS* 7:2 (2018): 121–140, https://necsus-ejms.org/earth-networks-the-human-surge-and-cognitive-mapping/.

31 Stephanie Dennison and Song Hwee Lim (eds.), *Remapping World Cinema: Identity, Culture and Politics in Film* (London and New York: Wallflower Press, 2006); Lúcia Nagib, Chris Perriam and Rajinder Dudrah (eds.), *Theorizing World Cinema* (London and New York: I.B. Tauris, 2012); Nataša Durovicová and Kathleen E. Newman (eds.), *World Cinemas, Transnational Perspectives* (London: Taylor & Francis, 2009); Rosalind Galt and Karl Schoonover (eds.), *Global Art Cinema: New Theories and Histories* (New York: Oxford University Press, 2010); Rob Stone et al. (eds.), *The Routledge Companion to World Cinema* (London and New York: Routledge, 2018); Elena Gorfinkel and Tami Williams (eds.), *Global Cinema Networks* (New Brunswick, Camden, and Newark, NJ. and London: Rutgers University Press, 2018).

barriers, travel across the globe and precipitate transcultural processes of hybridisation. Although it is hard to overestimate the importance of this scholarship in making film studies a more pluralistic and democratic discipline, it does lend credence to a concern raised by W. J. T. Mitchell already in 2007, when he noted that 'the general tendency has been to talk about the global *distribution* of images, their circulation in forms of mass media such as cinema, television, advertising, and the internet' rather than about 'images of the world and the global as such'. Mitchell proposed: 'We need to begin, then, by asking ourselves: How do we imagine, depict or know the global?'[32]

This book takes up this call. But it instead embraces the planetary as its organising descriptor as a way to provide a theoretical alternative to the circulation bias of world/global cinema and accordingly treat the world as a representational and physical entity in its own right.[33] In this, the book might seem to resonate with another current strand in film studies concerned with aesthetic and ethical processes of 'world-making'.[34] But its purchase on the concept must be primarily viewed in alignment with current ecological discussions of worlding, or 'cinema's powerful production of worlds in relation to *the* world', to cite Adrian J. Ivakhiv in his illuminating *Ecologies of the Moving Image* (2013).[35] Engaging in dialogue with the recent field of 'ecocinema' as a cinema that compels us 'to reflect upon what it means to inhabit this planet: that is, to be a member of the planetary ecosystem',

32 Mitchell, 'World Pictures', 50.

33 The term 'planetary cinema' has also been embraced by the duo Geocinema (Asia Bazdyrieva and Solveig Suess), although their focus and outputs (including moving-image experimental works) are significantly different from mine, considering as they do 'planetary-scale networks – cell phones, surveillance cameras, satellites, geosensors – as a vastly distributed cinematic apparatus'. See their website here: https://geocinema.network/ (last accessed 4 June 2021).

34 Daniel Yacavone, *Film Worlds: A Philosophical Aesthetics of Cinema* (New York: Columbia University Press, 2015); Karl Schoonover and Rosalind Galt, *Queer Cinema in the World* (Durham and London: Duke University Press, 2016). See also V. F. Perkins, 'Where's the World? The Horizon of Events in Movie Fiction' in John Gibbs and Douglas Pye (eds.) *Style and Meaning: Studies in the Detailed Analysis of Film* (Manchester and New York: Manchester University Press, 2005), 16–41.

35 Adrian J. Ivakhiv, *Ecologies of the Moving Image: Cinema, Affect, Nature* (Ontario: Wilfrid Laurier University Press, 2013), 8. Marchessault also explores many postwar world-building projects in relation to the planet as such in her excellent *Ecstatic Worlds*. See also Paula Willoquet-Maricondi (ed.), *Framing the World: Explorations in Ecocriticism and Film* (Charlottesville and London: University of Virginia Press, 2015). For a foundational collection on ecocinema, see Stephen Rust, Salma Monani and Sean Cubitt (eds.), *Ecocinema Theory and Practice* (New York and London: Routledge, 2013).

Planetary Cinema hopes to contribute to this conversation both in its focus on current films that literally attempt to imagine the planet itself and in its contention that we must historicise such a planetary impetus within a temporally larger media constellation.[36]

The works examined in the following pages include: IMAX films (*Blue Planet*, 1990; *A Beautiful Planet*, 2016, both directed by Toni Myers), the experimental film-essay *Medium Earth* (The Otolith Group, 2013) and the documentary *Earth* (*Erde*, Nikolaus Geyrhalter, 2019), all examined in chapter 1 in juxtaposition with the global panoramas of the past; contemporary BBC Earth series such as *Frozen Planet* (2011) and *Blue Planet* (2017), explored in chapter 2 within a natural-history cinematic lineage that includes the pioneering, early British films by Martin Duncan, Percy Smith and Cherry Kearton; collaborative documentary and web-based projects, including *The Global Remake: Man with a Movie Camera* (Perry Bard, 2007–2014), *Mass Ornament* (Natalie Bookchin, 2009) and *In Praise of Nothing* (Boris Mitić, 2016), placed in chapter 3 within a genealogy that includes the world symphonies *A Sixth Part of the World* (*Shestaya chast mira*, Dziga Vertov, 1926) and *Melody of the World* (*Melodie der Welt*, Walter Ruttmann, 1929), as well as Charles Urban's early cinema catalogues and stereocard collections; global documentaries inventorying the human face, such as *Samsara* (Ron Fricke, 2011) and *Human* (Yann Arthus-Bertrand, 2016), discussed in chapter 5 in relation to nineteenth-century photographic galleries of humanity; network films such as *Babel* (Alejandro González Iñárritu, 2006) and *The Human Surge* (*El auge del humano*, Eduardo Williams, 2016), examined in chapter 5 alongside D. W. Griffith's *Intolerance* (1916) as a precursor of the genre; and end-of-the-world projects such as *The Last Pictures* (Trevor Paglen, 2012) and *Homo Sapiens* (Nikolaus Geyrhalter, 2016), whose archival impetus is explored in chapter 6 in relation to Kahn's Archives de la Planète.

A few things are perhaps already clear from my corpus. First, the book combines a variety of genres, modes and texts, ranging from the popular to the experimental. These were selected on the basis of the degree to which they engage in a fruitful aesthetic conversation with earlier global imaginaries via the specific philosophical categories and concepts organising each chapter. As a result, many of the selected contemporary works, whether we think of the authors or companies behind them, emerge from the West, including Canada, the UK, the US, Austria and France. Yet there

36 Paula Willoquet-Maricondi (ed.), 'From Literary to Cinematic Ecocritcism' in *Framing the World*, 1–22; 10.

are some exceptions, and in many cases, the global circuits of financing and distribution upon which some of these current works rely complicate a facile correlation between nationality and worldview. For example, the 'British' duo The Otolith Group consists of a British-Ghanaian (Kodwo Eshun) and a British-Indian (Anjalika Sagar) national. Alejandro González Iñárritu is a filmmaker from Mexico, although most of his films, including *Babel*, are now largely made with US money and for a global market. Eduardo Williams is from Argentina, but his films are dependent upon the European film festival circuit. And this is not to mention that many of the texts selected compile images filmed all over the world, often to instantiate global conceptions of authorship, examples including *The Global Remake* and *In Praise of Nothing*, respectively conceived by the Canadian artist Perry Bard and the Serbian filmmaker Boris Mitić.

Second, I am not interested in how the Earth has been pictured from space or deployed in the science fiction film, which would be the topic of a different book.[37] Although some of my contemporary case studies do occasionally rely on the image of the planet from afar enclosed in one single frame, most of them, past and present, activate a synecdochal mechanism whereby individual frames, shots and films are conceptually and structurally tied to a wider whole that is the idea of the whole world. In so doing, they resonate with what both André Bazin and Stanley Cavell have theorised as lens-based media's ability to reproduce 'an image of the world' that implies and implicates the rest of the world (see chapter 6).[38] As Cavell notes, cinema's worlding is different from that of painting insofar as 'a photograph is *of* the world': 'When a photograph is cropped, the rest of the world is cut out. The implied presence of the rest of the world, and its explicit rejection, are as essential in the experience of a photograph as what it explicitly presents'.[39] Something similar is at stake in Bazin's 'myth of total cinema', which, while recognising that a reproduction of the world is ultimately impossible due to technical encumbrances, nevertheless concedes that cinema's mythical goal is the recreation of the 'world in its own image'.[40]

37 See, in this regard, Adrian Ivakhiv, 'The Age of the World Motion Picture: Cosmic Visions in the Post-Earthrise Era', in Stanley D. Brunn (ed.) *The Changing World Religion Map: Status, Literature and Challenges* (Dordrecht: Springer, 2015), 129–144.

38 André Bazin, *What Is Cinema?* volume 1, trans. Hugh Gray (Berkeley: University of California Press, 2005), 13; Stanley Cavell, *The World Viewed: Reflections on the Ontology of Film* (Cambridge, MA, and London: Harvard University Press, 1979), 20.

39 Cavell, *The World Viewed*, 24.

40 Bazin, *What is Cinema?*, 21.

To be sure, neither Bazin nor Cavell espoused an all-seeing and alienating vision of the world. On the contrary, if their thinking is relevant here, this is because they highlight the material locatedness and limited perspective of images as they are produced in the world and reproduce specific portions of the world. As Ivakhiv argues, cinema is 'geomorphic': films 'produce a *segmentation* or *fragmentation* of the world in that we are shown disconnected bits of world – pieces, images or glimpses that are woven together into a Cubist-like assemblage'.[41] But whereas some projects contemplated in this book tend to mask this fragmentation in an attempt to construct a smooth and undifferentiated picture of the world – or an enframed 'world picture', to cite again Heidegger – others acknowledge the impossibility of such a project by highlighting instead the fissures, cracks and dividing lines in their necessarily incomplete – or 'cropped', to use Cavell's term – rendition of the planet.[42] For, indeed, there are as many differences between an early-cinema programme promising to 'put the world before you' and an 'archive of the planet' personally funded by a millionaire, as there are between BBC Earth series and experimental documentaries – differences that range from the representational and the aesthetic all the way to the discursive and the ideological. I will return to some of these questions in the last section of this introduction; now it is time to provide a justification for this book's historical periodisation and its promiscuous mixing of a variety of media texts, genres and forms.

Multiple Media Worlds

Most chapters in the book juxtapose visual and audiovisual artefacts produced in different historical periods. The methodological reasoning for this comparative approach is borne out of two fields of enquiry, developed over the last decades, which this book hopes to combine into a third intellectual stream with the guiding help of philosophical concepts. These fields broadly pertain to globalisation and media history studies, and they have emphasised

41 Ivakhiv, *Ecologies of the Moving Image*, 74, emphasis in original.

42 I am not the first to make a connection between Heidegger's and Cavell's conceptions of worlding. See Brian Price, 'Heidegger and Cinema' in Temenuga Trifonova (ed.) *European Film Theory* (New York: Routledge, 2009), 108–121. However, whereas Price contends that the 'logic of film studies […] has only ever been synechdocal [*sic*]' as their 'use of "world"' in 'theories of the indexicality of the photographic image' hinge on 'the insistent relation of part to whole' (109), this book instead emerges from the belief that film studies has not been synecdochal enough when it comes to matters of worlding.

continuities between the late nineteenth century and our own time, even if in relation to different phenomena and not necessarily in conversation with one another. One of the aims of this book is to forge such a conversation.

Regarding globalisation, a growing number of studies is concerned with debunking the perception that it is a contemporary phenomenon whose beginning coincides with the end of the Cold War and the dawn of the digital revolution. These studies, coming as they do from an array of disciplines, aim instead to demonstrate that many of the structures, epistemologies, organisations and experiences associated with globalisation largely defined as a 'a concept [that] refers both to the compression of the world and intensification of consciousness of the world as a whole' – in Roland Robertson's much-cited definition – finds a number of technological, socio-economic, geopolitical and cultural precedents starting in the second half of the nineteenth century.[43] As historian Jürgen Osterhammel summarises in his suitably monumental *The Transformation of the World: A Global History of the Nineteenth Century* (2009), 'contemporary historians on the lookout for early traces of "globalization" are not the first to have discovered transnational, transcontinental, or transcultural elements in the nineteenth century, often described as the century of nationalism and the nation-state'.[44]

Thus, in their anthology *Nineteenth-Century Worlds: Global Formations Past and Present* (2008), Keith Hanley and Greg Kucich note the correspondences between what they 'call nineteenth-century "global formations" and the phenomenon of globalization today', a comparison of which, they propose, can 'illuminate some of the salient features, problematics, and arguable benefits of one the most significant early stages of globalization during the nineteenth century'.[45] Other accounts further show that seismic changes in international relations were cemented as the conjoined result of the rise of nation-states as sovereign political units, European imperialism, global industrialisation, economic integration, intercontinental migration, and transportation and communication networks – all justified and supported by the entwined, racist ideologies of 'progress' and 'civilisation'. As George Lawson and Barry Buzan contend: 'Not until the nineteenth century

43 Roland Robertson, *Globalization: Social Theory and Global Culture* (London, Thousand Oaks, CA, and New Delhi: SAGE Publications, 1992), 8.

44 Jürgen Osterhammel, *The Transformation of the World: A Global History of the Nineteenth Century* (Princeton and Oxford: Princeton University Press, 2009), xv.

45 Keith Hanley and Greg Kucich, 'Introduction: Global Formations and Recalcitrances' in Keith Hanley and Greg Kucich (eds.) *Nineteenth-Century Worlds: Global Formations Past and Present* (London and New York: Routledge, 2008), 1–16; 1–2.

did the world become a global system in which core states could quickly and decisively project the new mode of power around the world', a system that was 'both intensely connected and deeply divided'.[46]

Variously referred to as 'the second industrial revolution', 'the first globalization', 'the fin de siècle' or the 'long turn of the century', the periodicity of this historical moment, as with any historical periodisation, is to some extent artificial and varies depending on the focus of the study.[47] Yet by and large studies of nineteenth-century globalisation often agree on the period 1880–1919 to account for the peak of the transformations outlined above, or in some cases, the longer timeframe 1850–1930 to highlight that these transformations did not simply appear or disappear out of nowhere. And all of these studies concur that the operational center of such transformations was Europe, and more specifically Britain and France. Osterhammel sums up: 'The history of the nineteenth century was made in and by Europe' and no country before 'had projected their power to the farthest corner of the earth and had such a powerful cultural impact on "the Others" as Britain and France did in the nineteenth century'.[48]

This meant that if you happened to be in Britain or in France, and especially their capitals London and Paris, you would have been exposed to an effusive cultural globalism that was both part and parcel of these countries' geopolitical, economic and cultural influence worldwide. This is one of the reasons why these two countries, and especially Britain, constitute the core of this book when it comes to its historical focus, which coincides moreover with the periodicity mentioned above. Whereas the bulk of the chapters, especially those focusing on cinema, explores films and related projects within the period 1880–1919 as a 'long turn of the century', in chapters that deal more extensively with other mediums, such as the panorama and photography (chapters 1 and 4), there was the need to go further back in time, sometimes as early as the 1850s, in order to properly account for the gradual development and consolidation of a planetary imaginary. Conversely, my analysis of a cinema of things (chapter 3) stretches the timeframe to the late 1920s so as to consider Walter Ruttmann's *Melody of the World* and

46 Barry Buzan and George Lawson, *The Global Transformation: History, Modernity and the Making of International Relations* (Cambridge: Cambridge University Press, 2015), 2–3.
47 On 'the second industrial revolution', see Miriam R. Levin et al., *Urban Modernity: Cultural Innovation in the Second Industrial Revolution* (Cambridge, MA, and London: MIT Press, 2010); on the 'first globalisation', see Miguel Suárez Bosa (ed.), *Atlantic Ports and the First Globalisation c. 1850–1930* (Basingstoke: Palgrave Macmillan, 2014); on 'the fin de siècle' and the 'long turn of the century', see Osterhammel, *The Transformation of the World,* 58.
48 Osterhammel, *The Transformation of the World,* xx.

Dziga Vertov's *A Sixth Part of the World*: the former as the last vestiges of an interwar globalism and the latter as the cinematic formulation of a utopian global socialism. This being not a history book, however, it claims no intervention in issues of periodisation. Rather, my aim is to show that the social, economic and political changes described above could not have gone unexpressed in cultural and discursive realms and that they accordingly found manifestation in a number of media artefacts that both mirrored and contributed to an increased perception of the world as a wider, and increasingly interconnected, totality.

In some ways this was an optimistic story. To cite Miriam R. Levin, 'the future [was] generally portrayed as the linear result of scientific and technical progress – safe, increasingly prosperous, congenial and controllable'.[49] And that future was global: georamas, world exhibitions, photographic catalogues, and later, film programmes all, in one way or another, held the promise of a more organised, compartmentalised and controlled world.[50] Yet such a quest for management and order was itself the symptom of wider anxieties relating to a world that threatened to spin out of control as its parameters were enlarged due to economic and expansive imperatives, especially in Victorian Britain. Confidence and anxiety thus were, as Paul Young has shown in his study of the Great Exhibition (1851) and globalisation, often the two sides of the same global coin: 'given that the world was so clearly changing for the Victorians, and given that the Victorians were so clearly changing the world, was it possible to get this changing world into perspective?'[51]

However, this book claims that globalisation as a concept is also insufficient in itself to grasp the contours of this changing world. For, as I explore in chapter 1, the nineteenth century equally witnessed a *geologisation* of the Earth that stretched its timeline into a nonhuman past, with a sense of earthwide expansion thus felt on both a spatial and temporal level. The wildly popular framework of natural history, discussed in chapter 2, similarly entailed a broadening of earthly horizons into nonhuman realms, whereas Albert Kahn's Archives de la Planète, examined in chapter 6, was already informed by an acute sense of human-induced physical transformations on the planet thanks to its epistemological foundations in the discipline of

49 Miriam R. Levin, 'Dynamic Triad: City, Exposition, and Museum' in Levin et al. *Urban Modernity*, 1–12; 3.

50 On this quest to organise the world as an 'exhibit', see Timothy Mitchell, 'The World as Exhibition', *Comparative Studies in Society and History* 31:2 (1989): 217–236.

51 Paul Young, *Globalization and the Great Exhibition: The Victorian New World Order* (Basingstoke: Palgrave, 2009), 3.

'human geography'. It thus makes sense to attribute to this historical period the formation not only of a global but also of a 'planetary consciousness', as Joyce E. Chaplin has argued in her *Round about the Earth: Circumnavigation from Magellan to Orbit* (2012):

> Global is social – it implies the social relations that extend over the globe. In contrast, planetary is physical, implying the physical planet itself. Far more studies have focused on the former than the latter. This is because human-to-human interactions have been historians' major focus. Only recently have human relations with the non-human parts of nature have been put into dialogue with those human relationships; only recently have scholars begun to reread historical documents to discover our past sense of our place within nature.[52]

For Chaplin, the planetary consciousness that resulted from around-the-world travels in the late nineteenth century finds an equivalent in our time, though the sense of 'confidence' which defined that consciousness then, she maintains, has now been replaced by 'doubt', as 'the environmental costs of planetary domination have begun to haunt us'.[53] This book follows in these footsteps both in its embrace of the planetary as propelled by the current ecological emergency and as an organising and comparative descriptor that can allow us to reread historical documents with a 'planetary' lens, in my case media and film artefacts.

As a matter of fact, the idea that our contemporary media culture is comparable to that emerging in the late nineteenth century has been advanced by some of the most eminent film theorists of our time. Already in 1993, Miriam Hansen postulated a correlation between what she calls 'early cinema' and 'late cinema' in terms of 'parallels between pre-classical and post-classical forms of spectatorship, between early modern and postmodern forms of distraction and diversity'.[54] For Tom Gunning, 'the two ends of the Twentieth Century hail each other like long lost twins', both being eras of 'technological acceleration and transformation of the environment', and

52 Joyce E. Chaplin, *Round about the Earth: Circumnavigation from Magellan to Orbit* (New York: Simon and Schuster, 2012), xix. Before Chaplin, the term 'planetary consciousness' was coined by Mary Louise Pratt in relation to the European project of natural history, which I explore in chapter 2. See Pratt, *Imperial Eyes: Travel Writing and Transculturation* (New York and London: Routledge, 2008), 15–36.

53 Chaplin, *Round about the Earth*, xxi.

54 Miriam Hansen, 'Early Cinema, Late Cinema: Permutations of the Public Sphere', *Screen* 34:3 (1993): 197–210; 200.

both testifying to a 'voraciously competitive media environment' within which cinema finds itself.[55] Thomas Elsaesser has likewise noted that digital cinema has become 'the explicit reference point in the present from which to seek out precedents and parallels across a hundred-year span', given the 'equally rapid changes in the overall mediascape' of both periods.[56]

Inspired by, or in tune with, these ideas, a number of comparative methodologies have emerged with the intent of exploring cinema's intermingling with other medial and artistic entities, including 'media archaeology', 'media genealogy', 'parallax historiography' and 'intermediality'.[57] Their many differences notwithstanding, these approaches are keen on forging anti-canonical, anti-evolutionary, anti-chronological and anti-teleological film and media histories, often drawing on Benjamin and/or Foucault for inspiration and following in the footsteps of the New Film History approach launched in the now landmark 1978 FIAF Brighton Conference.[58] But they have also raised suspicions concerning issues of methodological promiscuity by downplaying medium specificity and contextual markers.[59]

As far as this book is concerned, it shares with these approaches a commitment to parallelisms and even anachronisms that can rekindle our understanding of film and media history away from linear determinisms

55 Tom Gunning, 'Re-Newing Old Technologies: Astonishment, Second Nature, and the Uncanny in Technology from the Previous Turn-of-the-Century' in David Thorburn and Henry Jenkins (eds.) *Rethinking Media Change: The Aesthetics of Transition* (Cambridge, MA, and London: MIT Press, 2003), 39–60; 42, 51; Gunning, 'Moving Away from the Index', 35–36.

56 Thomas Elsaesser, *Film History as Media Archaeology: Tracking Digital Cinema* (Amsterdam: Amsterdam University Press, 2016), 41, 354.

57 On 'media archaeology', see Erkki Huhtamo and Jussi Parikka (eds.), *Media Archaeology: Approaches, Applications, and Implications* (Berkeley, Los Angeles and London: University of California Press, 2011); Jussi Parikka, *What is Media Archaeology?* (Cambridge: Polity Press, 2012). On 'intermediality', see Jeffrey Geiger and Karin Littau (eds.), *Cinematicity in Media History* (Edinburgh: Edinburgh University Press, 2013); Ágnes Pethő, 'Intermediality in Film: A Historiography of Methodologies', *Acta Univ. Sapientiae, Film and Media Studies* 2 (2010): 39–72. On 'parallax historiography', see Catherine Russell, 'Parallax Historiography: The Flâneuse as Cyberfeminist' in Jennifer M. Bean and Diane Negra (eds.) *A Feminist Reader in Silent Cinema* (Durham, NC, and London: Duke University Press, 2002), 552–570; Paul Flaig and Katherine Groo (eds.), *New Silent Cinema* (New York and London: Routledge, 2016). On 'media genealogy', see Alexander Monea and Jeremy Packer, 'Media Genealogy and the Politics of Archaeology', *International Journal of Communication* 10 (2016): 3141–3159.

58 For the importance of New Film History for media archaeology and variants, see Wanda Strauven, 'Media Archaeology: Where Film History, Media Art, and New Media (Can) Meet' in Julia Noordegraaf et al. (eds.) *Preserving and Exhibiting Media Art: Challenges and Perspectives* (Amsterdam: Amsterdam University Press, 2013), 59–79; 61–63.

59 See Parikka, *What is Media Archaeology?*, 10; Thomas Elsaesser, 'Media Archaeology as Symptom', *New Review of Film and Television Studies* 14:2 (2016): 181–215.

and neat evolutionisms. And it follows Elsaesser in his contention that methodological unruliness can be averted via a 'more restricted focus that puts cinema tactically at the center while extending the scope of the medium in new directions'.[60] But *Planetary Cinema* also proposes a methodological intervention via its recruitment of specific philosophical categories – sublimity, wonder, things, faciality, network and disappearance – as a way to open up coherent conceptual pathways through which one can chart the continuities and disjunctures between planetary visions old and new. Engaging with the thinking of Alexander von Humboldt, Siegfried Kracauer, Hannah Arendt, Bruno Latour, Alain Badiou, Déborah Danowski and Eduardo Viveiros de Castro, among others, *Planetary Cinema* thus sits at the intersection of film/media history and theory/philosophy, and it claims, as a contribution to ecocinema, that we need this combined historical approach and expansive textual focus in order to understand the planetary in film and media.

At the same time, the book hopes to consolidate new sets of historical understanding on two main fronts. The first, as mentioned earlier, relates to its goal of challenging traditional historical genealogies that postulate the Apollo pictures of the Earth as the genesis for a planetary awareness (see also chapter 1). The second concerns the idea that 'the narrative of "modernity"', as Vanessa R. Schwartz and Jeannene M. Przyblyski summarise, 'has provided one of the most useful discursive frames for making sense of the relationship between visual experience and cultural hegemony in the nineteenth and twentieth centuries' through to 'current discussions of visual culture in the twenty-first century'.[61] This book starts from the premise that this narrative has not only been exhausted, but that modernity, to cite Buzan and Lawson, 'was a global process both in terms of its origins and outcome'.[62] I therefore take globality to be the ontological and epistemological precondition of modernity and not simply its backdrop, a figure-ground conceptual reversal which, I hope, can cast new light on nineteenth- and early twentieth-century media culture.

For, indeed, whereas Parikka is right to warn, via Elsaesser, that 'the multiple worlds of visual culture of the nineteenth century, with its "vaudeville, panoramas, dioramas, stereoscopic home entertainment, Hale's tours and

60 Elsaesser, *Film History as Media Archaeology*, 21.

61 Vanessa R. Schwartz and Jeannene M. Przyblyski, 'Visual Culture's History: Twenty-First Century Interdisplinarity and Its Nineteenth-Century Objects' in Vanessa R. Schwartz and Jeannene Przyblyski (eds.) *The Nineteenth-Century Visual Culture Reader* (London and New York: Routledge, 2004), 3–14; 8.

62 Buzan and Lawson, *The Global Transformation*, 7.

world fairs", [...] are a further good reminder of the dangers of homogenization', what is striking about these multiple worlds is that they often latched on to 'the world' as their discursive, conceptual and representational horizon.[63] In so doing, they both drew upon and contributed to globality as it was variously and differently formulated in a variety of domains and disciplines: geological, geographical, natural-historical, anthropological, cultural and technological. Today, planetary tropes and imagery are everywhere in our audiovisual landscape, no doubt as a response to an increasing awareness of the Earth's fragility in the context of an Anthropocene-induced uncertain future.

Although this book claims no intervention in Foucauldian studies, one way of understanding the pervasiveness of the global and the planetary in these two historical periods might be in terms of what Foucault conceptualises as historically situated discursive formations, which he termed 'epistemes'. Contra historiographic linearity, Foucault's archaeological method aims to uncover the breaks and ruptures between discourses as they manifest themselves at different moments in time. But as he hastens to add: 'To say that one discursive formation is substituted for another is not to say that a whole world of absolutely new objects, enunciations, concepts, and theoretical choices emerges fully armed and fully organized' and that others disappear completely in their wake. Rather, 'it is to say that a general transformation of relations has occurred, but that it does not necessarily alter all the elements', and in so doing, 'one can, on the basis of these new rules, describe and analyse phenomena of continuity, return, and repetition'.[64] For Elsaesser, a media history approach can only gain from this model, 'describing and reconnecting historical phenomena in a different conceptual space, either by positing distinct epistemes and discursive formations, or by a conjecture or a constellation that "makes new sense" explicitly from the point of view of the present'.[65]

Often likened to notions of progress, futurity and an unquenchable thirst to dominate the Earth's natural resources, the planetary of nineteenth-century discourses and figures may appear to be radically distinct from the planetary of our time, where uncertainty, doubt and anxiety about the future often prevail. This is to a large extent true. But as the following chapters will attest, there are also many 'phenomena of continuity, return,

63 Parikka, *What is Media Archaeology?*, 10.

64 Michel Foucault, *The Archaeology of Knowledge and the Discourse on Language*, trans. A. M. Sheridan Smith (New York: Vintage Books, 2010), 492.

65 Elsaesser, *Film History as Media Archaeology*, 34.

and repetition' across these two historical periods: whether we look at embryonic and proto-ecological conceptualisations of disappearance and human-induced change on a global scale in the late nineteenth century, or whether we look at how some colonial tropes, techniques and tools of globality have survived to this date – hence the necessity to disentangle the imperial from the global. By tracking the similarities and ruptures across these two periods, *Planetary Cinema* takes the environmental crisis and its ushering forth of the planetary as an opportunity to make new sense of the past in order to understand our present: hopefully this may lead us to a better future.

Towards the Planetary

The challenge facing us today is that, while planetary thinking is of the utmost urgency, the discourses that could provide models for this endeavour are implicated in troubling histories and lineages. Globalism is justifiably maligned in its networking of the world so as to maximise the flows of capital. Predicated on a supposed universality of all human beings inhabiting this Earth, Western discourses of universalism have often operated either as a negative principle that measured itself against those who were not considered humans, such as women and nonwhite peoples, or as a hierarchic system according to which some humans – white and male – were deemed far more intelligent and worthier than others. In this context, as Naomi Schor remarked in 1995: 'The dismantling of the universal is widely considered one of the founding gestures of twentieth-century thought'.[66] As the idea of someone who is at home anywhere on the planet, cosmopolitanism has also not gone unchallenged, accused of being a shorthand for the historical experience of the European male traveller who could tour around the world by trailing on the colonial networks of his country of origin.[67] As a result, global figurations and epistemologies are often likened to an 'imperial imaginary', 'imperial eyes', an 'imperial visuality', or an 'Apollonian vision' that sees expressly with the aim of surveying, demarcating and appropriating the world.[68]

66 Naomi Schor, *Bad Objects: Essays Popular and Unpopular* (Durham and London: Duke University Press, 1995), 3.

67 See Ursula K. Heise, *Imagining Extinction: the Cultural Meanings of Endangered Species* (Chicago and London: Chicago University Press, 2016), 225.

68 I am referring here to expressions from Pratt *Imperial Eyes*, Cosgrove *Apollo's Eye* and Mirzoeff *The Right to Look*. On the connection between the global and an 'imperial imaginary',

The question, then, becomes: can we disentangle these discourses from Eurocentric premises and presuppositions while retaining their collective thrust? For some thinkers, the answer is yes. Seeing 'the rise of gene-oriented or genomic constructions of "race"' as a 'welcome cue to free ourselves from the bonds of all raciology', Paul Gilroy has proposed the concept of 'planetary humanism' as one that offers 'the basis for belonging to one another and acting in concert'.[69] New cosmopolitanisms, as Paula Amad has shown, have equally aimed to 'reform the term from its connotations of elite, enlightened universalism [...] all the while remaining critical of global capital's negative effects'.[70] Universalism itself has been repackaged as a 'negative universalism', to cite Dipesh Chakrabarty, in order to account for the fact that the unprecedented environmental transformation currently on course 'poses for us a question of a human collectivity, an us, pointing to a figure of the universal that escapes our capacity to experience the world. It is more like a universal that arises from a shared sense of a catastrophe. It calls for a global approach to politics without the myth of a global identity'.[71]

But suspicions remain. Nicholas Mirzoeff, for example, has replied to Chakrabarty by stressing that 'while climate change certainly affects the entire planet, its impact is very different in different places, consistent with the usual indicators of wealth', which begs the question of 'how such universal history might be written'.[72] Mirzoeff's rebuke in turn draws on Denise Ferreira da Silva's important *Toward a Global Idea of Race* (2007), which defines raciality and globality as mutually constitutive and segregational onto-epistemological categories (see chapter 4). For Ferreira da Silva, Gilroy's 'planetary humanism' is 'the best example of the perverse effects' of a desire 'to recuperate the racial subaltern into an unbounded humanity'.[73]

There is no denying that racism – and for that matter the exclusion and discrimination of all minority groups – is very much alive and well in the world. But to simply reject planetary ontologies and epistemologies is, for me, not the solution, not only because it disregards the fact that these forms are

see Ella Shohat and Robert Stam, *Unthinking Eurocentrism: Multiculturalism and the Media* (London: Routledge, 1994), 100–136.

69 Paul Gilroy, *Against Race: Imagining Political Culture Beyond the Color Line* (Cambridge, MA: Belknap Press, 2000), 15.

70 Paula Amad, 'Experimental Cosmopolitanism: The Limits of Autour du Monde-ism in the Kahn Archive' in Trond Erik Bjorli and Kjetil Ansgar Jakobsen (eds.) *Cosmpolitics of the Camera: Albert Kahn's Archives of the Planet* (Bristol, UK, and Chicago, IL: Intellect, 2020), 133–154; 133.

71 Dipesh Chakrabarty, 'The Climate of History', *Critical Inquiry* 35:2 (2009): 197–222; 222.

72 Mirzoeff, 'It's Not the Anthropocene', 127.

73 Denise Ferreira da Silva, *Toward a Global Idea of Race* (Minneapolis and London: University of Minnesota Press, 2007), 8.

not univocal or immutable but also because it denies the pressing urgency and uniqueness of our particular moment, which immediately calls for collective solutions on the global scale. As a number of thinkers have also alerted us, including the philosopher Alain Badiou in his *Saint Paul: The Foundation of Universalism* (2003), aversion to the universal has resulted in extreme forms of identitarianism that subscribe to the logic of the market and reproduce a neoliberal ideology that trumps the individual over the collective.[74] To cite Amitav Ghosh, in his inspiring *The Great Derangement: Climate Change and the Unthinkable* (2016): 'The political is no longer about the commonweal or the "body politic" and the making of collective decisions', and yet climate change is an 'issue that concerns our collective survival'.[75]

To be clear: I am not proposing one should forget the ways in which universalisms and globalisms have acted in concert with exclusionary forces of power that persist to this day, nor am I saying that the environmental crisis affects everyone in equal measure. I therefore concur with Mirzoeff that 'while climate change certainly affects the entire planet, its impact is very different in different places'.[76] My problem with this line of argument is that the second part of this sentence is often seen as the justification to disregard the first part, as if they were mutually exclusive and as if we could afford to not think about the planet as a whole because some places are the first to bear the brunt of environmental disasters. As Eva-Lynn Jagoe rightly notes, the present ecological crisis, with its hurricanes and wildfires, 'spreads and subsumes differences in its urgency', hence the need 'to think collectively across racialized, gendered, and embodied identities'.[77] In other words, to acknowledge and come to terms with all the injustices that have been done in the name of the global and the universal still leaves us with a massive problem: one that is the size of the whole world.

At this point the reader may be asking himself or herself how a book that attempts to juxtapose nineteenth-century global imaginaries with contemporary

74 Alain Badiou, *Saint Paul: The Foundation of Universalism*, trans. Ray Brassier (Stanford, CA: Stanford University Press, 2003), 10–11. See also Eva Lynn-Jagoe, 'Jumping the Break: Wildfire and the Logic of Separation', *Discourse* 40:2 (2018): 231–251; 235, 237. Lynn-Jagoes's article, which reflects on the 2017 Banff Research in Culture (BRiC) she co-organised, generated a response from its participants. See Rosa Aiello et al., 'Staying with the Breaks: Dispropriating the Universal: A Response to Eva-Lynn Jagoe', *Discourse* 41:1 (Winter 2019): 167–175. Lynn-Jagoe's response to this response was published in the same issue of *Discourse*, titled 'Broken? Notes toward 2067', 176–180.

75 Amitav Ghosh, *The Great Derangement: Climate Change and the Unthinkable* (Chicago: University of Chicago Press, 2016), 125–127.

76 Mirzoeff, 'It's Not the Anthropocene', 127.

77 Lynn-Jagoe, 'Jumping the Break', 237, 247.

ones can contribute to this debate. After all, are not many of these imaginaries responsible for the suspicion attached to global figures and ideas? As Cosgrove argues, current 'one-world' and 'whole-earth' discourses – the former related to the idea of globalisation, the latter to the global environmental crisis – 'inherit the most persistent and contradictory feature of the Western global imagination, its sense of global mission', and must 'therefore be treated with scepticism in the light of the genealogy of Apollonian vision'.[78] I do not deny this problematic genealogy: in fact, many chapters in this book prove that universalism and globality were corrupted in their supposedly egalitarian premises (chapters 4 and 5). My point is that by confronting earlier global imaginaries and constructions we can be better equipped to explore their legacies as they persist in our time. For, indeed, as Bruno Latour observes in his *Down to Earth: Politics in the New Climactic Regime* (2018), while Europe 'can no longer claim to dictate the world order', it 'can offer an example of what it means to rediscover inhabitable ground'.[79] He goes on:

> After all, it is indeed Europe that claims to have invented the Globe, in the sense of space captured by the instruments of cartography. A system of coordinates so powerful – too powerful – that it makes it possible to record, preserve, and store the multiplicity of life forms. This is the first representation of a common world: simplified, of course, but common; ethnocentric, of course, but common; objectivizing, of course, but common.[80]

As this book argues, this global invention equally necessitated other instruments for its construction and delineation: media artefacts that both imaged and imagined, produced and reproduced the world. However contentious in their simplified, ethnocentric and objectifying dimensions, these artefacts may offer useful resources, or at the very least they ought to be thoroughly assessed as we confront the urgent need to build a common world.

The worlding examined in this book is therefore not one evenly distributed across the globe, but one that largely arises and survives in the West, with all the implications in terms of positionality and power relations that this location entails. There is no doubt that we need, now more than ever, different ways of conceptualising a shared planet, and Amerindian,

78 Cosgrove, *Apollo's Eye*, 265–266.

79 Bruno Latour, *Down to Earth: Politics in the New Climactic Regime*, trans. Catherine Porter (Cambridge: Polity Press, 2018), 101.

80 Ibid.

diasporic and non-Western cultures and communities can certainly help us with the task of imagining different worlds, or at least imagining this world differently.[81] But this would be a different study. For no book, not even, or perhaps especially, a book about the world, could claim to include worldviews from all parts of the world. Whereas it is my hope that the present work will spark different conversations and take media world-making in new directions, *Planetary Cinema* is primarily devoted to exploring the formation of a Western planetary consciousness in the second half of the nineteenth century and its significance and repercussions for our contemporary moment.

However, this is not to say that the texts and forms discussed in the following pages form a monolithic bloc. On the one hand, as I hope to show, nineteenth-century global visions and projects can be imperial, but not unilaterally. Here, I follow in the dialectical footsteps of Bruce Robbins, Jay Winter, Janine Marchessault, Tanya Agathocleous and Brooke Belisle, all of whom have recognised, in relation to different techniques and projects, that utopian energies can underpin conceptions of globality and sometimes uneasily co-exist with expansionist ideologies.[82] As Robbins summarises, any perspective from any angle is subject to contradictions and inconsistencies, meaning that in and of itself 'the global scale is not ethically and politically distinct from other, smaller scales'.[83] As I explore in chapter 1, Alexander von Humboldt's breathtaking holistic worldview, forged in the mid-nineteenth century, is remarkably useful for many of the planetary issues that afflict us now and cannot be reduced to an overdetermined imperial rhetoric. Marx's global utopianism, which was formulated as a response to the worldwide encroachment of Western capitalism, is similarly an essential framework to consider when exploring Vertov's *A Sixth Part of the World* and his dreams of a world cinema – one that has been given a new lease of life by Perry Bard's *The Global Remake*, as I discuss in chapter 3. As outlined in the same chapter,

81 On the contribution of Amerindian cultures to conceptions of worlding (and the end of the world), see Danowski and Viveiros de Castro, *The Ends of the World*, 61–78. See also Salma Monani and Joni Adamson (eds.) *Ecocriticism and Indigenous Studies: From Earth to Cosmos* (New York: Routledge, 2016); Davi Kopenawa Yanomami and Bruce Albert, *The Falling Sky: Words of a Yanomami Shaman*, trans. Nicholas Elliott and Alison Dundy (Cambridge, MA, and London: Harvard University Press, 2013).

82 Agathocleous, *Urban Realism and the Cosmopolitan Imagination*; Bruce Richards, *Feeling Global: Internationalism in Distress* (New York and London: New York University Press, 1999), 7; Marchessault, *Ecstatic Worlds*; Belisle, 'Whole World Within Reach', 130; Jay Winter, *Dreams of Peace and Freedom: Utopian Moments in the 20th Century* (New Haven, CT, and London: Yale University Press, 2006).

83 Robbins, *Feeling Global*, 5.

a thinker such as Kracauer, not unlike Benjamin, championed the need to reflect on the affects and effects of global media culture through a direct confrontation with, rather than a dismissal of, their sensory-perceptual reorganisation of the world.

On the other hand, many recent audiovisual works examined in the following chapters eschew the traps of what Martin Roberts defines as 'coffee-table globalism' by engaging critically with the treacherous proclivities of the global.[84] Natalie Bookchin's multichannel *Mass Ornament* (chapter 3), for example, is not only made up of amateur videos she sources from the Internet but is itself an astute reflection on a world of proliferating images. The oeuvre of the Austrian filmmaker Nikolaus Geyrhalter is also replete with attempts to document the exclusionary forces of globalisation and the destruction of the Earth, as exemplified by his *Earth* (chapter 1) and *Homo Sapiens* (chapter 6). Similarly, The Otolith Group's *Medium Earth* (chapter 1) and Eduardo Williams's *The Human Surge* (chapter 5) ponder over earthly forces that both underpin and exceed humanity in the context of an unequal Anthropocene and an unequal globalisation. Taken together, these works affirm the existence of an interconnected world that includes the human and the nonhuman, while never succumbing to a glossy, homogenising globalism. They likewise eschew myopic localism and point to the necessity of a visualisation of the planet as a singular whole that necessitates attention and care for the future.

This book is therefore both critical and affirmative. Without overlooking the problematic histories of world visions, it also includes works and projects that advance generative and transformative world thinking. *Planetary Cinema*'s main goal is to examine, historicise and theorise the role of cinema and related media in both shaping and responding to a planetary consciousness. It emerges from the belief that we must confront figures and figurations of the planetary in all their complexities and intricacies so that we can better understand our world in order to change it.

The Chapters

The undergirding concept of chapter 1, 'Sublime Earth', is the sublime, which is explored with recourse to Alexander von Humboldt's planetary philosophy as outlined in his magnum opus *Cosmos: Sketch of a Physical Description*

84 Martin Roberts, '*Baraka*: World Cinema and the Global Culture Industry', *Cinema Journal* 37:3 (1998): 62–82; 66.

of the Earth (1845–1862). While tracing the connections between a sublime aesthetics and the panorama, the chapter pays particular attention to its sister medium, the georama, including a consideration of its ambivalences in the context of imperialism, globalisation and an awakened geological consciousness in the second part of the nineteenth century. These considerations provide the conceptual substratum for the chapter's contemporary focus, which first turns to the IMAX Earth films *Blue Planet* (1990) and *A Beautiful Planet* (2016) as panoramic visions that conflate the natural and the technological sublimes into a distanced and distancing view of the planet. The last two sections of the chapter look at two films that counter such a view through a visualising of the Earth as ground and soil: whereas *Medium Earth* (2013) reinstitutes the sublime geological stirrings of the Earth's strata through a focus on the cracks and fissures on the planetary surface, the documentary *Earth* (2019) turns its camera to extractive activities carving out holes and vaults in the inner structure of our planet.

Wonder is the organising philosophical idea connecting two different eras in chapter 2, 'The Unseen World Across the World'. The concept is first explored in relation to natural-history visual culture in Victorian Britain and then two early-cinema strands that revitalised that culture: the popular science genre, as seen in the microscopic and time-lapse visualisations of organic phenomena masterminded by Martin Duncan and Percy Smith; and the global expedition genre, as seen in Cherry Kearton's quest to capture unperturbed wildlife across the world. I argue that these first forays into natural history filmmaking are crucially relevant for an understanding of current BBC Earth series, such as *Frozen Planet* (2011) and *Blue Planet* (2017). These series enact a technological rekindling of a never-before-seen aesthetic that paradoxically necessitates the idea that some things may never be seen again due to the speed at which the environmental crisis advances. I suggest that this paradox can be understood as a struggle on these series' part to keep their foundations intact within a natural history paradigm in the age of the Anthropocene, according to which a distinction between human and natural histories is no longer tenable.

Cinema's capacious ability to record and catalogue anything and everything is the focus of chapter 3, 'The Universal Equality of Things'. Adopting Kracauer's reflections on photography and cinema's rapport with 'things' as a methodological guide, the chapter is divided into three sections. The first looks at around-the-world stereocard boxes, as well as the early-cinema catalogues of UK-based entrepreneur Charles Urban, as foundational and rudimentary efforts to inventory the staggering wealth of subjects and objects captured by photography and cinema. The second section turns to what I term the 'world symphony' genre in a reference to the city symphony film, looking specifically at Vertov's *A Sixth Part of the World* (1926) and Ruttmann's *Melody of the World* (1929) as comparable montage efforts to rearrange the catalogue into a more tightly integrated whole. The last section looks at contemporary web-based and -sourced projects dealing with the avalanche of images flooding the Internet, including Perry Bard's *The Global Remake: The Man with a Movie Camera* (2007–2014), Natalie Bookchin's *Mass Ornament* (2009), and the YouTube documentary *Life in a Day* (2011). Despite their divergences, these projects are all concerned with assuaging the loss of meaning identified with the contingent; and in this context, I conclude, the parodic global symphony *In Praise of Nothing* (2017) gains in significance due to its stated if ultimately unsuccessful attempt to embrace the no-thingness of the world.

Chapter 4, 'The Face of the World', explores the trope of faciality in relation to extinction discourse. The chapter first examines the advent of photography and the way it was recruited to the typological project of classifying the faces of humanity according to nineteenth-century biological theories of race. I show how the head-on facial portraiture, ranging from carte de visite collections through to anthropometric photography, lent visual concreteness to the idea that supposedly inferior races were to vanish thanks to their presumed inability to catch up with 'civilisation' and 'progress'. I then examine the longevity of this trope by looking at two recent documentaries, *Samsara* (2011) and *Human* (2015), that give a new lease of life to the facing face as the global signifier of a soon-to-be perishing human life. Although both films try to distance themselves from evolutionary human hierarchies, their programmatic adherence to the codes of nineteenth-century still portraiture, I suggest, reveals the persistence of colonial modes of looking in contemporary galleries of humanity.

In chapter 5, 'A Networked Humanity', the discourse of film as universal language is investigated in relation to the network narrative. I contend that D. W. Griffith's infamous quest to transmute such a discourse into a quadripartite textual structure in *Intolerance* (1916) constitutes an illustrious precursor of the global network narrative genre. This is not only because of its world-historical mixing of four autonomous stories but also because of its contemporary narrative, in which the networks of modernity – roads, railways and telephone signals – prove crucial for the film's happy ending. This analysis is substantiated by my reading of Alejandro González Iñárritu's *Babel* (2006), a film that bears remarkable resemblance to *Intolerance* in its unashamed belief in the idea of cinema as a universal language – one that is correspondingly supported by a multinarrative textual design and visualised through a focus on corporeal suffering as the lingua franca of a helpless humanity. The chapter closes with a consideration of Eduardo Williams's small-budget, roaming *The Human Surge* (2016), a film in which totalising ideas of humanity and globality are also mapped onto a networked narrative configuration. Here, however, miserabilism plays no part and digital online networks are not only debunked in terms of their universal pretences but they are also relativised in a wider system of nonhuman networks that materially constitute the Earth itself.

How lens-based media such as photography and film have confronted the task of recording disappearance, often via ruinous imagery, is the subject of the last chapter, 'A Disappearing Planet'. I first explore Albert Kahn's colossal Archives de la Planète (1909–1931), a multimedia project whose raison-d'être was the capturing of disappearing realities owing to a sweeping global modernisation. Kahn's Archive is the most eloquent example of the historical value accorded to indexical images as they promise the preservation of lives and lifeworlds for the future. Yet, as I also show, disappearance appears in another, equally significant guise in the Archives de la Planète, which contains not only one of the most exhaustive collections of warfare destruction on French soil but also films of natural calamities, thus cementing a conception of planetary time as both irreversible variability and unpredictable rupture. These considerations provide the backdrop for my analysis of contemporary works interested in archiving the planet for a nonhuman future. Looking in particular at Geyrhalter's hybrid fiction-documentary *Homo Sapiens* (2016) and Trevor Paglen's *The Last Pictures* (2012) project, I explore their attachment to the indexical image as the radicalisation of a world-archiving enterprise, sometimes as a way to sidestep the inexorable force of terrestrial time.

The films, projects and artefacts explored in this book testify to the uses and abuses to which cinema and related media have been put as they bring into view the lands, peoples, nonhuman animals and entities that make up this world. Often revelatory, sometimes shameful, these planetary visions remind us that there is beauty and horror in this world, but ultimately that we still have a world – and that it is our only world.

1 Sublime Earth

Abstract
The undergirding concept of this chapter is the sublime, which is first explored with recourse to Alexander von Humboldt's nineteenth-century planetary philosophy and in relation to the panoramic medium and variations. These considerations provide the conceptual bedrock for the chapter's contemporary focus, which explores the IMAX Earth films *Blue Planet* (1990) and *A Beautiful Planet* (2016) as panoramic visions that conflate the natural and technological sublimes into a distanced view of the planet. The last two sections look at films that visualise the Earth as ground and soil: whereas *Medium Earth* (2013) reinstitutes the sublime geological stirrings of the Earth's strata, the documentary *Earth* (2019) turns its camera to extractive activities carving out vaults in the planet's inner structure.

Keywords: sublime, panorama, Alexander von Humboldt, IMAX, *Medium Earth, Earth*

General views enlarge the bounds of our intellectual existence, and while we ourselves may be living in retirement they place us in communication with the whole globe.
Alexander von Humboldt

Though pictures of the blue planet had been taken in the preceding decades by lunar probes, satellites and rockets, it was the image of a half-shadowed Earth rising in the distance from a lunar landscape that first caused a stir. Taken by astronaut William Anders during the Apollo 8 mission in December 1968 and soon nicknamed Earthrise, its impact was not quite so fully anticipated. The official target of the mission was the moon, not the Earth, so much so that when the three astronauts – Anders, Frank Borman

de Luca, T. *Planetary Cinema. Film, Media and the Earth.* Amsterdam: Amsterdam University Press, 2022
DOI: 10.5117/9789463729628_CH01

and James Lovell – were back on terrestrial ground, they all recalled their astonishment at seeing the planet from its outer orbit. Borman noted: 'It was the only thing in space that had any color to it'. And Lovell mused: 'From out there, it really is "one world"'.[1] Several Apollo missions followed suit, all bringing back their Earth pictures. Yet it was not until the Apollo 17 mission in December 1972, the last of that space programme, that another photograph would cause as much of an impact. Taken by Harrison Schmitt, the Blue Marble showed a frontal view of the Earth, with no parts in shadow and the African continent highly prominent. It would become one of the most widely disseminated photographs in human history.

In his illuminating history of the Apollo pictures, Robert Poole has shown the contradictory meanings these photographs negotiated as they precipitated or were appropriated by a plethora of theories, publications, movements and events in the late 1960s and into the 1970s. Products of cold war nationalism on the one hand and an astrofuturist discourse on the other, the pictures turned public attention back to the Earth in line with a rising concern for the environment. Poole writes: 'An "eco-renaissance" took place during the Apollo years of 1968–72, framed almost exactly by the "Earthrise" and "Blue marble" photographs four years apart. These were the years of the legendary *Whole Earth Catalog*, Friends of the Earth, and Earth Day'.[2] Though the space programme was the apex of American technological expansionism, these Earth images also acted as a catalyst for holistic treatises, including Buckminster Fuller's 'Spaceship Earth' and James Lovelock's Gaia hypothesis, both indebted to systems thinking and cybernetic theory. As Lovelock recalled in 1979, the real gift of the space programme was that 'for the first time in human history we have had a chance to look at the Earth from space'.[3] His idea was that our planet is 'a single living entity', which he named Gaia with reference to the Greek goddess of the Earth: a self-regulating organism 'involving the Earth's biosphere, atmosphere, oceans and soil'.[4] Nearly a decade later, the sights of the Earth still reverberated. Thus, in his 1987 book *The Overview Effect*, Frank White noted that, thanks to the Apollo pictures, we can see 'that the Earth is a

1 Cited in Robert Poole, *Earthrise: How Man First Saw the Earth* (New Haven, CT: Yale University Press, 2008), 2.

2 Ibid., 13.

3 James Lovelock, *Gaia: A New Look at Life on Earth* (Oxford: Oxford University Press, 2000), 8.

4 Ibid., 10.

whole system and humanity one of many interdependent species calling the planet home'.[5]

Not everyone was enthusiastic about the overview, however. Already in 1958, Hannah Arendt had drawn attention to the 'uncomfortable military and political circumstances attending' the launch of the satellite Sputnik into space in 1957, 'an earth-born object made by man' that was admitted to the 'sublime company' of the 'heavenly bodies'.[6] For Arendt, this event epitomised a modern science that 'owes its great triumphs to having looked upon and treated earth-bound nature from a truly universal viewpoint, that is, from an Archimedean standpoint taken, wilfully and explicitly, outside the Earth'.[7] She concluded: 'Had the scientist [...] raised questions such as *What is the nature of man and what should be his stature?* [...] he would never have arrived where modern science stands today', for the answers to these questions 'would have acted as definitions and hence as limitations of his efforts'.[8]

Arendt's reflections on the stature of man must be seen alongside an array of shape-shifting perceptions of the human before the Earth and the cosmos, all of which acquired differences in scale and dimensionality depending on how and from where one looked. Sometimes these conflicting views were felt to be experienced at the same time, as when the astronaut Michael Collins recalled that from space the planet appeared like a 'tiny pea, pretty and blue', yet that did not make him feel like a 'giant' but, on the contrary, 'I felt very, very small'.[9] A few years later, the scientist Carl Sagan observed: 'The Apollo pictures of the whole Earth conveyed to multitudes something well known to astronomers: [...] humans are inconsequential'.[10] Yet, for Sagan, the 'frame-filling' Blue Marble still sent the wrong message. He thus came up with the idea of having another picture taken from the NASA deep space probe Voyager in 1990, this time 'from a hundred thousand times farther away' in order to 'help in the continuing process of revealing to ourselves our true circumstance and condition', namely that we inhabit

5 Frank White, *The Overview Effect: Space Exploration and Human Evolution* (Boston: Houghton Mifflin Company, 1987), 75.

6 Hannah Arendt, *The Human Condition* (Chicago and London: University of Chicago Press, 1998), 1.

7 Ibid., 11.

8 Hannah Arendt, 'Man's Conquest of Space', *The American Scholar* 32:4 (Autumn 1963): 527–540; 528, emphasis in original.

9 Cited in Denis Cosgrove, *Apollo's Eye: A Cartographic Genealogy of the Earth in the Western Imagination* (Baltimore, MD, and London: Johns Hopkins University Press, 2001), 260.

10 Carl Sagan, *Pale Blue Dot: A Vision of the Human Future in Space* (New York: Ballantine Books, 1994), 34.

a planet which is 'a mere point in a vast encompassing Cosmos', or a Pale Blue Dot, as the picture was to be named.[11]

The sublime is a useful philosophical framework for thinking about the scalar transfiguration of the Earth from the wholeness of the Blue Marble into a Pale Blue Dot and the dialectics between discourses of human mastery and insignificance that these photographs unleashed. Put to the service of self-aggrandising techno-imperialist discourses on the one hand, these pictures acted as proof of the diminutiveness of humanity on the planet within an intergalactic cosmos on the other. Sublimity often denotes an aesthetic experience that combines exhilaration and terror before something so overpowering or magnificent that the mind has difficulty grasping it. Centuries old though it is, the concept gained particular force in the late eighteenth century thanks to Edmund Burke and Immanuel Kant, both of whom reflected on natural objects before which the human imagination is disturbed and enlarged, including mountains, ocean storms and earthquakes. Yet, as an ever-evolving category, the sublime has acquired a myriad of conceptual guises through the twentieth and into the twenty-first centuries, including discourses of the 'technological sublime' invoked in the space race.[12] More recently, it has been central in environmental debates in relation to a 'toxic sublime' aesthetics through which human-modelled landscapes are framed as simultaneously awe-inspiring and terrifying.[13]

While engaging with these concepts, the aim of this chapter is to probe what I will term a 'planetary sublime' as tied to the idea and figuration of the whole world. As a celestial body of vast, inconceivable geo-cosmic magnitude, the Earth fits into traditional descriptions of the sublime object, as proved by the discourses and responses generated by the Apollo pictures. In line with the aims of this book, however, my goal is to counter the historical exclusivism commonly attributed to a planetary consciousness during the space age. Though the sublime sight of the whole Earth was unavailable in the nineteenth century, its imaginary was already widely inscribed in

11 Ibid., 34.

12 For a comprehensive account of new sublimes, see Temenuga Trifonova, 'Editor's Introduction', in Temenuga Trifonova (ed.) *Contemporary Visual Culture and the Sublime* (New York and London: Routledge, 2018), 1–41.

13 See, for example, Jennifer Peeples, 'Toxic Sublime: Imaging Contaminated Landscapes', *Environmental Communication: A Journal of Nature and Culture* 5:4 (2011): 373–392; T.R. Kover, 'Are the Oil Sands Sublime?: Edward Burtynsky and the Vicissitudes of the Sublime' in Robert Boschman and Mario Trono (eds.) *Found in Alberta: Environmental Themes for the Anthropocene* (Ontario: Wilfrid Laurier University Press, 2014), 125–148.

visual culture and thought in the form of horizon-expanding panoramas, giant globes and planetary treatises, including Alexander von Humboldt's breathtaking five-volume *Cosmos: Sketch of a Physical Description of the Universe* (1845–1862).

Remarkably, Humboldt formulated in *Cosmos* an all-enfolding, secular overview of the planet as a whole made up of intricately interdependent lifeworlds, prefiguring the Gaia hypothesis by more than 150 years ('Gäa' was one of the titles contemplated by him for his project).[14] He alerted his readers about the potential damages caused by unchecked progress, and his holistic thinking still shapes ecological ideas to this day. As his biographer Andrea Wulf argues: 'Humboldt gave us our concept of nature itself. The irony is that Humboldt's views have become so self-evident that we have largely forgotten the man behind them'.[15] Together with Laura Dassow Walls's passionately argued *The Passage to Cosmos* (2009), Wulf's *The Invention of Nature* (2014) forms part of a recent trend to rehabilitate Humboldt for our environmentally precarious times.[16] Once relegated to an 'imperial gaze' and dismissed as esoteric, Humboldt has re-emerged in these new accounts as someone who harboured deeply egalitarian, proto-ecological and anti-colonialist attitudes.[17]

This chapter modestly hopes to contribute to this Humboldtian revival by showing that his writings hold a crucial importance for the concept of the sublime when it comes to entwined perceptions of the natural world and the whole world. Humboldt's fate has been such that even in an otherwise illuminating recent book devoted to the environmental ethics of the romantic sublime, his name is never mentioned.[18] Inspired though he was by his compatriot Immanuel Kant, Humboldt's take on

14 Andrea Wulf, *The Invention of Nature: The Adventures of Alexander von Humboldt, the Lost Hero of Science* (London: John Murray, 2015), 9.

15 Ibid., 9.

16 New translations and reprints of Humboldt's writings have also been lauded for their importance for ecological thinking. See, for example, the introduction to Alexander von Humboldt and Aimé Bonpland, *Essay on the Geography of Plants*, ed. Stephen T. Jackson, trans. Sylvie Romanowski (Chicago and London: University of Chicago Press, 2009).

17 Alexander von Humboldt, *Cosmos: Sketch of a Physical Description of the Universe*, volume 1, ed. Edward Sabine (New York: Cambridge University Press, 2010 [1848]), 355–356. On Humboldt's 'imperial gaze', see Mary Louise Pratt, *Imperial Eyes: Travel Writing and Transculturation* (New York and London: Routledge, 2018), 109–140. On Humboldt's liberal and anti-imperialist views, see Wulf, *The Invention of Nature*, 3; Laura Dassow Walls, *The Passage to Cosmos: Alexander von Humboldt and the Shaping of America* (Chicago and London: University of Chicago Press, 2009), 19–22.

18 Emily Brady, *The Sublime in Modern Philosophy: Aesthetics, Ethics, and Nature* (New York: Cambridge University Press, 2013).

the sublime was also unique in its richly effusive description of nature and the cosmos. More importantly, unlike Kant, for whom sublimity was strictly identified with natural forces, Humboldt championed mediatised sublime experiences in the form of 360-degree panoramas, which held in his view the potential of inciting the love of nature and a communion with the whole.

In this chapter Humboldt will thus serve as a guide with whom we will traverse nineteenth-century global cultural history, and through to our time. We start with a consideration of his planetary ideas in relation to the concept of the sublime and then move on to examine how such ideas aesthetically intersected with contemporaneous mediums such as panoramas and georamas. As my analysis will show, although these forms were no doubt aligned with an imperial imaginary, they also acted as vessels for utopian ideas as a way to make sense of a planet that expanded not only geographically but also temporally into nonhuman trajectories. In other words, global panoramas emerged as uniquely suited mediums to negotiate the sublime mixture of awe and anxiety generated by processes of globalisation and geological formulations of the Earth.

We then travel forward in time so as to examine contemporary forms in which the panorama and the overview have been rekindled in relation to renewed conceptions of planetary sublimity in the age of the Anthropocene. My selection of case studies is purposefully diverse so as to demonstrate the pervasiveness of these concepts across a range of mediums, modes and genres. One of these is IMAX, a descendent of the georama that, as seen in films like *Blue Planet* (1990) and *A Beautiful Planet* (2016), can channel the technological sublime of the whole-planet view into a narrative that elevates humanity as a telluric force. The last sections turn to the film essay *Medium Earth* (2013), by The Otolith Group, and the documentary *Earth* (*Erde*, 2019), by Nikolaus Geyrhalter. While invoking concepts and figures of the sublime, I argue that these films provide a corrective to sublime anthropocentrism by resisting or countering the supremacy of the view from above through an emphasis on the geological autonomy of the Earth from down below.

By employing the sublime as a lens with which to cut through a variety of literary, visual and audiovisual forms and texts across time, my hope is to show not only the continual transmutation of the concept as it reflects our ever-changing world but also its continuing relevance for a better understanding of our Earth-bound condition in times of environmental collapse.

Humboldt's Panoramas

On 23 June 1802, the remarkably erudite explorer Alexander von Humboldt (1769–1859) climbed the Chimborazo, situated in today's Ecuador and then believed to be the highest mountain in the world. That climb was the culmination of a three-year journey through Latin America on which Humboldt, then 33 years old, had embarked with a companion, the young French scientist Aimé Bonpland. Born in Berlin into an aristocratic Prussian family, Humboldt had since an early age demonstrated an insatiable curiosity towards all things, a love for nature and a longing to visit far-away lands. It was not, however, until the death of his stern mother in 1796 that he found himself able to act upon such a longing, and not before a number of failed attempts to sail from many ports in Europe, then convulsed by the Napoleonic wars. Financed entirely at his own expense, Humboldt's voyage finally came about thanks to the Spanish crown, which authorised his request to travel to its colonies.

That voyage would change the way Humboldt saw the world. It would also make the world see him. Upon his return to Paris in 1804, Humboldt devoted himself to publishing an account of his adventures and the findings of his travels through Spanish America. His mission was to spread natural and scientific knowledge, which he believed had the power to change the world for the better. Combining travel description, scientific rigour and aesthetic enjoyment of nature, these publications became part of *Voyage to the Equinoctial Regions of the New Continent*, a mammoth literary project that eventually came to encompass 34 volumes compiled over 20 years and including: *Essay on the Geography of Plants* (published in German and later in French, in 1807), *Views of Nature* (published in German and French in 1808), and the unfinished seven-volume *Personal Narrative of Travels to the Equinoctial Regions of the New Continent during the Years 1799–1804* (published in French from 1814 to 1831).[19]

Already in his *Essay on the Geography of Plants*, Humboldt had formulated his distinctive approach to nature not as a classificatory experiment content with grouping specimens morphologically, as with the popular form of natural history discussed in the next chapter, but rather as a deeply felt and textured mode of description that brought forth the sensuous dimension of the natural world in all its interlinked intricacies. Humboldt advocated for an all-embracing view of nature that was materialised in novel ways of conceiving it visually. For example, the centerpiece of *Essay on the Geography*

19 Wulf, *The Invention of Nature*, 505–507.

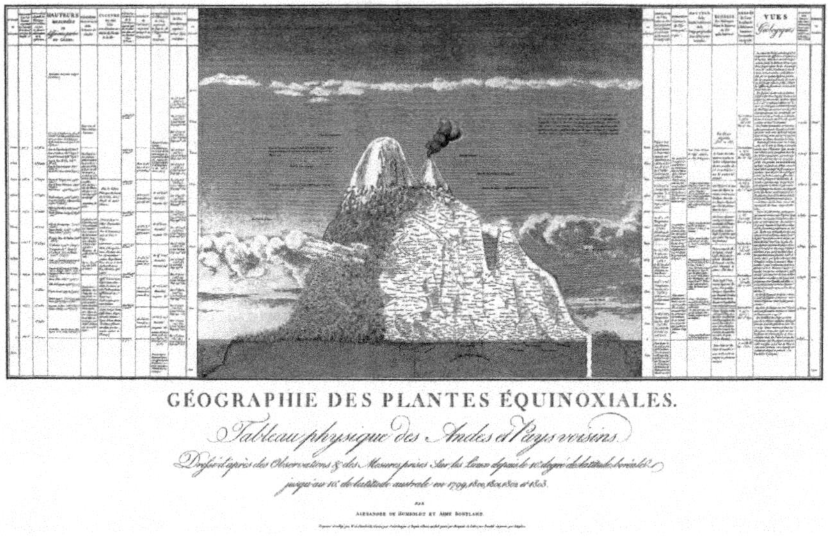

Figure 1.1 Humboldt's tableau evinces his vision of an interconnected Earth. Engraving from *Essay on the Geography of Plants* (1807).

of Plants was an engraving titled *Physical Tableau of the Andes and Neighbouring Countries* (figure 1.1). This depicted a cross-sectional profile of the Chimborazo, populated with the names of plants, flanked by columns containing information about air temperature, geological phenomena, barometric pressure and even the intensity of the sky's azure in relation to altitude and other regions on the planet. The aim, as Humboldt put it, was to connect 'together all the phenomena and productions on the surface of the earth'.[20]

As a network of interrelated phenomena, the Chimborazo tableau was the crystallisation of a *Naturgemälde*, 'an untranslatable German term that', as Wulf explains, 'can mean a "painting of nature" but which also implies a sense of unity or wholeness'.[21] For Humboldt, this sense of wholeness was especially felt when the world was seen from great heights, at the summit of mountains, as when, atop the Chimborazo, he saw 'the earth as one great living organism where everything was connected'.[22] It followed that humans

20 Humboldt and Bonpland, *Essay on the Geography of Plants*, 79.

21 Wulf, *The Invention of Nature*, 101.

22 Ibid., 2. See also Peter J. Bowler, 'Climb Chimborazo and See the World', *Science* (4 Oct 2002), available at https://science.sciencemag.org/content/298/5591/63 (accessed 20 August 2019); Bernard Debarbieux, 'The Various Figures of Mountains in Humboldt's Science and Rhetoric', *Cybergeo: European Journal of Geography* (2012), article 618, available at http://journals.openedition.org/cybergeo/25488 (accessed 20 August 2019).

should be careful not to destabilise this delicate web, as he observed in relation to the disastrous effects of deforestation in the colonial plantations at Lake Valencia in Venezuela as early as 1800, making him, as Wulf observes, 'the first scientist to talk about harmful human-induced climate change'.[23] As Humboldt would note decades later: 'Man can only act upon nature, and appropriate her forces to his use, by comprehending her laws, and knowing those forces in relative value and measure'.[24]

Although secular holism provided the aesthetic and conceptual substratum of Humboldt's philosophy of nature throughout his life, this philosophy only found its full-blown maturation in *Cosmos: Sketch of a Physical Description of the Universe*, his five-volume magnum opus published between 1845 and 1862. By that time, Humboldt had already acquired global fame as a writer, humanist, scientist, explorer and naturalist, which allowed him to count on a network of collaborators. Partly based on his Berlin lectures on physical geography (1827–1828), the first volume of *Cosmos* was an instant bestseller, with over 20,000 copies of the German edition sold in a couple of months, and translations in English, Dutch, Italian, French, Danish, Polish, Swedish, Spanish, Russian and Hungarian following suit.[25] Though it eventually sprawled into many volumes, *Cosmos* was initially conceived as a two-part project, with the first devoted to a description of the celestial system, the Earth's geology and global vegetation distribution, and the second to humanity's framing of the natural world by way of poetry, landscape painting and the sciences. As Humboldt saw it, his vision of the world was no 'mere encyclopedic aggregation' of phenomena but rather, a knowledge enterprise that considered 'partial facts only in relation to the whole'.[26]

As an effort to visually enlarge the world, *Cosmos* works on two interrelated levels. On the one hand, Humboldt conceptualises his holistic gaze through recourse to figures of altitude as a metaphorical strategy to connect descriptive and enumerative epistemologies. He writes: when 'the mind of man attempts to subject to itself the world of physical phenomena', he feels 'raised to a height from whence, as he glances round the far horizon, details disappear, and groups of masses are alone beheld [...] as by an effect of aerial perspective'.[27] This imagined suspended perspective allows 'different regions of the earth [to be] comprehended in one glance, and placed under

23 Wulf, *The Invention of Nature*, 6.

24 Humboldt, *Cosmos*, vol. 1, 37.

25 Wulf, *The Invention of Nature*, 290.

26 Humboldt, *Cosmos*, vol. 1, 41.

27 Ibid., 67.

the dominion of intellectual combination' – a 'wider horizon' that 'renders it possible to embrace, at one view, with greater clearness, a wider field of knowledge'.[28] On the other hand, these overviews are more than figures of speech. For, as Humboldt also notes, it is '[w]hen we behold a plain bounded by the horizon' that 'an impression of the existence of comprehensive and permanent laws governing the phenomena of the universe' is 'revealed to the mind'.[29] This 'presentiment of order and law' is 'derived from the contrast of the narrow limits of our being with that image of infinity, which every where reveals itself in the starry heavens, in the boundless plain, or in the indistinct horizon of the ocean'.[30]

However, Humboldt's cosmology was not restricted to the all-embracing view, whether in its phenomenological or figurative mode. Organised around three fields of enquiry, the first volume of *Cosmos* develops a treatise along the lines of a continuously differentiating and descending perspective on Earth that sees it from a cosmic distance, hovers over its terrestrial surface and burrows its deep strata:

> I propose to begin with the depths of space and the remotest nebulae, and thence gradually to descend through the starry region to which our solar system belongs, to the consideration of the terrestrial spheroid with its aerial and liquid coverings, its form, its temperature and magnetic tension, and the fullness of organic life expanding and moving over its surface under the vivifying influence of light.[31]

For Humboldt, then, it was not enough to see the planet from a distance. Scanning the life-support systems on its surface and digging deep into its structure was just as essential in order to gain insight into its history, since 'the globe on which we live also reveals the knowledge of an earlier state: the strata of sedimentary rocks, which compose a large portion of its crust, present to us earlier forms of organic life'.[32]

Although Humboldt's planetary topology was genuinely awe-inspiring, it did not, of course, hang in a vacuum, inspired as it was by ideas of Immanuel Kant and his close friend Johan Wolfgang von Goethe. From Kant, who was attempting to find a third way between rationalism and empiricism

28 Ibid., 18, 55, 31.
29 Ibid., 6.
30 Ibid., 6.
31 Ibid., 67.
32 Ibid., 58.

in his transcendental philosophy, Humboldt inherited both an interest in physical geography and the belief that human consciousness acted as a subjective filter in the perceptual apprehension of reality – or, as Humboldt evocatively put it, the idea that 'the external world and our ideas and feelings melt into each other'.[33] The 'Cosmos', as 'a presentiment of the true harmony and order of the universe', was ultimately a melding of the natural world with the human mind.[34] As Walls argues, Humboldt 'was forging something new, a truly Kantian synthesis, a dynamic concept of nature that would correct the excesses and blindnesses' of German idealism on the one hand and French positivism on the other.[35] At the same time, Goethe's rejection of a Cartesian, mechanistic view of the universe in favour of an approach that attended 'to the wholeness of nature, to the transformation of its parts, and to the aesthetic impression made by nature on human beings resonated deeply with Humboldt'.[36] And, like Humboldt, Goethe gave pride of place to the horizon as seen from great heights, describing his experience atop Strasbourg Cathedral thus: 'Such a fresh glance into a new land in which we are to abide for a time has still the peculiarity, both pleasant and foreboding, that the whole lies before us like an unwritten tablet'.[37]

When regarded from 'a higher point of view', as it were, the predilection for the horizon one finds in Humboldt and Goethe becomes less particular. As Stephan Oettermann has shown, the horizon constituted a '*see*-change' in visual perception and culture at the time, with the word denoting 'not so much a particular kind of line in art or mathematics as a sensory experience', especially in a German context.[38] Oettermann goes on: 'in their letters, memoirs, and literary works at the beginning of the nineteenth century, Germans in great numbers reported climbing towers'; then, when 'towers were not enough to satisfy the craving, people took to climbing mountains'.[39] And when the novelty of mountain climbing had worn off, the invention of hot-air balloons offered the promise of new thrills and higher sights.[40]

33 Ibid., 64.

34 Ibid., 4.

35 Walls, *The Passage to Cosmos*, 33.

36 Ibid., 31.

37 Cited in Stephan Oettermann, *The Panorama: History of a Mass Medium*, trans. Deborah Lucas Schneider (New York: Zone Books, 1997), 9.

38 Oettermann, *The Panorama*, 8, emphasis in original.

39 Ibid., 9, 11.

40 And yet when Humboldt climbed the Chimborazo at an altitude of 19,413 feet in 1802, he was still surpassing balloonists in Europe. See Wulf, *The Invention of Nature*, 100.

Such a craze for the horizon, for sights of immensity and boundless vistas, was both part and parcel of the prevailing concept of the sublime, according to which the experience of nature's vastness expanded and agitated the human imagination in its inability to grasp overwhelming phenomena. Although its conceptual lineage stretched back in time, the sublime received a boost with Edmund Burke's *A Philosophical Enquiry into the Origin of Our Ideas of the Sublime and Beautiful* in 1757, followed by Kant's *Observations on the Feeling of the Beautiful and Sublime* (1764–1765) and *Critique of Judgement* (1790). For both Burke and Kant, the sublime experience is attained provided the subject is at a safe distance. This sense of security acts as a spatial and/or temporal filter through which danger and anxiety can be partially circumvented and nature experienced in its sublimity, leading to moral discernment. Similarly, for both thinkers the power of the sublime is in inverse proportion to our knowledge of what causes it. Kant writes: to 'call the sight of the ocean sublime', we must get rid of 'all kinds of knowledge' that regard the ocean as a 'vast kingdom of aquatic creatures' in order to take it 'as poets do, merely by what strikes the eye', as 'an abyss threatening to overwhelm everything'.[41]

However, whereas for Burke the sublime engulfs the subject, as 'we shrink into the minuteness of our own nature and are in a sense annihilated', for Kant, the self emerges victorious, since we find, when confronted with the imaginative inadequacy to grasp phenomena too immense or powerful, the discovery in us of a 'faculty of resistance of a quite different kind, which gives us courage to measure ourselves against the apparent almightiness of nature', namely: reason.[42] In this sense, the Kantian sublime – whether incited by a vast natural sight (the mathematical sublime) or terrifying occurrences (the dynamic sublime) – exists in the mind of the observer, which feels, through reason, 'the proper sublimity of its destination, in comparison with nature itself'.[43]

Much has been made of the environmental implications of this aspect of Kant's sublime in its elevation of the rational self over nature, something I will return to and complicate later in this chapter. For now, let us note that Humboldt attempted to solve some of these impasses. He refuted the notion 'that our ignorance of natural things is the principal source of [...] the feeling

41 Immanuel Kant, *Critique of Judgement*, trans. J.H. Bernard, E-book (Digireads.com, 2012), loc. 2870; 2878.

42 Edmund Burke, *A Philosophical Enquiry into the Origin of Our Ideas of the Sublime and the Beautiful* (Mineola, NY: Dover Publications, 2008), 49; Kant, *Critique of Judgement*, loc. 2700.

43 Kant, *Critique of Judgement*, loc. 2710

of the sublime', and likewise did not qualify it against the superiority or inferiority of the rational self.[44] Instead, for Humboldt, 'the study of general natural knowledge' could not only intensify sublimity but also make us 'enter into a more intimate communion with the external world', as a result of which the mind and the whole harmoniously commingled rather than wrestled with each other.[45]

Whether or not the Irishman Robert Baker was familiar with these ideas, the sublime is a remarkably suitable framework for examining his popular invention. Initially baptised as 'La Nature à coup d'oeil' (French for 'nature at a glance') but soon changing to 'panorama' (Greek for 'see all'), Baker's creation opened in 1794 in London's entertainment district, Leicester Square. Indebted to a landscape visual tradition, the panorama was a large-scale, 360-degree frameless painted surface placed inside a dimly lit rotunda, with natural light let in through an upper aperture. The cylindrical painting was contemplated from the vantage point of a centralised and elevated platform. The content varied from military battles and imperial imagery with propagandistic purposes (particularly after 1815) to 'views' of far-flung places that offered a surrogate for travel. These occasionally did include natural phenomena, such as rivers, waterfalls and mountain summits, but cities were the dominant subject, with visitors able to 'travel' to Constantinople (1803), Rome (1804), Grand Cairo (1810), Messina (1812) and Paris (1814) for a modicum of money.

Unlike landscape painting, however, the panorama abolished the boundaries of the frame and a single-perspective system, thereby enveloping the observer while implying a world that extended beyond one's visual grasp. This was achieved through a painstaking reproduction of cityscapes beyond which the line of horizon was highlighted: the intent was to make the spectator doubt for a second his or her actual location. As one booklet for a New York–based panorama of Paris put it: 'The picture is simply and exactly the reflex of Nature itself. The illusion of spreading landscape and outreaching vista is as perfect as if the spectator were standing in Trinity steeple, gazing out upon the undulating scenery of New Jersey'.[46] By conflating the actual location in which the panorama was housed with the one depicted in the painted surface, this remark evinces the form's discursive reliance on the globally transporting powers of its realism. As Erkki Huhtamo summarises:

44 Humboldt, *Cosmos*, vol. 1, 20.

45 Ibid., 36.

46 *The Great Siege of Paris 1870–71*, promotional booklet, held at Bill Douglas Cinema Museum, University of Exeter. Item number: 12877.

'Although it was not wired in the sense of broadcasting or the Internet, [the panorama] was capable of teleporting its audience to another location, and dissolving the boundary between local existence and global vision'.[47]

For some, such as the English poet William Wordsworth, such illusionism was condemnable, amounting to commodified, alienating 'imitations' of the 'absolute presence of reality'.[48] Others, such as Humboldt, praised the panorama on the basis of its democratic potential to offer sublimity and natural knowledge for all. As he saw it, the experience 'in Barker's panoramas, by the aid of Prevost and Daguerre, [may] be converted into a kind of substitute for wanderings in various climates', since 'in a panorama, the spectator, enclosed in a magic circle and withdrawn from all disturbing realities, may the more readily imagine himself surrounded on all sides by nature in another clime'.[49] While regretting that panoramas had chiefly favoured cities as their visual content, Humboldt expressed his wish for scenic vistas of 'wild luxuriance and beauty', even better if based on photographs for accuracy, in order to 'enhance the love of the study of nature'.[50] He concluded:

> if large panoramic buildings, containing a succession of such landscapes, belonging to different geographical latitudes and different zones of elevation, were erected in our cities, and [...] thrown freely open to the people, it would be a powerful means of rendering the sublime grandeur of the creation more widely known and felt. The comprehension of a natural whole, the feeling of the unity and harmony of the Cosmos, will become at once more vivid and more generally diffused.[51]

Thanks to its immersive properties, the panorama acted for Humboldt not only as a way to condense the global within the local, but also as a sublime site where one could sense the dynamic unity of the whole, much like a *Naturgemälde*. Unlike Kant and Burke, then, Humboldt believed that the sublime could be achieved through *mediation* as part of a utopian project to make the invigorating experience 'freely open to the people'.

47 Erkki Huhtamo, *Illusions in Motion: Media Archaeology of the Moving Panorama and Related Spectacles* (Cambridge, MA, and London: MIT Press, 2013), 5.

48 Cited in Tanya Agathocleous, 'Wordsworth at the Panoramas: The Sublime Spectacle of the World', *Genre* 36:3–4 (2003): 295–316.

49 Alexander von Humboldt, *Cosmos: Sketch of a Physical Description of the Universe*, volume 2, ed. Edward Sabine (New York: Cambridge University Press, 2010 [1848]), 90.

50 Ibid., 91.

51 Ibid.

To examine the panoramic experience with the conceptual aid of the sublime enables a more nuanced account of the ambivalences and contradictions the form negotiated as one of the first mass mediums of the nineteenth century. Prevailing accounts often stress its Foucauldian, power-inflected mode of vision as enabled by the elevated platform, which presumably gave spectators the potent, omniscient feeling of surveying the landscape from a safe distance, not unlike a military commander and in line with Jeremy Bentham's contemporaneous idea of the panopticon as a surveillance technology.[52] According to this perspective, the panorama fostered a sense of ocular mastery in tune with an 'imperial visuality' that appropriated landscapes for disciplinary, military and regulatory purposes.[53] Yet, as contemporaneous descriptions also tell us, the gigantic size and circular format of the panorama overwhelmed spectators and divested them of control, producing disorienting sensory experiences and even unpleasant physical sensations such as nausea and vertigo.[54] Indeed, as William Galperin notes, the panorama denied the 'stable and controlling subject-position' identified with classical perspective, which complicates a one-sided conception of the form in relation to architectures of visual power and demands instead that we see it in terms of an experience that was *both* empowering and disorienting.[55]

It seems to me that Tanya Agathocleous has advanced a more nuanced understanding of the panorama when she notes that it 'was a deeply ambivalent formation, complex enough to be read as both conservative and progressive; adaptable enough to serve as imperial propaganda-tool or as a champion of cosmopolitan democracy', including (as with Humboldt) utopian discourses.[56] Promising the exacting reproduction of specific sites, the panorama simultaneously gestured towards the whole world through the horizon, engendering what Agathocleous terms an 'urban sublime' in the context of an acutely felt globalisation. She writes: 'Panoramas were global not only because they depicted a range of cities from around the world but

52 See, for example, Alison Griffiths, *Shivers Down Your Spine: Cinema, Museums, and the Immersive View* (New York: Columbia University Press, 2013), 77; Bernard Comment, *The Panorama* (London: Reaktion Books, 1999), 136; Nicholas Mirzoeff, *The Right to Look: A Counterhistory of Visuality* (Durham, NC, and London: Duke University Press, 2011), 17; Angela Miller, 'The Panorama, the Cinema, and the Emergence of the Spectacular', *Wide Angle* 18:2 (1996): 34–69; 35–36.

53 Mirzoeff, *The Right to Look*, 17.

54 Oettermann, *The Panorama*, 12.

55 William Galperin, *The Return of the Visible in British Romanticism* (Baltimore, MD: Johns Hopkins University Press, 1993), 40.

56 Agathocleous, 'Wordsworth at the Panoramas', 302.

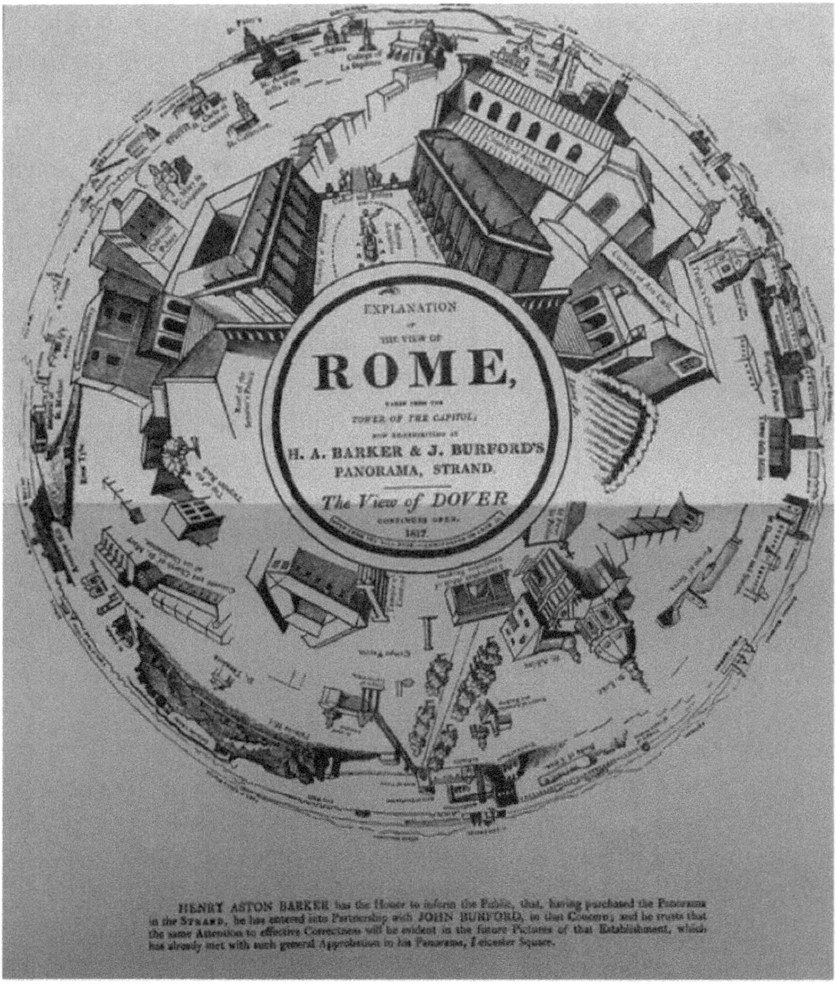

Figure 1.2 A panoramic key of Rome evinces the global aspirations of the medium. Courtesy of the Bill Douglas Cinema Museum, University of Exeter.

because, in their emphasis on the infinite and all-encompassing extension of the horizon, they situated the landscapes in the world as a whole'.[57]

This situating of the local within the global was literalised in the panoramic keys distributed to patrons, and their transformation across the first decades of the panorama is worth noting. Initially these keys were drawn with an anamorphic technique that translated the 360-degree format into a two-dimensional circular figure around which monuments and sights were arranged, and within which the centrality of the spectator

57 Ibid., 305.

was implicitly or explicitly demarcated. Gradually, however, these circular drawings acquired three-dimensional, global contours (figure 1.2). Looking at Baker's circular keys from 1794 to 1816, Denise Blake Oleksijczuk notes how the 'design shifts from locating the spectator within a flat, immersive landscape that ends at the limit of the horizon to an elevated view at the pinnacle of the globe that extends the space encompassed by pushing the horizon back into the distance'.[58] By implying a global vastness that stretched into infinity, the panorama thus provided the figurative and phenomenological means through which to confront and make sense of a rapidly expanding world, offering sublime views of the whole ridden with ideological contradictions.

Humboldt's Globes

If the panorama implied a boundless world, some of its variations, such as the georama, depicted the entire world as a bounded globe. Originally created by Charles Delanglard, who submitted his project for approval at the Geographical Society of Paris in 1822, the georama first opened its doors in 1825 at the Boulevards des Capucines. Modelled though it was on the panorama, the georama introduced important changes. The spectator walked into a spherical rather than cylindrical rotunda: a hollow globe measuring 120 feet in circumference and around 40–50 feet in diameter. Rather than a painted landscape, the attraction displayed a world map, with two circular stairs spiralling their way up from the Antarctic to the Arctic and three viewing platforms in between. While the landmasses were made of opaque paper, the water was of a transparent material that reacted to the natural light outside. In some respects, the georama was the enlarged version of globes used as geographical instruments, yet here not only was the visitor inside the sphere but the map of the world was 'outside in', that is, painted on a concave rather than convex surface and available to be contemplated from down below. As Jean-Marc Besse notes, this inversion was championed by Delanglard as the project's main originality, since neither the 'world map nor the convex sphere allow for the entire surface to be embraced at one

58 Denise Blake Oleksijczuk, *The First Panoramas: Visions of British Imperialism* (Minneapolis: University of Minnesota Press, 2011), 128. From 1815, the anamorphic drawing was replaced by horizontal prints that sacrificed circularity for perspectival illusionism, which Oleksijczuk attributes to a desire to facilitate engagement with the visual content, especially as these began to serve nationalism.

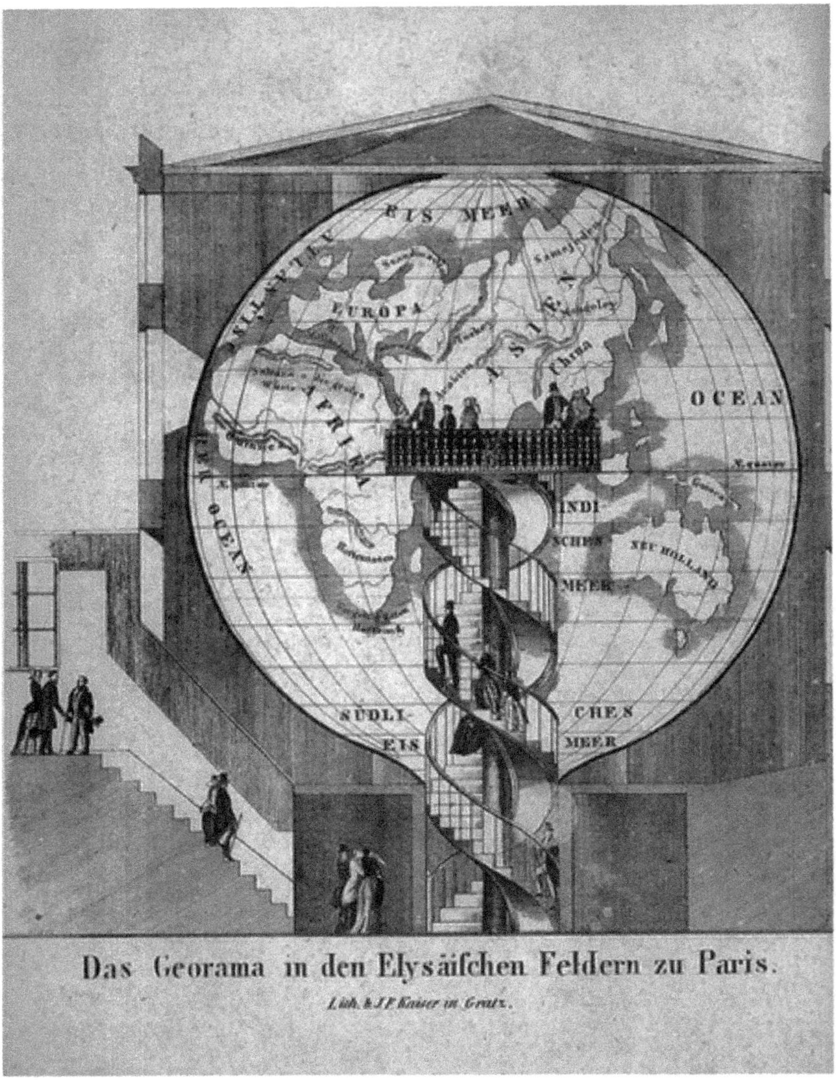

Figure 1.3 A cross-section of Charles-Auguste Guérin's georama, which opened its doors in 1844.

single glance'.[59] In response to this conundrum, the georama attempted to forge 'a global, and so to speak immediate, view of the totality of the surface of the Earth', allowing one 'to directly see the whole'.[60]

59 Jean-Marc Besse, *Face au monde: Atlas, jardins, géoramas* (Paris: Desclée de Brouwer, 2003), 199. My description of the georama is indebted to Besse's work.
60 Ibid., 214.

Himself a member of the Geographical Society of Paris, where he lived at the time, Humboldt reportedly expressed his admiration for the attraction. In the words of another society member, he had thought 'ingenious' the idea of placing 'the spectator in a situation where he could embrace in one single glance all countries in the globe and the immensity of seas, and bring closer and compare the different countries by a sort of immediate intuition'.[61] Delanglard's attraction was relatively short-lived, closing its doors in 1833. Yet it proved inspirational, serving as the model for another georama designed by Charles-Auguste Guérin in Paris (1844), which measured around 105 feet in circumference and contained one spiral staircase leading on to a single platform situated at the height of the Equator (figure 1.3), and a few years later, James Wyld's famous Great Globe (1851) in London.

Wyld was the son of a founding member of the Royal Geographical Society, from whom he inherited a map-making firm in 1836, and became Royal Geographer to Queen Victoria and Prince Albert in 1839. Inspired by the news of the first Great Exhibition of the Works of Industry of All Nations, or the Great Exhibition, he came up with the idea of erecting his own georama. The initial plan, rejected by the Royal Commissioners, was to house the Globe inside the Crystal Palace, the glass building in Hyde Park where the worldly goods and inventions were to be displayed. Eventually Wyld rented a spot in Leicester Square, finding support in the district's venue proprietors, who welcomed the addition of a more instructive attraction in its seedy quarters. The idea was for the opening of the Globe to coincide with the Great Exhibition in order to capitalise on the number of visitors expected to flock to the capital.[62]

To the surprise of many, and after a few incidents including a fire, the giant sphere, officially titled Wyld's Model of the Earth, was completed on time thanks to round-the-clock construction work. It opened in June 1851 and swiftly became the second most-visited attraction in London, behind only the Great Exhibition itself. More impressive than previous georamas, the Globe was painted blue on the outside, with silver stars, and measured 188 feet in circumference and 60 feet in diameter. Lectures were delivered every hour, with Wyld's maps, atlases and globes sold in the circular corridor around the sphere, which visitors had to traverse. Inside, a series of staircases led to four ascending platforms (figure 1.4). The concave map

61 A. Jullien, cited in Besse, *Face au monde*, 203.

62 This description is indebted to Bernard Lightman, 'Spectacle in Leicester Square: James Wyld's Great Globe' in Joe Kember, John Plunkett and Jill A. Sullivan (eds.) *Popular Exhibitions, Science and Showmanship, 1840–1910* (London: Pickering & Chatto), 19–39.

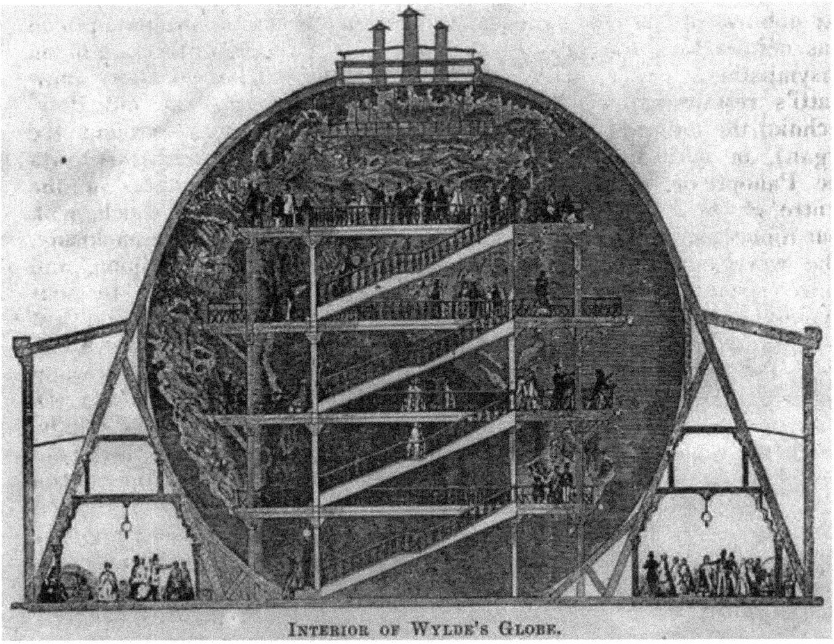

Figure 1.4 A cross-section of James Wyld's Globe (1851), printed in *The Illustrated London News*, 7 June 1851.

was three-dimensional, made up of plaster casts and geographical features differentiated by colours and materials, as Rick Altick describes: tufts of cotton wool simulated the eruption of volcanos with 'red-painted peaks; snow-covered mountains were represented by roughly modeled masses of a glittering white substance; deserts were painted in a tawny color, the oceans blue, the fertile areas green'.[63]

In the Globe's opening year, patrons also received a substantial booklet titled *Notes to Accompany Mr Wyld's Model of the Earth*, comprising an introduction signed by Wyld and four chapters. Such a comprehensive textual accompaniment was not unusual. As Ralph O'Connor notes, 'texts were integral to early-nineteenth-century visual culture, and not just pretexts for it', with guidebooks providing directives for the viewer's phenomenological experience and the intellectual framework within which attractions were to be understood.[64] As far as the Globe's brochure is concerned, it firmly

63 Richard D. Altick, *The Shows of London* (Cambridge, MA and London: Belknap Press, 1978), 465.

64 Ralph O. Connor, *The Earth on Show: Fossils and the Poetics of Popular Science, 1802–1856* (Chicago and London: University of Chicago Press, 2007), 287.

subscribed to imperial ideologies according to which the impressive size of the world as contemplated inside the sphere was lauded in proportionate measure to the achievements of the British Empire. Paul Young has shown that the Great Exhibition served to promote Britain's 'expansionist drive and to justify it to a metropolitan audience' in terms of a 'Victorian imperial mission'.[65] A similar discursive impetus can be observed in the accompanying notes, in which the Globe is positioned as an opportunity to celebrate the 'political and moral preponderance of the Germanic race', and to help 'judge accurately [...] the extent of our progress' as 'the chief seat of a race to which the largest portion of the earth's surface – perhaps, also, the largest portion of the earth's population – belong'.[66]

Nevertheless, like panoramas, georamas accommodated conflicting discourses, including utopian and educational ones. A case in point is the unrealised globe conceived by the anarchist geographer Élisée Reclus for the Paris 1900 exhibition. In a paper titled 'A Great Globe' presented at the Royal Geographical Society in 1898, Reclus stressed the need for spherical renditions of the Earth 'where every man will find himself at home', without 'any privilege for race or nationality'.[67] As globalisation was made felt from the newspaper 'that brings us news from all parts of the world' to the fact everyone now has 'friends across both oceans', 'the moment has come for us to have grand representations of our common home'.[68] Reclus's reference to globalising processes is significant, for in some ways the Great Globe built decades earlier served to mirror and assuage anxieties identified with such processes. As Young notes, although nineteenth-century globalisation generated a sense of confidence in boosting imperial aspirations, it also constituted a disorienting experience in light of 'the overwhelming nature of the expansive imperatives and economic transformations that characterized the period'.[69] The profusion of global projects and figures at the time, from universal expositions through to walk-in allegorical globes that accommodated 'contradictory economic, social, cultural and philosophical agendas', promised in this context to alleviate anxieties by making the world 'whole again'.[70]

65 Paul Young, *Globalization and the Great Exhibition: The Victorian New World Order* (Basingstoke, UK: Palgrave, 2009), 9–10.

66 *Notes to Accompany Mr Wyld's Model of the Earth* (London: Blackburn & Burt, 1851), xvii, xviii, xii. Item held at The British Library.

67 Élisée Reclus, 'A Great Globe', *The Geographical Journal* 12:4 (1898): 401–406; 406.

68 Reclus, 'A Great Globe', 406.

69 Young, *Globalization and the Great Exhibition*, 2.

70 Ibid., 6, 3.

Although Wyld's Globe was hailed, in the brochure, as revealing that 'artificial distinctions' between east and west and northern and southern hemispheres 'cease to be of importance', it was not invested with the same utopian energies as the one envisioned by Reclus.[71] But the same booklet does show the extent to which the Globe entertained educational ambitions in an avowed reference to Humboldt's planetary treatise, providing combined and scientific knowledge on varied aspects of physical geography, botany, geology and astronomy. As with Delanglard's georama, the Globe's vision was lauded on the basis of its superiority to maps and convex globes, since 'whatever may be the magnitude of the map or the globe over which the eye ranges, the earth's ball, or spherical shape, prevents the whole of its surface from being comprehended at a single glance'.[72] In response, the Globe depicted 'the whole extent, figure, magnitude, and multifarious features of the world we live in, as if it were one vast plain'.[73]

The rhetorical transmutation of the globe into a vast plain reiterates the panoramic roots of the georama, even though the latter offered no line of horizon stretching into infinity but, rather, a bounded world inside out. Yet, judging by contemporary accounts, georamas also elicited sensory experiences that posed a threat to the subject in line with panoramic sublimity. Take this 1851 article, titled 'A Journey Round the Globe', from the satirical *Punch* periodical:

> It is most agreeable to stand in the centre of the Earth, and to see yourself surrounded by oceans and continents, – first, to feast of a bit of land, and then to drink in with your eyes a whole Atlantic-full of water. Drink as much as you will, you cannot take all the water in. You dread lest the waters should close in around you, and swallow you up like a cork in the middle of a water-butt. You cling to the railings for support; but the sight of land cheers you the next moment. All the World is before you; you have only to choose where to go to.[74]

The references to feelings of 'dread' as the attraction threatens to 'close in' and 'swallow' patrons who then must 'cling to the railings for support', confirm that the georama was construed as an overwhelming experience before the physical immensity of the Earth. Spatial magnitude was then

71 *Model of the Earth*, 3.

72 Ibid., 1.

73 Ibid., 2.

74 'A Journey Round the Globe', *Punch, or the London Charivari* XXI (1851), 4.

given a temporal spin in the accompanying brochure, which defined itself as a holistic geography that 'embraces the whole natural history of the earth' and 'the substances of which it is composed, the changes it has undergone'.[75] 'The great Humboldt' was accordingly duly cited, whether to substantiate the claims advanced via recourse to his observations on topics as diverse as land formation and the internal structure of the Earth or, indeed, as a way to position the 'Great Model Globe' as a welcome and more up-to-date companion to *Cosmos*, by that time already a bestseller in its second edition.[76]

The influence of *Cosmos* on the Globe is also discernible in the four-chapter structure of the booklet, which emulates, with a few variations, Humboldt's roving planetary perspective, as summarised in the opening of the penultimate chapter:

> In the foregoing chapters, a general survey has been taken of the surface of the globe [...] as if existing apart from the rest of the universe; and then extending our views, we contemplated it as a part of that planetary system of which the great seat of light is the centre. From these exalted heights of nature, we must now come down to a more contracted scene – the interior of the earth itself.[77]

The Great Globe's guidebook therefore shows how Humboldt's ideas had travelled some way from his more specialised and scientific writings to become the conceptual and structural inspiration for such a popular attraction. In fact, after the Great Exhibition drew to a close, Humboldt himself became involved in a project that aimed to branch the Globe out into a sprawling educational museum, to be aptly named 'The Cosmos Institute'.[78] Though this never saw the light of the day, the Globe spectators, like *Cosmos* readers, were asked to conceive of the world in terms of a vision that was not content with scanning its planetary surface and which accordingly delved deep into its historical layers. Whereas the brochure emphasised that 'it is clear that the surface of the globe must be under a continuous process of change' and that 'the earth has been at various periods violently convulsed', 'the throwing out and heaping up of volcanic rocky matter' was

75 *Model of the Earth*, 1.
76 Ibid., xiii, 3.
77 Ibid., 19.
78 'The Cosmos Institute', *London Illustrated News* (1853), 180.

materialised in the three-dimensional rendering of the globe, in which, as noted, volcanoes were depicted in full activity.[79]

In a way, the idea of contemplating the Earth's surface as if from its interior conflated cosmic and geological visions of the planet. This was not lost on satirists at the time. The aforementioned *Punch* article thus noted, in relation 'to the peculiar formation of the interior' of the attraction:

> Mr. Wyld has made a grand discovery. He has satisfactorily proved that the interior of the Globe is not filled with gases, according to AGASSIZ; or with fire, according to BURNET; neither has he filled it, like FOURIER, with water, as if the Globe were nothing better than a globe of gold fish. No; Mr. WYLD has lately shown us that the interior of the Globe is occupied by immense strata of staircases![80]

The importance of allusion to geological concepts, theories, and practitioners such as Louis Agassiz, Charles Burnet and others, cannot be overestimated. It implies that some readers would have been familiar with those, while reminding us that, in addition to a globalisation, the nineteenth century witnessed a geologisation of the Earth. Charles Lyell's seminal *Principles of Geology*, originally published 1830–1833 and a founding geological text that influenced Darwin's evolutionary biology, had by that time proposed a stratigraphic reading of the Earth's deep time that exploded previous temporal parameters and chronologies.[81] In this context, as O'Connor has shown, 'the new science of geology was publicized in spectacular and theatrical forms which enabled it to gain the cultural authority it enjoys today'.[82]

While not a geologically themed attraction strictly speaking, the Globe nonetheless contributed to a geohistory that was 'far longer and stranger than the story of a literal six-day Creation'.[83] This it did by not only offering an outside view of the planet as if from its inside, but also in its textual and pictorial description of planetary phenomena in continuous variability and uncontrollable change, in line with contemporaneous geological theories and planetary treatises. The sublime experience it offered was therefore

79 *Model of the Earth*,20–21.

80 'A Journey Round the Globe', 4; emphasis in original.

81 For a recent critique of Lyell's geology on the basis of its racist presuppositions, see Kathryn Yusoff, *A Billion Black Anthropocenes or None* (Minneapolis: University of Minnesota Press, 2018), esp. 74–85.

82 O'Connor, *The Earth on Show*, 2.

83 Ibid., 3.

based on the idea of a world that expanded along both spatial and temporal axes: the *longue durée* of a turbulent planet against which human existence was sublimely inconsequential. To quote Walls: 'The wide, wide globe was becoming a deep, deep planet'.[84]

IMAX Whole-Earth

Brooke Belisle has shown that nineteenth-century walk-in spheres, by offering 'an embodied encounter with the spectacle of the globe', can be situated within a media archaeology of immersive world maps that finds contemporary incarnations in audiovisual artefacts as diverse as the VR-installation Reality Deck and the digital globe Google Earth.[85] As she notes: not unlike the georama, the 'concatenation of images sutured together to form the smooth globe of Google Earth [...] articulates an updated panoramic strategy of digital representation, a way in which parts are algorithmically made to cohere into the appearance of a "total view"'.[86] The visual appeal of Google Earth no doubt derives from this synecdochal interplay, one that allows viewers to spin a 3D planet and then zoom in to panoramic, 360-degree street views. For Ursula K. Heise, it is this 'ability to display the whole planet as well as the minute details of particular places in such a way that the user can zoom from one to the other and focus on different parts of information' that makes the geo-browser the vessel for what she terms a 'de-territorialised eco-cosmopolitanism'.[87] Perhaps Humboldt would have approved of Google Earth.

At the same time, the georama can be productively linked to giant-screen immersive experiences such as IMAX, which is often situated in media genealogies that include the panorama as an important precursor.[88] More

84 Laura Dassow Walls, 'Natural History in the Anthropocene' in John Parham and Louise Westling (eds.) *A Global History of Literature and the Environment* (Cambridge: Cambridge University Press, 2016), 187–200; 191.

85 Brooke Belisle, 'Nature at a Glance: Immersive Maps from Panoramic to Digital', *Early Popular Visual Culture* 13:4 (2015): 313–335; 317.

86 Ibid., 330.

87 Ursula K. Heise, *Sense of Place and Sense of Planet: The Environmental Imagination of the Global* (New York: Oxford University Press, 2009), 67.

88 See, for example, Charles R. Acland, 'Imax Technology and the Tourist Gaze', *Cultural Studies* 12:3 (1998): 429–445; Oliver Grau, *Virtual Art: From Illusion to Immersion*, trans. Gloria Custance (Cambridge, MA, and London: MIT Press, 2003); Martin Roberts, 'Baraka: World Cinema and the Global Culture Industry', *Cinema Journal* 37: 3 (1998): 62–82; Griffiths, *Shivers Down Your Spine*, esp. 37–113.

specifically, in this section I want to argue that a comparison between nineteenth-century whole-world spectacles and IMAX makes sense when we look at the prevalence of what I will term the IMAX Earth film. As seen in *Blue Planet* (1990) and *A Beautiful Planet* (2016), the genre has been consistent through IMAX's history and gives pride of place to images of the Earth from space from within a discourse of ecological sustainability. Whereas sublimity was built into global panoramas and georamas as a means to negotiate globalising processes, imperial expansionism and the world's geohistory, in these films the sublime overview is often recruited as the means by which to grapple with the planetary scale of anthropogenic changes. In so doing, I suggest, *Blue Planet* and *A Beautiful Planet* participate in a sublime discourse that conflates its technological and natural variations into a self-aggrandising idea of human sublimity.

Conceived and developed in Canada between two world exhibitions (Expo 67 Montreal and Expo 70 Osaka) by Graeme Ferguson, Roman Kroitor, Robert Kerr and William C. Shaw, IMAX successfully managed to secure a place outside universal fairs thanks to, among other things, partnerships with educational museums and theme parks. At the time of writing, IMAX owns more than 1,000 purpose-built theatres in 66 countries, most of which are equipped with a giant screen typically measuring six to eight stories in height.[89] Unlike the wraparound panorama, the IMAX screen is vertical and frontal, though it similarly aims to overwhelm the viewer's senses through image size, surround stereo system and, frequently, 3D technologies. In particular, variations such as the IMAX Dome, with 180-degree half-spherical domed screens below which spectators are trapped in inclined seats, especially recall the georama in its promise of extreme immersion.

The largeness of the high-definition IMAX image was originally achieved by running 15/70 mm film format through the camera horizontally rather than vertically, requiring films to be shot with purposefully designed cameras if they were to be screened at IMAX theatres. Since 2002, however, any film can be shown at IMAX venues when subjected to a Digital Media Remastering (DMR) enlargement process as part of what has been deemed a dramatic rebranding.[90] Nevertheless, since IMAX must still cater for science and natural history museum demographics, the educational genres

89 Griffiths, *Shivers Down Your Spine*, 95. See also: https://www.imax.com/content/corporate-information (accessed 2 August 17).

90 Beja Margitházi, 'See More, Think Big: The IMAX Brand Before and After the Digital Remastering' in Ágnes Petho (ed.) *Film in the Post-Media Age* (Newcastle upon Tyne: Cambridge Scholars, 2012), 143–158.

that made it popular, often organised around the idea of historical and/ or geographical travel, are still prevalent.[91] In these genres, the IMAX experience is often branded a superior form of vision both on a qualitative and quantitative level: the former through the 'unparalleled brightness' of its 'awe-inspiring' images and the latter via its claim to take 'spectators to inaccessible, far-off foreign places', which, as Oliver Grau notes, closely follows 'a pattern prefigured by the panorama'.[92]

That IMAX sees itself as a visual repository of the planet is confirmed by DVD box sets associated with its name, including the four-film *IMAX Earth* (2012) and the 10-film *Mother Earth IMAX* (2014), both comprising films that are each set in a different natural environment.[93] At the same time, the Earth itself constitutes an IMAX genre since its genesis, thereby evincing the planetary ambitions of this panoramic medium. This is exemplified by titles such as *Man Belongs to Earth* (1974), *Living Planet* (1979), *World Coaster* (1981), *Chronos* (1985), *Only the Earth* (1989), *Blue Planet* (1990), *The Secret Life on Earth* (1993), *SOS Planet* (2003), *Sacred Planet* (2004), and, more recently, *The Earth Wins* (2013) and *A Beautiful Planet* (2016). Granted, these films differ in important respects. Whereas some compile aerial shots of landscapes filmed across the globe in the form of a visual catalogue, others privilege images of the Earth from space in their textual constitution. In turn, the latter films resonate with the IMAX space genre produced in association with the Smithsonian's National Air and Space Museum.[94] Yet, differences aside, all of these films participate in some measure in environmental discourse, which is moreover a crucial staple of IMAX's brand image.

Today IMAX bears the logo 'Together for a Better Planet', with the corporation detailing on its website its involvement in global sustainability enterprises. These include the 'IMAX Big Picture' initiative, in partnership with the United Nations Environment Programme (UNEP), which involves a series of IMAX film screenings so as 'to leverage [IMAX's] legacy of immersive experiences and storytelling to change our environment and improve all

91 Tana Wollen, 'The Bigger the Better: From Cinemascope to IMAX' in Philip Hayward and Tana Wollen (eds.) *Future Visions: New Technologies of the Screen* (London: BFI, 1993), 10–30; 21–22.

92 See https://www.imax.com/content/imax-difference (accessed 3 March 2018); Grau, *Virtual Art*, 161.

93 While *IMAX Earth* contains four different films – *Ring of Fire* (volcanos), *Search for the Gray Sharks* (deep sea), *The Greatest Places* (deserts), and *Tropical Rainforest* – *Earth IMAX Mother* contains 10 individual titles, including *Africa*, *Alaska*, *Australia* and the *Great North*.

94 Allison Patricia Whitney, *The Eye of Daedalus: A History and Theory of IMAX Cinema* (PhD thesis, University of Chicago, 2005), 59. My thanks to Whitney for sharing her thesis with me.

life on earth'.[95] Whether the films themselves are deemed ecologically effective, however, is another story. Charles R. Acland, for example, has coined the term 'postcard environmentalism' to refer to what he examines as a discrepancy between the imposing presence and clarity of the IMAX image, with its 'vibrant colours and fine details', and its purported conservationist intentions.[96] Looking specifically at *The Living Sea* (1994), 'an IMAX documentary about the majesty and fragility of the world's oceans', Acland contends that it is 'notable less for its eco-friendly message than for the schools of multicoloured fish [which] appear to be literally brought before you'.[97]

Acland's remarks tacitly allude to the sublime experience that is frequently associated with IMAX. Indeed, as Allison P. Whitney has shown, in addition to the prevalence of sublime subject matter in the films – ocean depths, mountain ranges, outer space – the IMAX 'viewing experience' is touted as evoking 'the sublime in and of itself' through rhetorical recourse to the 'vastness' of the screen, the 'difficulty' of image capture and the 'brightness' of the image, all Burkean categories.[98] Yet IMAX sublimity must also be understood in relation to a 'technological sublime', one that is actualised both in the content of films celebrating technology (also a dominant IMAX genre) and the foregrounding of its own technology as 'imposing, powerful, and even frightening'.[99] Whitney notes: 'the fact that technological marvels are deemed an appropriate subject for IMAX films stems not only from the medium's capacity to represent sublime objects, but also to generate, in this case, an experience of the technological sublime'.[100]

If for Burke and Kant the sublime was to be found in natural forces, and its experience mingled terror with pleasure, the technological sublime often subtracts fear in what becomes a more straightforward experience of astonishment at humanmade achievements. According to David E. Nye, 'The technological sublime is an integral part of contemporary consciousness, and its emergence and exfoliation into several distinct forms during the past two centuries is inscribed within public life'.[101] Charting the history of this sublime mode in the US, as materialised in successive accomplishments such

95 See https://www.imax.com/content/corporate-social-responsibility (accessed 23 December 17).
96 Acland, 'Imax Technology', 429.
97 Ibid., 429–430.
98 Whitney, *The Eye of Daedalus*, 152.
99 Ibid., 176.
100 Ibid., 178.
101 David E. Nye, *American Technological Sublime* (Cambridge, MA: MIT Press, 1994), xiii.

as railroads, bridges and electricity, Nye singles out the 1969 Apollo mission as a milestone of technological sublimity in its combination of 'irresistible power, magnificence, complexity, and a journey into the infinite reaches of space', all combined with more than a dose of patriotism.[102]

Although Nye does not cite the pictures produced during the Apollo missions, such pictures, as we saw earlier, became the locus of conflicting visions of human power and insignificance on a shape-shifting planet floating against the infinity of space. IMAX often taps into these figures and discourses while highlighting its own technology as sublime in its Earth films, as I will now explore in relation to *Blue Planet* (1990). My choosing of this film is not arbitrary. Although IMAX films had featured images of the Earth before, such as *The Dream is Alive* (1985), *Blue Planet* is striking from today's perspective in that it prefigures Anthropocene visualities and narratives in significant ways. When placed beside the more recent *A Beautiful Planet* (2016), as we shall see, *Blue Planet* also demonstrates that an IMAX Earth aesthetics has consistently extolled the space overview as a supreme form of vision, which, paradoxically, proves myopic in its one-dimensionality.

Produced in association with NASA, *Blue Planet* was written, directed and narrated by Toni Myers, who has to this date made numerous space films for IMAX. The film opens with the oral recordings and footage of the moon landing from the Apollo 11, 12 and 16 missions, as Myers recounts how, as the astronauts looked back into space, 'the Earth stood out [as a] tiny oasis'. These visual documents are followed by the film's own image of the planet's curvature as the camera tilts down and the Earth gradually fills the screen, with Myers noting: 'to look at our Earth from the outside is to discover an entirely new planet'. From the outset, *Blue Planet* thus situates itself in relation to the discourses surrounding the Apollo images, promising an experience of visual revelation – the discovery 'of an entirely new planet' – that is indissoluble from space-viewing – 'our Earth from the outside'.

Granted, *Blue Planet* does not remain in space all the time but 'drops down to the surface of the Earth' every now and then, as the voiceover reiterates. For example, when the camera hovers over the African continent and 'a strange red lake' becomes discernible (figure 1.5), Myers notes the need 'to observe this environment more closely', which acts as the cue to a low aerial view of Lake Natron. Later, the formation of a hurricane from space gives way to ground-level amateur footage of storm destruction and then to another space view, now with no clouds over the Earth. Another sequence

102 Ibid., 246.

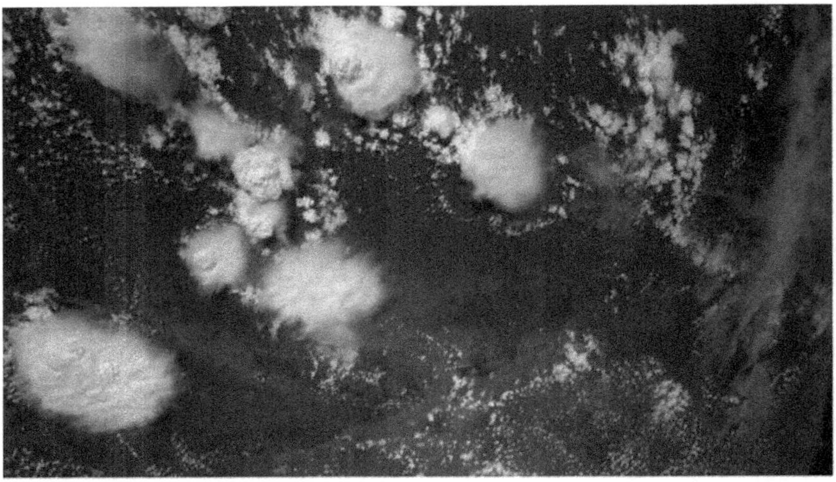

Figure 1.5 The supremacy of the view from above in IMAX *Blue Planet* (1990). Courtesy of IMAX.

takes the viewer, via montage, from rifts visible on the surface of the planet to lava chimneys at the bottom of the sea, then back to ground-level images of a volcano, and up into the Earth's orbit in order to show us that 'the smoke [can be seen] all the way from space'. While these sequences would thus seem to adopt a Humboldtian vision in its rootless exploration of the Earth, it is telling that they all end up in space so as to underline the supremacy of the view from above, which is a thematic, visual and structuring principle of the film.

Blue Planet is divided into two parts. The first focuses on the interconnectedness of the Earth's life-support systems and the unpredictable behaviour of 'our turbulent planet'. Speaking of the ice ages, Myers muses: 'We don't know why these local changes occur, but we do know that the Earth's climate as a whole has changed over much longer periods'. The Earth is further attributed agency through a focus on 'nature's powerful whims', such as earthquakes and thunderstorms. While the extended time devoted to these sublime phenomena seems at first justified by their thrill effect, these scenes serve the function of hinting at recent climate changes, which are explored in the film's second part. Here, *Blue Planet* shifts its tone by stressing how human action has become a powerful entity comparable to natural forces, though it is unclear whether the viewer should feel awe or terror at this momentous transfiguration. Myers notes, ambivalently: 'now a new force, as threatening as any in nature, has begun to change the Earth. *We* are that force. To our ancestors only a few years ago, the forests, oceans and skies seemed vast and almost limitless but all that has changed'.

Through recourse to sublime vocabulary and iconography, *Blue Planet* confirms its figuring of a 'human sublimity', one that may be correlated with a Kantian sublime as the human rises up over the natural world. Crucially, however, human superiority is here not the result of the self-willed mental exercise of reason over the mightiness of nature, as Kant maintained. Rather, the idea is that *humanity is now a sublime force in and of itself,* thus prefiguring a dominant Anthropocene narrative which, as Christophe Bonneuil and Jean-Baptiste Fressoz summarise, 'magnifies the irruption of human action as a telluric force'.[103]

Importantly, this sublime Anthropos enters into the field of visibility thanks to the sublimity of space technology. As Myers observes: 'It is only now we can see it from space that we realise the *magnitude* of what *we* are doing to the Earth', a statement that is followed by a shot of smoke hovering over the planet's surface. The cause of this smoke, however, is no longer the active volcano depicted earlier in the film but, rather, a forest in flames cleared for farming by humans. The voiceover reiterates: 'only from space can you see how much it is burning'. Meanwhile, satellites emblazoned with the US flag are praised for their ability to prevent or detect environmental changes, including the amount of energy we consume (illustrated in the film with composite imagery) and the hole in the ozone layer. Although *Blue Planet* never mentions the military and surveillance uses on the basis of which satellites were first built and launched into space, its focus on satellite monitoring alludes to the crossroads between environmentalism and the military-industrial apparatus.[104] The inclusion of images from the space missions at the film's start, themselves the direct result of cold war nationalism, further reinforces the thorny geopolitical implications of the film's miscellaneous imagery.

By combining footage from the Apollo missions, simulations, and images filmed by astronauts specifically for the film, *Blue Planet* fashions itself as a teleological survey of planetary visualisation. And it places its own vision at the vanguard of this sublime teleology. In fact, the perceived supremacy of the view from afar still prevails in IMAX Earth films, as confirmed by the 2016 release of *A Beautiful Planet,* also directed by Toni Myers, produced in conjunction with NASA and filmed in 4K. Narrated by Hollywood A-list star Jennifer Lawrence and filmed by astronaut crews

103 Christophe Bonneuil and Jean-Baptiste Fressoz, *The Shock of the Anthropocene: The Earth, History and Us*, trans. David Fernbach (London: Verso, 2017), 85.
104 See Allison Carruth & Robert P. Marzec, 'Environmental Visualization in the Anthropocene: Technologies, Aesthetics, Ethics', *Public Culture* 26:2 (2014): 205–211; 205.

from the International Space Station, *A Beautiful Planet* was launched to coincide with IMAX's partnership with the United Nations as part of its aforementioned 'Big Picture' project. Focused partly on the everyday routine of astronauts from Italy, Japan, Russia and the US, *A Beautiful Planet* has a more cosmopolitan, lighter feel than *Blue Planet.* A large portion of the film, for example, is dedicated to showing the curiosities of spaceship daily life, from showers without water through the difficulties of putting on a space suit. Yet, as with *Blue Planet*, the premise of *A Beautiful Planet* is to reveal, as Lawrence emphasises, 'what we here on the ground can't see so clearly: what is happening to our Earth and how we are changing it'.

One could go so far as to say that *A Beautiful Planet* radicalises an IMAX overview aesthetics. For, whereas *Blue Planet* still makes recourse to ground-level images in order to laud its space-bound views by contrast, *A Beautiful Planet* dispenses with terrestrial sights almost entirely (with the exception of two scenes). Its ending is also significant. Lawrence remarks that astronomers have found a new solar system, 'Kepler 186', that contains 'a remarkable five Earth-sized planets', one of which, the outermost Kepler 186f, 'is just the right distance from its star for water and life to exist'. She muses: 'Could this be another Earth?'. The implication here seems to be that, in the gloomiest of scenarios, a habitable planet may be on the horizon, even though the film ends by noting, cheerfully, that 'we can do great things together' and that 'if we do our part [...] our Earth will always be a beautiful planet'.

Like *Blue Planet*, which concludes that 'we can undo the damage that we have caused' over whale vocalisations, *A Beautiful Planet* thus ends on a hopeful if rather rushed note. Both films therefore reinforce a prevailing Anthropocene narrative according to which the whole of humanity is not only attributed ecological culpability but it is also assumed to continue to have mastery and control, 'but now in the service of saving, rather than destroying the planet', as Kelly Oliver puts it.[105] The paradox of this narrative is that while it presupposes the more-than-human history of the Earth, it re-inscribes human exceptionalism with a vengeance. By the same token, the foregrounding of the view from above to the detriment of other views reinforces a pervasive environmental idea that 'it is only possible to understand and manage the planet's problems properly by looking at it from space, in a kind of "de-Earthed" vision'.[106] While satellites and probes

105 Kelly Oliver, *Earth & World: Philosophy after the Apollo Missions* (New York: Columbia University Press, 2015), 24.
106 Bonneuil and Fressoz, *The Shock of the Anthropocene*, 63.

can of course help us monitor and predict ecological disturbances, one may wonder, with Bruno Latour (and, for that matter, Arendt), whether the 'nature-as-universe' dictum that '*to know is to know from outside*' has not made us 'see less and less of what is happening on Earth'.[107] Indeed, if there is one lesson that Humboldt can teach us today, it is that we must activate and interconnect different visions in order to grasp the energetic interdependence of planetary processes. Let us now look at two films that respond to this imperative by treating the world not so much (or only) as a floating planet that can be monitored from afar, but as a deep terrestrial soil that writes its own, eventful and volatile, history.

Who Does the Earth Think It Is?

The London-based duo Anjalika Sagar and Kodwo Eshun, or The Otolith Group, is known for creating research-based works that straddle the boundaries between filmmaking and installation pieces, experimental cinema and audiovisual essay, fiction and documentary. This includes a series of political essay films they made between 2003 and 2009, known as the *Otolith Trilogy*, and more recently a number of moving-image works that grapple with ecological questions and their entanglement with human and nonhuman histories.

Commissioned by the Los Angeles media art center REDCAT and the Haus der Kulturen der Welt in Berlin as part of the Anthropocene Project, their single-channel *Medium Earth* (2013) is no exception to this rule. Filmed on location in Yucca Valley, the Joshua Tree National Park and Los Angeles, *Medium Earth*, true to the concept of the Anthropocene, attempts to show the interplay between human and geological temporalities. However, the film resists self-aggrandising human narratives by suggesting that ultimately it is the Earth itself that has the upper hand as the agent of its own history. Through an exploration of geological instabilities, the film, as per its title, frames the planet itself as a medium whose signs, signals and disturbances we would do well to attend to. Its attunement to the Earth is translated into circular panoramas whose tactile surveying of rock surfaces acts as a gateway into a deep geohistory; in this sense, *Medium Earth* adds a new spin to panoramic sublimity both in form and content.

107 Bruno Latour, *Down to Earth: Politics in the New Climactic Regime* (Cambridge: Polity Press, 2018), 67, 70.

Although there are no humans to be seen in the film, it depicts anthropogenic meddling with natural landscapes through recurrent shots of subterranean car parks and highways cutting through mountainsides. As Eshun explains: 'Vast chunks of California were "reclaimed" from the desert. That's why *Medium Earth* features so many parking garages'.[108] But no sooner do these human-modelled environments appear than they are shown to be entirely susceptible to the Earth's own tectonic doings.

Take for example the first sequence, which strings together static images filmed in empty underground car parks, some of which are many levels below the planet's surface. Then a shot showing a fissure on the ground snaking its way across the concrete floor prompts the camera to tilt down slowly and close in on the crack as if its aim is to penetrate deeper. The following sequence, as if acting upon this quest, is filmed amidst gigantic rocks, with the camera slowly rising from the depths and revealing as it moves up only a patch of the sky and tree branches blowing with the wind. Throughout the film, the camera closely inspects a number of cracks winding their way through deserts, floors and freeways, indexical traces of a planetary calligraphy that asserts the volition of the tectonic as an empirical fact. No matter how many vaults humans may carve out in its structure, *Medium Earth* suggests, the Earth may still impinge its will. As if responding to Latour's call for a 'terrestrial' mode of seeing that digs 'deep down into the Earth with its thousand folds', the recurrent low-angle shots taken amidst rocks and through which tiny glimpses of the sky are only just visible, appear in this context as the reverse mirror image of space overviews in their demand that we see our planet from an earthly, underside standpoint.[109]

This reversal of perspective finds its counterpart in the many circular panoramas employed in the film. Shot at both ground and underground level, these pans convey their attachment to the Earth as a historiographic soil rather than an object one coolly examines from a distance. In one of these, a slow-moving camera, placed at a high angle and from a short distance, films the scratched surface of a desert. In another, 180-degree shot, the initially stationary camera frames two busy highways flanked by mountainsides to the noises of cars flashing past. Then, as the camera begins moving, an eerie soundtrack featuring muffled detonations and atmospheric static, by Thomas Köner, is superimposed on the audio. The

108 'The Otolith Group: *The Radiant* and *Medium Earth*', *Conversation Pieces* (blog), June 2 2015, https://agratza.wordpress.com/2015/06/05/the-otolith-group-the-radiant-and-medium-earth-2/ (accessed May 29 2020)
109 Latour, *Down to Earth*, 81.

gentle handheld camera continues panning, moving past the highways and on to the ragged mountainside. It patiently surveys the mountain's crevices and folds while contouring their outlines, to then return to the other side of the highway. By interspersing its tactile embrace of undulating sedimentary structures with sped-up cars on a busy highway, all the while respecting the physical contiguity of these entities via a panoramic shot, *Medium Earth* discloses the material coexistence and interaction of geological and human times.

That the circular pan is utilised here to the service of a grazing gaze challenges the omniscient and all-surveying vision usually attributed to the panoramic medium. If nineteenth-century panoramas promised the sublime experience of an expansive world beyond the horizon and if the space overview takes these horizons to sidereal heights, *Medium Earth* subverts this visuality from within by employing panning shots as a proximate view of earthly matter. Granted, the film features a few shots of wide expanses in line with a landscape visuality. Yet the majority of panoramas in the film preclude the distance that is often viewed as the prerequisite for the all-embracing experience, giving us instead haptic images of the Earth as a way to open up and access its more-than-human history. As Humboldt once noted, rocks 'give to the eye of the geologist a peculiar kind of animation to the landscape, acting on his imagination as traditional monuments of an earlier world. Their form is their history'.[110]

This is made especially visible in one 360-degree shot that deploys an oscillating rather than steady camerawork. The handheld camera surveys boulder outlines while mimicking the shape of these geological formations, with the spatial movement of the shot thus dictated by, and enfolded onto, the geo-temporalised form of its content. The idea of a stirring planet in continuous and convulsive tectonic movement is then underlined by the soundtrack, as the softly spoken voiceover (Sagar herself) notes: 'Under the sedimentary strata, thrust faults move in the dense thickets'. Later, she asks, as the camera moves with the movement of boulder crops: 'Who does the Earth think it is? We are sinking below its crust, into the plates colliding, forming new mountains'.

Taken together, these statements and images suggest a living planetary landmass that defies prediction let alone human intervention. They recall Nigel Clark's contention, in his book *Inhuman Nature: Sociable Life on a Dynamic Planet* (2011), that we must confront 'how the world actually works' beyond and above anthropogenic change if we are to produce correct

110 Humboldt, *Cosmos*, vol. 1, 58.

answers 'to the question of what sort of planet and cosmos we inhabit and what kinds of imperatives arise out of this inhabitation'.[111] As Clark notes, while Kant's assertion of a self-willing subject is often deemed the philosophical source behind a human-nature separation, his formulation of the sublime in fact responded to a very specific event – the 1755 Lisbon earthquake – that haunted the philosopher, as he recognised that 'the temporal and spatial dominion of our species was disturbingly inconsequential when viewed in the context of the earth's eventful history or the vastness of interplanetary space'.[112] In a similar move, Emily Brady has suggested that Kant values 'nature for the challenges it presents to us, as something that is difficult for us to face, against which morality provides the resources needed to cope'.[113] In this sense, the sublime triggers 'an awareness of an independence from the power of nature while at the same time judging nature to be powerful'.[114]

From this perspective, the sublime can be deployed as an aesthetic tool to recognise rather than evade the idea of a turbulent Earth whose instabilities we must acknowledge and respect. By focusing on the equally threatening and fascinating fault lines of an ever-shifting planet, *Medium Earth* makes visible a sublime history of the Earth as a medium via the indexical medium of film.

Unearthing the Earth

Nikolaus Geyrhalter's documentary *Earth* (*Erde*, 2019) is also concerned with the terrestriality of the Earth. The focus here, however, falls more heavily on how the soil has been reclaimed by human enterprises and turned into mud. No stranger to issues of ecological import, Geyrhalter has over the last two decades directed the lens of his camera (he is often the cinematographer of his films) to the impact of globalisation on the transformation of the environment, including: *Our Daily Bread* (2005), centered on the high-tech food industry; and *Homo Sapiens* (2016), which, as discussed in chapter 6, catalogues abandoned ruinous sites from different parts of the world.

Earth is firmly situated in Anthropocenic terrain. As its opening statement announces: 'Every day, 60 million tons of surface soil are moved by

111 Nigel Clark, *Inhuman Nature: Sociable Life on a Dynamic Planet* (London: Sage, 2011), xviii.
112 Ibid., xii.
113 Brady, *The Sublime in Modern Philosophy*, 85.
114 Ibid., 87.

rivers, wind and other forces of nature. Humans move 156 million tons of rock and soil per day, making humankind the most decisive geological factor of our time'. The film is divided into seven episodes, each of which depicts a specific location: a construction site in San Fernando Valley, California; the Brenner base tunnel between Austria and Italy; a coal strip mine in Gyöngyös, Hungary; a marble quarry in Carrara, Italy; a copper mine in Minas de Riotinto, Spain; a former salt mine-turned-storage site for radioactive waste in Wolfenbüttel, Germany; and the oil sands in Fort McKay, Canada.

From this description one might conclude that *Earth* is another tale of how humans have changed the Earth into an *Eaarth*, to cite Bill McKibben's neologism to account for the uncanniness of our altered planet.[115] This is no doubt true, yet I want to suggest in the remainder of this chapter that the film counters a facile narrative of human superiority via internalised dualities in both its textual and structural constitution. While the first opposition is crystallised around the juxtaposition of overhead/panoramic and underground/tactile views, the second comes into fruition in the final episode of the film, which relates an animistic outlook on the Earth that sits in contrast with the preceding developmental perspective. These overlaying dichotomies, I suggest, produce a multi-perspectival view that does justice to the complexity of human-world interactions *and* the sublime autonomy of our planet.

Each section of *Earth* is introduced by an intertitle superimposed over satellite images of the depicted location. As the credits tell us, these are from European Space Imaging, a German company that brands itself as the 'leading provider' of 'pictures of Earth, taken from space'.[116] In effect, the imagery produced by European Space Imaging is designed to help the very activities the film scrutinises: construction, mining and resource extraction. In *Earth*, however, they are appropriated to amaze the viewer with the astonishing vastness and deepness of human intervention: the gigantic craters and gaping holes we see from on high in the images of the Brenner tunnel construction, the Carrara marble extraction area (figure 1.6) and the Minas de Riotinto site – all turned into undulating, surprising shapes and forms through the strangely flat angle of the aerial vertical shots, which convey a machinic, one may say 'extra-terrestrial' quality.

115 Bill McKibben, *Eaarth: Making a Life on a Tough New Planet* (New York: Henry Holt and Company, 2010), 24.

116 See https://www.euspaceimaging.com/ (accessed 27 November 2019)

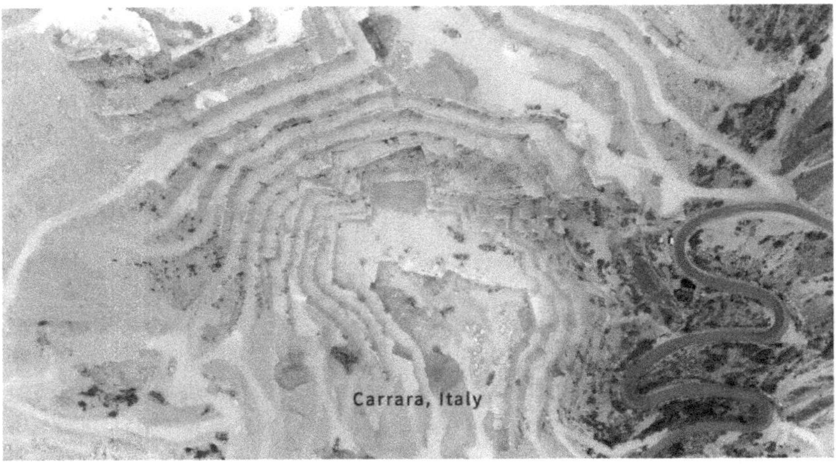

Figure 1.6 The satellite overview in *Earth* (2019). Courtesy of NGF.

Traditional panoramas also make appearances in *Earth*: Geyrhalter admits that he likes 'open space in the image, an open horizon'.[117] Some of these images emphasise the insignificance of industrial sites. An example is the film's opening shot where a barren deserted field in California stretches as far as the eye can see, with the construction site scarcely perceptible on the horizon. Other views, conversely, underline the enormity of human-induced intervention. A case in point is the sequence starting with a short-distance shot of workers in an archaeological site in Minas de Riotinto, followed by a panoramic image that relativises this site and these workers, now mere dots in the frame, within a gargantuan open-pit mine. Another image shows the extent to which the extraction of Carrara marble on the mountainsides has seemingly turned into a morphological feature of these valleys (figure 1.10).

To some extent, these overhead and panoramic views may be understood in relation to what is now referred to as the 'toxic sublime', often invoked in discussions of the work of Edward Burtynsky and Yann Arthus-Bertrand. Whereas Burtynsky manufactures eye-popping, 50x60 inch photographs of nickel mines, recycling sites and scrap dumbs, Arthus-Bertrand (discussed in chapter 4) privileges the flyover view to make visible the destruction of nature, with his hit *Home* (2009) filmed entirely from helicopters. As a new aesthetic incarnation of sublimity, the toxic sublime would seem to differ from both the natural and technological sublimes: the former because it

117 Geyrhalter, cited in Scott MacDonald, *The Sublimity of Document: Cinema as Diorama* (New York: Oxford University Press, 2019), 129.

frames human-tainted rather than natural landscapes, the latter because it does not unequivocally celebrate industrial achievement. Rather, as Jennifer Peeples theorises, the toxic sublime is defined by a constitutional ambiguity, countering a sense of awe 'with alarm for the immensity of destruction one witnesses'.[118]

This ambiguity has been the topic of heated debates in recent years, with many deeming it reproachable on the basis of its aestheticisation of ecological catastrophe.[119] Although the criticism that the toxic sublime has been 'gentrified' cannot be discounted, my problem with this view is that it presumes aesthetics to be antithetical to ethics, when, as Jacques Rancière has taught us, the politics of aesthetics resides in the way it shatters the normative 'presupposition [...] that some subjects are suitable for artistic presentation; while others are not'.[120] In any case, there is not one but many toxic sublimes, variously related to questions of medium specificity and aesthetic choices. Geyrhalter's stationary, geometric shots are much more sobering and clinical when compared with the quasi-abstract plasticity of Burtynsky's photographs or Arthus-Bertrand's soaring imagery and soundtrack. Unearthing before our eyes activities that are placed too conveniently out of sight, *Earth* instead subscribes to what Scott MacDonald terms 'the sublimity of document' in relation to filmmakers (Geyrhalter included) who use the 'camera's facility with documenting reality' to 'provide panoramas of sites/sights [they] feel we need to be aware of'.[121]

More importantly, the overviews and panoramas opening each section prove to be preludes to a dropping down to the surface and an excavation into earthly strata. Unlike *Home*, which never once lands its feet on the ground, *Earth* favours instead what could be defined as a Humboldtian,

118 Peeples, 'Toxic Sublime', 380.

119 See, for example, Zoë Druick, 'Environmentalism from the Sky?', *Docalogue*, https://docalogue.com/october-anthropocene-the-human-epoch/ (accessed June 8 2020); Zoë Druick, '"A Wide-Angle View of the Fragile Earth": Capitalist Aesthetics in the Work of Yann Arthus-Bertrand', *Open Cultural Studies* 2 (2018): 396–405; T. J. Demos, *Against the Anthropocene: Visual Culture and Environment Today* (Berlin: Stenberg Press, 2017).

120 Jacques Rancière, *The Future of the Image*, trans. Gregory Eliott (London and New York: Verso, 2009), 118. It must be noted, however, that Rancière is wary of the concept of sublime as incarnated in its postmodern, Lyotardian variation. See Stephen Zepke, *Sublime Art: Towards an Aesthetics of the Future* (Edinburgh: Edinburgh University Press, 2019), esp. 202–240. On the gentrification of the toxic sublime, see Clint Burnham, 'Photography from Benjamin to Žižek, via the Petrochemical Sublime of Edward Burtynsky' in Sheena Wilson, Adam Carlson and Imre Szeman (eds.) *Petrocultures: Oil, Politics, Culture* (Montreal: McGill-Queen's University Press, 2017), 458–475; 409.

121 MacDonald, *The Sublimity of Document*, 13, 14.

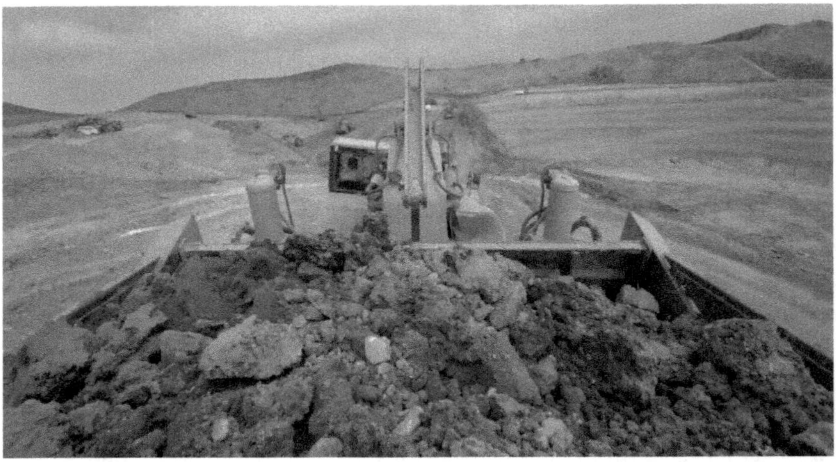

Figure 1.7 The tactile earth/Earth in *Earth* (2019). Courtesy of NGF.

pan-visual aesthetic that combines a myriad of figurative and literal angles: terrestrial and extra-terrestrial, human and nonhuman, external and internal, superficial and subterranean. For example, the film continually registers in durational shots the soil being manoeuvred, burrowed and shovelled by bulldozers, scoopers and bucket-wheel machines. At other times, viewers get to experience the perspective of these machines with cameras attached to their front or back, such as a scooper picking up and dumping the soil, underlined in all its earthiness (figure 1.7). Establishing shots filmed at ground level further give us views of machines scuttling around over vast terrains of grey soil.

And then *Earth* delves deeper. We see excavators opening up vaults in subterranean sites; take a ride on tunnel conveyor belts designed to transport goods; follow workers in elevators going down 490 metres below the surface; and hear and see dynamite imploding the earth in seismically disturbing shots. In the film, then, anthropogenic impact is given materiality not only through high and open views but also via tactile shots and long takes that register in real time and up close the sheer matter-of-factness of destruction as machines and detonators burst the earth/Earth open in order to extricate its sought-after materials.

The visual movement the film proposes is therefore not one projecting from the ground into outer space as a means of obtaining a universal perspective, but, on the contrary, one that descends to and burrows into the Earth as a means of gaining insight into the violent remodelling of its layers. Like a georama of sorts, *Earth* performs a balancing act between internal and external perspectives on the planet, an oscillating vision underlined by the workers themselves. For example, one notes that his job enables 'a connection to the depths of the Earth' and that, consequently, you 'feel a bit like an astronaut because you are the first human to come across this spot'. Another exclaims that mining is 'like travelling into the past' as you encounter substrata formed over millions of years. Of course, the irony here is that the film can only show a subterranean Earth, and consequently its history, insofar as it inspects what extractive activities have done to the planet: the humongous vaults, craters and holes gouged out for profit. But it is through a focus on such activities that the film is able to throw light on the Earth's rock-forming *longue durée*.

One particular sequence presses home the geological time-travel theme. This takes place in a Hungarian natural history museum, the only time the film ventures into a cultural institution. A female worker introduces an exhibition on the remnants of standing swamp cypresses that were found preserved, in their original form, in the coal strip mine of Bükkábrány. Initially visible in a large-scale, puzzle-like photograph on the wall behind the worker, chunks of the fossilised trees make an appearance in the following shot on the right-hand side of a diorama, with the coal mine painted as a background (figure 1.8). On the left-hand side, as the worker explains, 'we can look seven million years into the past [as] the environment might have looked like back then', with replica trees and vegetation in the foreground, and a painted forest in the background. The diorama's two-part arrangement, with one depicting the present and the other a nonhuman past, makes visible a structuring principle of the film. It resonates with landscape shots that conflate in the same frame two temporal registers, one belonging to the Earth's geological past and the other to its geo-modified present. A shot of the marble quarry eating away at the mountainsides is a case in point (figure 1.9), with human 'progress' made visible as a disturbingly literal encroachment upon the natural world.

Uncharacteristically for a Geyrhalter film, *Earth* is replete with interviews, which add welcoming, down-to-earth perspectives to its emphasis on the nonhuman and the machinic. The director's questions are often audible, many related to the ecological footprint of mining. Some workers revel in the sublimity of their jobs. One boasts he 'literally moves mountains' for a living, another that 'once we get here is not nature anymore', and yet another that 'what we're doing here is nothing short of a masterpiece'. However, the film is never judgmental of its subjects, and indeed, *Earth* includes a multiplicity of voices in line with its heterogeneous visual track. One worker, for example, remarks that 'extracting anything from the soil is a really violent process'. Another admits that 'We don't like to ruin the mountains

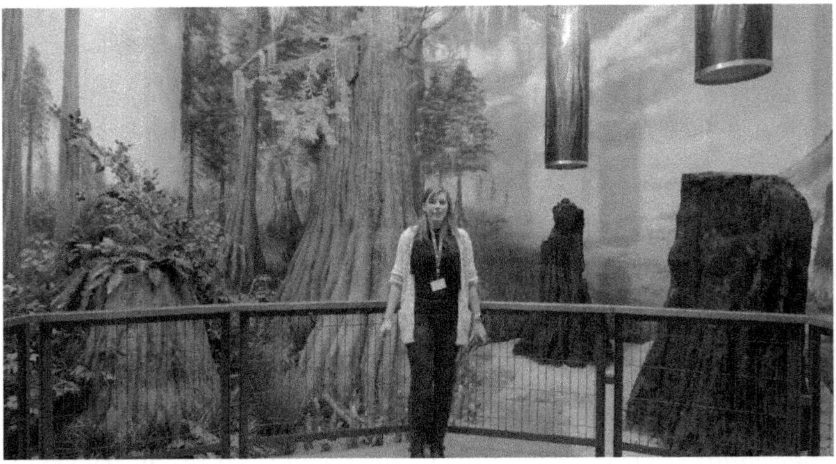

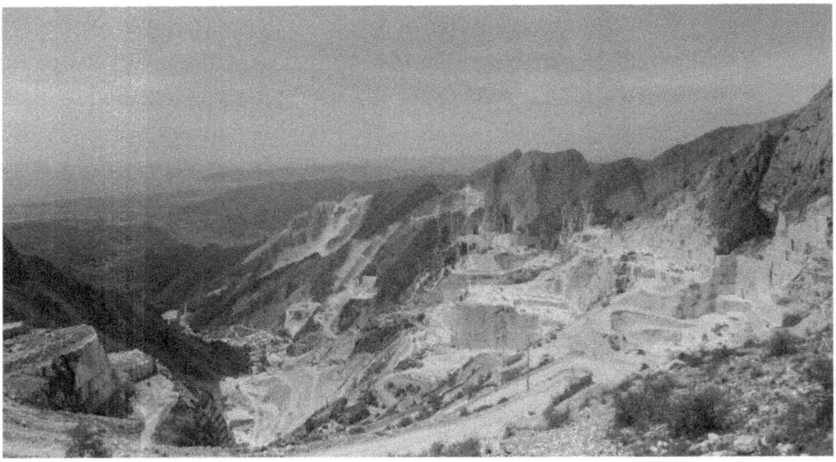

Figures 1.8 & 1.9 A diorama conflating human and nonhuman temporal frames in *Earth* resonates with panoramic images showing human destruction encroaching upon the natural world. Courtesy of NGF.

but that's what we are doing'. Phrases such as 'If I don't do it somebody else is going to do it', 'We can't stop transporting goods', and 'Copper is essential to our life with electricity' punctuate the testimonies. Although *Earth* does not follow the paths of these minerals from extraction to consumption, these testimonies implicate the viewer by highlighting one's dependence on the materials we see being excavated onscreen – all the way down to the material components of laptops on which the film might be watched.

If *Medium Earth* depicts the Earth as a geological medium, *Earth* instead unearths what Jussi Parikka terms 'a geology of media' to refer to the chemicals, metals and materials of which media technologies are made. As Parikka argues, 'instead of networking, we need to remember the importance of copper or optical fiber for such forms of communication; instead of a blunt discussion of "the digital", we need to pick it apart and remember [its] mineral durations'.[122] *Earth*'s revelatory dimension consists in showing the way in which the extrication of materials needed to support modern life enacts a destructive incision into the temporal folds of the planet, while refusing easy answers. Though the word capitalism is never uttered in the film, these workers see themselves and their jobs as trapped within the extractivist models of production that enable and support this socio-economic system. In the film, there is no sublime humanity that may revert this state of affairs as if on a whim but, rather, an acknowledgment of the overwhelming complexity of the task that awaits us precisely because the earthly resources on which we are dependent are finite. To paraphrase one engineer in the film: 'What's the alternative?'.

Although *Earth* refrains from providing a fully formed answer to this question, it gestures, in its final section, towards cultures and traditions that offer alternative ways to engage with the fertility of the planet. Here, the film does not record the workers on the oil sands in Fort McKay but two First Nations Canadian people from the Dene group who live in the surrounding area. One of these is Jean L'Hommecourt, an activist who has drawn attention to the damage that the extractive oil-industry has inflicted on her community, including food contamination. Whereas the decision to follow these people seems to have been dictated by the fact that entrance to the oil sands was denied to the film crew, there is a dramatic shift in perspective in concrete and figurative terms. The distant panoramas gain in significance by virtue of their being the only possible views of the prohibited site. Rather than drilling machines and job chores, the viewer is afforded an insight into a different relationship with the soil away from domination and possession. As L'Hommecourt relates to the camera: 'For me and my culture, being a Dene means people of the land, so we are of the earth, and we need the earth to survive, to exist as a human being, and in our culture we believe every element of Earth has a Spirit'. She goes on: 'Earth means life. Simple. It provides everything for human life to exist. Without all of this around us, we wouldn't be able to survive'. Driving around the sands, her friend remarks: 'They're taking to [the Earth's] flesh. Big wounds [...] They're saying they are going to reclaim but nobody can reclaim the world'.

Déborah Danowski and Eduardo Viveiros de Castro have noted that, while the belief systems of the 'autochthonous peoples of the American

122 Jussi Parikka, *A Geology of Media* (Minneapolis: University of Minnesota Press, 2015), 4.

continent' are often 'accused of naïve localism, primitivism, irrationalism', they constitute 'one of the possible chances, in fact, of a *subsistence of the future*'.[123] One could go so far as to say that these belief systems can help us in the project of revitalising 'the natural sublime as an *environmental sublime*', which, as Brady notes, 'can come to symbolize something beyond human control, and thereby teach us humility and respect'.[124] L'Hommecourt and her friend's beliefs encapsulate and enrich this 'humbling sublime' by treating the Earth as an animistic entity that inspires admiration, deference and even fear. Tellingly, Jean defines herself not as someone on but *of* the Earth, which is regarded as a nourishing soil that engenders life rather than a mineralogical reserve. Her statement recalls Arendt's insistence that the 'earth is the very quintessence of human condition, and earthly nature, for all we know, may be unique in the universe in providing human beings with a habitat in which they can move and breath without effort and without artifice'.[125] By leaving the First Nations Canadians to its final section, *Earth* sets up an opposition between an Earth that is cherished for the humility it commands in certain humans and another that is reduced to calculation, demarcation and appropriation.

Bruno Latour has noted that to grapple with the terrestrial soil in terms of 'the attentive care that it requires' is diametrically opposed to the idea of 'a plot of ground that a development or real estate project has just grabbed. The ground, the soil, in this sense, cannot be appropriated. One belongs to it; it belongs to no one'.[126] It is this generative conception of earthly belonging that L'Hommecourt advocates, as opposed to a developmental view of the ground. But whereas the latter is given concrete visual form throughout *Earth*, the Dene's vitalist and fertile conception of planetary existence constitutes the film's conspicuous visual absence.[127] It is an image that is never formed, left for the viewer's imagination, a negative structuring principle against which piles of grey soil must be qualified in retrospect.

123 Déborah Danowski and Eduardo Viveiros de Castro, *The Ends of the World*, trans. Rodrigo Nunes (Cambridge: Polity Press, 2017), 121, 123, emphasis in original. Latour makes a similar point in *Down to Earth*, 74.

124 Brady, *The Sublime in Modern Philosophy*, 183, 202, emphasis in original.

125 Arendt, *The Human Condition*, 2.

126 Latour, *Down to Earth*, 92.

127 This is not to suggest that indigenous cultures do not change nature, which is never really separate from human interference, but simply that theirs are more resourceful models of interaction with the natural world. See in this respect Ursula K. Heise, *Imagining Extinction: The Cultural Meanings of Endangered Species* (Chicago and London: University of Chicago Press, 2016), 9–10.

And here we are back full circle to Humboldt, who already in 1845 had alerted us to the 'danger' of nations that know 'no pause in unceasing movement, development, and production'.[128] Unchecked progress, he maintained, 'must be averted by the earnest cultivation of natural knowledge. Man can only act upon nature, and appropriate her forces to his use, by comprehending her laws, and knowing those forces in relative value and measure', otherwise there could be 'the partial diminution, and perhaps the final annihilation, of [...] resources'.[129] *Earth* is the unsettling but necessary evidence that this is now a global reality.

128 Humboldt, *Cosmos*, vol. 1, 37. It was for this very reason that Humboldt was fascinated by the indigenous cultures he had the chance of encountering while travelling through Spanish America. See Lee Schweninger, 'A Return to Nature's Order: Indigenous Peoples and the Politics of Alexander von Humboldt's *Political Essay on the Kingdom of New Spain*', *ELOHI* (2016), 87–102; see also Wulf, *The Invention of Nature*, 80–81.
129 Humboldt, *Cosmos*, vol. 1, 37.

2 The Unseen World Across the World

Abstract

This chapter adopts wonder as its organising philosophical idea. The concept is first examined in relation to natural-history visual culture in Victorian Britain and the popular science and global expedition early-cinema genres. I argue that these first forays into revealing the 'unseen' world of nature are crucially relevant for an understanding of contemporary BBC Earth series. These series enact a technological rekindling of a never-before-seen aesthetic that paradoxically relies on the idea that some things may never be seen again due to the current environmental crisis. I suggest that this paradox can be understood as a struggle on these series' part to keep their foundations intact within a natural history paradigm in the age of the Anthropocene, according to which a distinction between human and natural histories is no longer tenable.

Keywords: wonder, natural history, *The Unseen World*, never-before-seen, *Frozen Planet*, *Blue Planet*

If we are to write a history of wonder we must write a natural history.
John Onians

At the beginning of the BBC documentary series *Blue Planet II* (2016), David Attenborough states: 'The oceans, seemingly limitless, invoke in us a sense of awe and wonder, and also, sometimes fear. They cover 70% of the surface of the planet, and yet they are still the least explored'. These remarks evince that, not unlike the works discussed in the last chapter, the show's visual aesthetics are designed to elicit a sublime experience that combines awe and fear in relation to boundless oceanic expanses. Yet in this chapter I want to turn attention to that other concept in Attenborough's statement – wonder – in order to explore its connection with the ability of recording technologies to reveal the unseen in BBC natural history series.

de Luca, T. *Planetary Cinema. Film, Media and the Earth.* Amsterdam: Amsterdam University Press, 2022
DOI: 10.5117/9789463729628_CH02

Although similar in some respects, sublime and wonder elicit different phenomenological responses and pertain to distinct if sometimes overlapping philosophical lineages.[1] As Emily Brady explains: 'Both experiences provide limits through perceptual, immediate experiences of a range of natural objects'. However, whereas for sublimity 'it will be great things that belittle us', for wonder 'it can be something large but also small, such as a delicate spider's web', with the latter concept denoting 'a broader range of qualities that do not shock or frighten, but rather amaze, stirring intellectual curiosity rather than fear'.[2]

In his book *Wonder, the Rainbow and the Aesthetics of Rare Experiences* (1998), Philip Fisher theorises wonder as 'the deep surprise of the visual', one that is dependent upon 'first sight and first experience': 'Wonder is the outcome of the fact that we see the world [...] The object must be unexpectedly, instantaneously seen for the first time'.[3] It follows for Fisher that neither memory nor the ordinary can ever play a part in wonder, which is exclusively attached to the amazement at seeing things never before seen. Over the last two centuries, wonder has been especially bound up with scientific and technological revelationism. In his *The Age of Wonder* (2008), Richard Holmes expounds on early nineteenth-century 'Romantic science' as one that participated in a 'process of revelation' of nature, 'allowing man not merely to extend his senses passively – using the telescope, the microscope, the barometer – but to intervene actively, using the voltaic battery, the electrical generator, the scalpel or the air pump'.[4] Tom Gunning, for his part, defines the end of that century as a 'period of cultural wonder' par excellence, 'an era of technological acceleration and transformation of the environment' that finds echoes and parallels in our own time.[5] Yet, for Gunning, 'the "newness" of new technology, its capacity to dazzle us, is always in some sense the product of discourses surrounding it', that is,

1 For an illuminating philosophical history of wonder in relation to the sublime, see Genevieve Lloyd, *Reclaiming Wonder: After the Sublime* (Edinburgh: Edinburgh University Press, 2018).

2 Emily Brady, *The Sublime in Modern Philosophy: Aesthetics, Ethics and Nature* (New York: Cambridge University Press, 2013), 197–198.

3 Philip Fisher, *Wonder, The Rainbow and the Aesthetics of Rare Experiences* (Cambridge, MA, and London: Harvard University Press, 1998), 30, 19, 17.

4 Richard Holmes, *The Age of Wonder: How the Romantic Generation Discovered the Beauty and Terror of Science* (London: HarperPress, 2008), 17.

5 Tom Gunning, 'Re-Newing Old Technologies: Astonishment, Second Nature, and the Uncanny in Technology from the Previous Turn-of-the-Century' in David Thorburn and Henry Jenkins (eds.) *Rethinking Media Change: The Aesthetics of Transition* (Cambridge, MA, and London: MIT Press, 2003), 39–60; 42.

a discourse of technological wonder is often 'not only of innovation, but precisely one of novelty, maximizing the dazzling experience of the new'.[6]

These ideas provide a solid epistemological foundation for an understanding of contemporary BBC natural history series, which, I argue in this chapter, are not simply about opening our eyes to the wonders of the natural world, but more precisely about the wonders of technologies that capture natural wonders in a manner 'never before seen' (a recurrent tagline). Here, a discourse of planetary revelation, tirelessly reiterated in textual and extratextual materials, is formulated through a vainglorious aesthetic spinning around two axes: a quantitative axis, according to which new technologies of reproduction can show previously unthinkable reaches of the planet; and a qualitative axis, premised upon the high-definition clarity of the image. Taking the viewer on a flyover tour across the planet's surface (*Planet Earth*, 2006; 2016), deep down in the oceans (*Blue Planet*, 2001; 2017) and to the ends of the Earth in the polar regions (*Frozen Planet*, 2011), these series promise to inspect all corners of the planet while relying on aerial views, time-lapse and microcinematography to disclose images otherwise imperceptible or unavailable to the human eye.

At the same time, choosing the concept of wonder to account for these documentaries allows us to discern remarkable conceptual and stylistic links with early natural history cinema in Britain and its adherence to a wondrous aesthetic. Whether considered in terms of its popular science or global expedition genres, natural history film was eager to show, through an exhibitionistic display of techniques and technologies, *The Unseen World* and *Wild Life across the World* – these being respectively the titles of an 1903 early film programme by Francis Martin Duncan and an 1913 photography book by Cherry Kearton, both of whom, alongside Percy Smith and others, emerged as key figures in the consolidation of nature filmmaking. Indebted to the amateur cultural matrix of Victorian natural history, itself tied to postulates of aesthetic surprise and delight, these filmmakers revitalised natural wonder via the medium's motion properties and capacity to move around the world. They therefore offer a valuable framework to explore how contemporary global visions recycle a sense of marvel at nature, demanding that we take 'natural history' not merely as a label designating the BBC unit or its series but as a clearly defined tradition within which these series exist and which in many ways constrain and dictate their visualisation of nature.[7]

6 Ibid., 42, 43.

7 On the importance of natural history for an understanding of these series, see Michael Jeffries, 'BBC Natural History versus Science Paradigms', *Science as Culture* 12:4 (2003): 527–545;

Although a history of natural wonder stretches as far back as the twelfth century and gained currency with early-modern cabinets of curiosities, it was only in the eighteenth century that natural history began to acquire the global and popular contours with which it is associated today.[8] A turning point was Carl Linnaeus's *The System of Nature* (*Systema Nature*, 1735), which, as Mary Louise Pratt notes, inaugurated a specifically European sense of 'planetary consciousness' derived from 'the knowledge-production project of natural history'.[9] If the panoramas discussed in the last chapter implied or depicted the world through realistic paintings of cities and cartographic devices that intimated the exterior expanse of the globe, natural history mapped out 'the internal "contents" of those land and water masses whose spread made up the surface of the planet', ransacking 'every visible square, or even cubic, inch of the earth's surface'.[10]

A century later natural history had become immensely popular. This was so especially in Victorian Britain, where professional naturalists shared the field with amateur practitioners on whom they relied for the gathering and examination of insects, plants, fossils and rocks.[11] Popularisers, such as Reverend John G. Wood, published compendia, children's books, weekly serials and encyclopedias, concomitant with the appearance of herbariums and natural history museums featuring monumental displays of fossils, taxidermy and dioramas.[12] Whereas illustrated publications offered themselves as guides through which readers could see the natural world anew or with the help of microscopic technology, museum displays ordered global nature and its specimens in dazzling arrangements. The act

Jean-Baptiste Gouyon, 'The BBC Natural History Unit: Instituting Natural History Film-Making in Britain', *History of Science* 49:4 (2011): 425–451; Eleanor Louson, 'Taking Spectacle Seriously: Wildlife Film and the Legacy of Natural History Display', *Science in Context* 31:1 (2018): 15–38.

8 See Lorraine Daston and Katharine Park, *Wonders and the Order of Nature, 1150–1750* (New York: Zone Books, 1998); John Onians, 'I wonder ... A Short History of Amazement' in John Onians (ed.) *Sight and Insight: Essays on Art and Culture in Honour of E.H. Gombrich at 85* (London: Phaidon Press, 1994), 11–33.

9 Mary Louise Pratt, *Imperial Eyes: Travel Writing and Transculturation*, second edition (London and New York: Routledge, 2008), 24.

10 Ibid., 30.

11 See Jean-Marc Drouin and Bernadette Bensaude-Vincent, 'Nature for the People' in N. Jardine, J. A. Second and E. C. Spary (eds.) *Cultures of Natural History* (Cambridge: Cambridge University Press, 1996), 408–425; 417.

12 These included *The Boy's Own Book of Natural History* (1860), *Animal Traits and Characteristics* (1860) and *Common Objects of the Microscope* (1861). See Bernard Lightman, '"The Voices of Nature": Popularizing Victorian Science' in Bernard Lightman (ed.) *Victorian Science in Context* (Chicago: University of Chicago Press, 1997), 187–211. On natural history museums, see Louson, 'Taking Spectacle Seriously', 26–31.

of seeing nature had become deeply mediated: engravings, microscopes, magnifying glasses and glass cases acted as mediums through which the natural world was organised, exhibited and revealed as a realm separate from and for the enjoyment of humanity.

Natural history thus enacted a widening of the field of the visible into micro and macro perspectives, conjoining the very near and the very far. On the one hand, it mobilised the entire planet as a field for taxonomy and evaluation. As Laura Dassow Walls notes: 'railroads and steamships carried globetrotting naturalists into its farthest corners, whence they returned with books and collections to urban centers, where, at last, the farthest and the nearest could be assembled and compared, named and catalogued'.[13] On the other hand, it functioned as a phenomenology through which, according to Barbara T. Gates, people 'learned to reorder nature, retraining their eyes to look as never before to witness what was around them in their everyday worlds'.[14] Nature was deemed aesthetically rewarding for the visual senses as well as a source of moral renewal: it followed that its wilderness needed to remain separate from the human domain if these benefits were to be preserved.

Although we should be careful not to establish an unequivocal link between the perception of an endangered nature and the popularisation of natural history, there is little doubt that the latter gained force in tandem with growing fears related to large-scale changes wrought upon the natural world by industrialisation and modernisation. This is proved by the mushrooming of the conservation movement in many countries, including, in the UK: the Commons Preservation Society (1865), the Society for the Protection of Birds (1889) and the National Footpaths Preservation Society (1899).[15] As Ursula K. Heise argues: 'The enclosure of the commons, the construction of railroads, and deforestation worried nineteenth-century advocates for nature in Europe and North America', with these environmental forebears relying on:

> the idea that modern society has degraded a natural world that used to be beautiful, harmonious and self-sustaining and that might disappear

13 Laura Dassow Walls, 'Natural History in the Anthropocene' in John Parham and Louise Westling (eds.) *A Global History of Literature and the Environment* (Cambridge: Cambridge University Press, 2016), 187–200; 191.

14 Barbara T. Gates, 'Ordering Nature: Revisioning Victorian Science Culture' in *Victorian Science in Context*, 179–186; 182.

15 John Bevis, *The Keartons: Inventing Nature Photography* (Axminster, UK: Uniformbooks, 2016), 48, 148.

> completely if modern humans do not change their way of life [...] In these stories, the awareness of nature's beauty and value is intimately linked to a foreboding sense of its looming destruction.[16]

At the same time, the chasm between natural history and what is today known as the life sciences was growing wider as cinema dawned, with laboratory experimentation and analysis supplanting the older tenets of cataloguing, typology and display.[17] As a knowledge-building enterprise tied to an aesthetics of wonder, however, natural history did not so much retain its cultural importance as become wholly reenergised with the advent of film thanks to its ability to render the world in motion.

In the first two sections of this chapter we will turn to a few individuals who, over the course of the first decades of the twentieth century, were instrumental in the consolidation of natural history film in the UK. I argue that cinema's technological appendages and novelties revitalised the wonders of natural history through the medium's capacity to show hitherto imperceptible, microscopic and unperturbed life forms on both local and global levels, sometimes against the embryonic idea of nature's decline. The last two sections turn to contemporary natural history television series in order to explore their technological rekindling of a wondrous aesthetic. My focus throughout will be largely on extratextual – promotional, institutional and journalistic – accompaniments and materials, many of which were retrieved through archival research. Such an extratextual focus is partly a necessity, given that the majority of the early films mentioned here no longer survive. Yet there is also a conceptual reason for this approach, since, following Gunning, my aim is to show that wonder's technologised incarnations necessitate a discursive boost to augment their sensory-perceptual impact.

Much has been made of the ambivalent environmental credentials of BBC natural history series in their perceived reluctance to acknowledge human presence and intervention in pristine settings. My hope is to demonstrate that this hesitancy can be understood as a struggle on their part to keep their foundations intact within a natural history paradigm in the age of the Anthropocene, according to which a distinction between human and natural realms is no longer tenable.

16 Ursula K. Heise, *Imagining Extinction: The Cultural Meanings of Endangered Species* (Chicago and London: University of Chicago Press, 2016), 7.

17 That said, as Lynn K. Nyhart has shown, a separation between natural history and the 'new' biology was far from clear-cut or definitive. See Lynn Nyhart, 'Natural History and the "New" Biology' in *Cultures of Natural History*, 426–443.

Unseen Worlds

In his book *The Nature of Natural History*, originally published in 1950, Marston Bates defines the field as 'not equivalent to biology. Biology is the study of life. Natural history is the study of animals and plants – of organisms'.[18] The emergence of nature filmmaking in Britain, as masterminded by Francis Martin Duncan and Percy Smith, is often seen as perpetuating this version of natural history against the new biological sciences and laboratory experimentation. As Oliver Gaycken notes: 'Duncan and Smith participated in a mode of scientific practice that was becoming marginalized, as the modern disciplines of biology and zoology distinguished themselves from the prior "natural history" model of knowledge about nature'.[19] Timothy Boon similarly observes that the 'Urban-Duncan films' are closest to an 'observational and descriptive natural-historical approach to nature'.[20] In this section I want to complicate this view by showing that nature films were also deemed aligned with 'the study of life' thanks to the ability of microcinematography and time-lapse to turn inanimate into animate matter. Cinematic vitalism, in this context, often necessitated natural history's association with stasis and death as a negative pole against which cinema's added wonder effects could be positively valued in comparison.

The credentials of Francis Martin Duncan within the domain of natural history are easily verifiable. His father, Peter Martin Duncan, was a fellow of the Zoological and Linnean societies, member of the Microscopical Society (where he served as president, 1881–1883), and editor of Cassell's *Natural History* (1876–1882) illustrated encyclopedia. Following in his father's footsteps, Duncan authored many natural history books, including *Some Birds of the London Zoo* (1900) and *Some Curious Plants* (1900), though his specialty was nature photographs, some of which appeared in later editions of Cassell's *Natural History* and his own *First Steps in Photo-Micrography: A Hand-Book for Novices* (1902).

Film was the next logical step. A member of London's Quekett Microscopical Club, Duncan debuted in the cinema with a programme for the Charles Urban Trading Company, which, as discussed in chapter 3, specialised in nonfiction educational films. Titled *The Unseen World: A Series of Microscopic*

18 Marston Bates, *The Nature of Natural History* (Princeton: Princeton University Press, 1990 [1950]), 7.

19 Oliver Gaycken, *Devices of Curiosity: Early Cinema and Popular Science* (New York: Oxford University Press, 2015), 55.

20 Timothy Boon, *Films of Fact: A History of Science and Documentary Films and Television* (London and New York: Wallflower Press, 2008), 10.

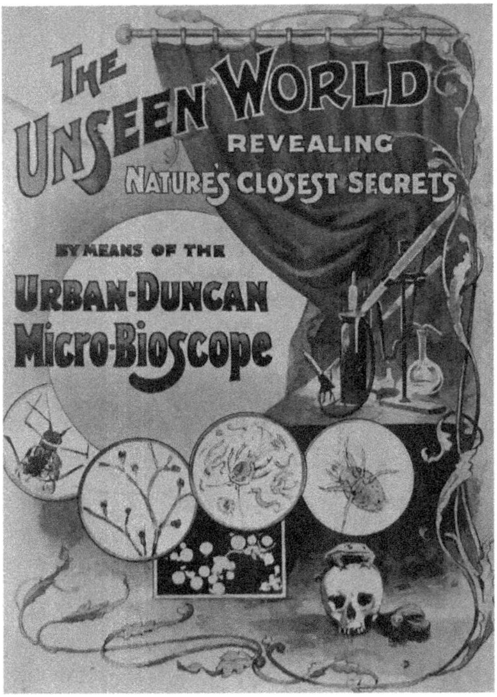

Figure 2.1 Promotional poster for the Urban-Duncan's *The Unseen World* (1903), printed in the Urban 1903 catalogue.

Studies by Means of the Urban-Duncan Micro-Bioscope, the 20-minute show premiered at London's Alhambra theatre in August 1903 and prided itself on conveying the wonders of science to a popular audience, combining instruction and amusement. Its first promotional poster (figure 2.1), reprinted in Urban's 1903 catalogue, shows four circular frames containing insects and a plant as though visualised through light-field photomicrography, just above a rectangular image mimicking the dark-field technique, and all against the background of laboratory utensils and a partially drawn curtain. As Gaycken points out, this illustration situated *The Unseen World* within the nineteenth-century tradition of lantern shows where lecturers projected and commented upon microscopic images of organisms; indeed, Duncan himself acted as a lecturer in the Alhambra show.[21]

The revelation of the invisible was not restricted to magnification, however. Although the Alhambra's first programme and the majority of films no longer survive, a quick glance at the 1903 catalogue, as well as the reviews and previews of the programme's premiere, confirm that *The*

21 Gaycken, *Devices of Curiosity*, 23.

Unseen World combined both close-range views of animals *and* subjects filmed with the help of a microscope. Another slide (figure 2.2) – featured in the 1903 and 1905 catalogues as an introductory image to 'Marine Studies' – further shows that the 'unseen' extended to underwater imagery. It depicts a microscope sitting at the bottom of the sea behind two glass plates, with a scuba diver standing next to it on the right holding a lantern whose beam of light illuminates some fishes, and above the diver an octopus within a circular frame recalling a lanternslide.

The image of a microscope under the water thus indicates that, for the *Urban-Duncan Micro-Bioscope*, microcinematography and images of marine life were on the same ontological footing when it came to their promise of visual revelation – even though this image was also misleading since none of the aquatic subjects were filmed by Duncan via microcinematography, nor were they shot underwater but through aquarium glasses. Yet, as evidence suggests, spectators and critics clearly differentiated the microscopic from the close-up images. One review thus noted: 'The views of [marine life] thrown on the screen are of such a size that every detail can be seen from any part of the theatre. *In addition*, a number of pictures enlarged from the microscope were shown last night'.[22]

Boon has examined the way in which science and nature filmmaking participated in the 'aesthetic of astonishment' famously theorised by Tom Gunning as one that 'directly solicits spectator attention, inciting visual curiosity, and supplying pleasure through an exciting spectacle [...] that is of interest in itself'.[23] In a sense, for Gunning, *all* films in the early years of cinema, from travelogues to humorous gags, provoked 'some sense of wonder or surprise [...] if only at the illusion of motion', that is, at the medium's ability to present, for the first time, images of the world 'through a technologically mediated experience of space and movement'.[24] The surprise at seeing onscreen movement, however, could not last forever: as Fisher notes, the 'rapid wearing out of the new is also part of the aesthetics of wonder'.[25]

22 'The Alhambra', *The Globe* (20 October 1903), reprinted in 'We Put the World Before You' (1903), Charles Urban catalogue, available online at https://archive.org/details/weputworldbeforeoounse/page/n6, 80, my emphasis (last accessed 30 December 2020).

23 Boon, *Films of Fact*, 13–14; Tom Gunning, 'The Cinema of Attraction[s]: Early Film, Its Spectator and the Avant-Garde' in Wanda Strauven (ed.) *The Cinema of Attractions Reloaded* (Amsterdam: Amsterdam University Press, 2006), 381–388; 384.

24 Tom Gunning, 'An Aesthetic of Astonishment: Early Film and the (In)Credulous Spectator' in Linda Williams (ed.) *Viewing Positions: Ways of Seeing Film* (New Brunswick, NJ: Rutgers University Press, 1995), 114–133; 125, 126.

25 Fisher, *Wonder*, 6.

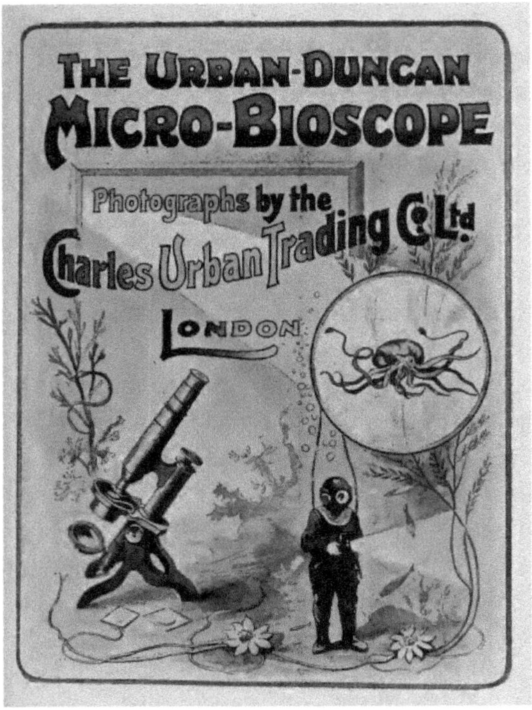

Figure 2.2 A poster for the *Urban-Duncan Micro-Bioscope* shows that its promise of visual revelation extended to underwater imagery. Image printed in the Urban 1903 catalogue.

Microcinematography was in this context recruited as a technique that could revitalise the marvels of cinematic motion, as proved by this 1903 review: 'A few years ago it was thought sufficiently wonderful to show the picture of a frog jumping. Go to the Alhambra this week and you may see upon the screen the blood circulating in that same frog's foot'.[26]

Significantly, this statement chimes with the Urban catalogue's summaries of *The Circulation of Protoplasm in the Waterweed* (1903) and *Circulation of the Blood in The Frog's Foot* (1903). The latter is described as showing 'a frog swimming in a tank', then the 'frog's webbed foot' and then 'the veins in the web of a frog's foot'.[27] The 'zooming in' effect intimated here reveals

26 'Daily Telegraph (18 August 1903)', 9, reprinted in 'Reprints of Press Articles Re: The Urban-Duncan Micro-Bioscope' (September 2013), 3–16, Charles Urban Papers, URB-10, National Science and Media Museum, Bradford, UK. For a comprehensive account of these reviews see Gaycken, *Devices of Curiosity*, chapter 1, 15–53.
27 'We Put the World Before You', 85.

that differently dimensioned images were highlighted not only in relation to various films within programmes but also within the textual space of the films themselves as a way to mitigate the potential illegibility of the magnified image and accordingly increase its wondrous effect. This is confirmed by *Cheese Mites* (1903), the most famous of *The Unseen World* films, and one of the few to survive. It starts with a medium profile shot of Duncan eating his lunch while using a magnifying glass to inspect a piece of Stilton, which offers the cue for the enlarged image of the titular mites. The recourse to an optical device within the diegesis to facilitate the transition between shots of differing scale was not an uncommon strategy in early cinema, famous examples including *Grandma's Reading Glass* (G.A. Smith, 1900) and *As Seen Through a Telescope* (G.A. Smith, 1900). But *Cheese Mites* offers a telling example not only because no magnifying glass could achieve the magnification of a microscope, but because Duncan's wide-eyed response to the amplified sight would have functioned, as Gaycken notes, as 'a self-conscious performance of experiencing perceptual novelty' that was to mirror the spectator's own.[28]

In fact, a promotional photograph recurrent in the catalogues and in the press showed Duncan, in a laboratory, handling organisms under a microscope. Moreover, the Urban catalogue recommended films of naturalists at work to act as transitions for microscopic views, including *The Naturalist at Work with the Microscope* (1903) and *The Naturalist Fishing for Specimens* (1903), both of which reiterate the Urban-Duncan films' roots in the field of natural history.[29] Indeed, while Derek Bousé contends that the term 'natural history film' begins to 'show up in trade journals around 1913', this was certainly not the case in the UK, where Duncan's films were often set against contemporaneous natural-history practices and representations as a strategy to heighten their wonder effect by way of comparison.[30]

Natural history's mediated ocularcentrism is well documented: from improved woodcut illustrations and engravings in literary publications through to dioramas of taxidermic specimens and spectacular displays of fossils and the like in natural history museums, the subject matter relies on technical mediation and the powers of the visual to inspire wonder.[31] The Urban collection of films was appraised in the press in relation to this

28 Gaycken, *Devices of Curiosity*, 33.

29 'We Put the World Before You', 86, 87.

30 Derek Bousé, *Wildlife Films* (Philadelphia, PA: University of Pennsylvania Press, 2000), 37. See for example Percy Smith's quote in article referenced in footnote 40.

31 Drouin and Bensaude-Vincent, 'Nature for the People', 410; Louson, 'Taking Spectacle Seriously', 26–31.

aesthetic and cultural matrix, be it in terms of its higher dose of entertainment when compared to 'curiously-antiquated encyclopaedias';[32] its inclusion of moving sea life that '"Little Mary" would treat with scant consideration' (an allusion to Mary Anning, a British collector of fossils who had died in 1847); or as a way to underline its superiority to taxidermy. Thus, in a 1903 *Daily Telegraph* article, the manager of the Palace Theatre was quoted as saying: 'I see no other way of picturing life in being. You can have stuffed animals in any position you like, but the public don't care for them'.[33] *The Bioscope* editor likewise remarked in 1911: 'The ordinary [photographic] camera can do a great deal but *life* is not there; for aught we know the animals might be dead, and the picture produced by the aid of a skilful taxidermist and a background'.[34]

In contrast with the immobility of natural history visual exhibits and illustrations, the beings on display in the Urban-Duncan films were lauded in terms of an unrivalled kinetic vitalism. A cursory glance at the catalogues of the time is enough to discern the hyperbolic language recruited: 'The pelicans rush from their pen [...] tossing their heads, jostling each other' (*Feeding the Pelicans*, 1903); 'Great water beetles rush backwards and forwards jostling and knocking each other about' (*Pond Life*, 1903); 'the full-grown newt [...] is seen actively swimming about' (*The Newt*, 1903).[35] The same is true of the microscopic films, which, according to the catalogues, depicted 'the restless stream of protoplasm' (*The Circulation of Protoplasm in the Waterweed*, 1903); the 'constant and varied motion' of the freshwater infusorian (*The Freshwater Infusorian*, 1903); and the 'restless unceasing movement' of typhoid germs that 'have been photographed ALIVE' (*Typhoid Bacteria*, 1903).[36]

Motion and life were further equated in these films' critical reception, which underlined that, in addition to magnification, 'Mr Urban found a way to photograph [animals] alive and in movement'.[37] One review mused: 'The chief impression [*The Unseen World* films] leave behind is an extraordinary

32 'Popular Science at the Alhambra', Reprints of Press Articles, 15.

33 'Nature on the Stage', *Daily Telegraph* (21 October 1903), reprinted in 'We Put the World Before You', 78–80; 78.

34 The Editor, 'Education by Cinematograph; Being a Review of the World already Accomplished in this Direction by Scientists throughout the World, and Appealing to the Government of this Country to Open their Eyes to the Educational Value of Moving Pictures', *The Bioscope* 225 (2 February 1911), 5, emphasis in original.

35 'We Put the World Before You', 62, 65.

36 Ibid., 85–86.

37 'New Camera Wonders', Reprints of Press Articles, 3.

sense of movement. Rest, repose, stillness, seem absolutely opposed to the principles of Nature'.[38] By disclosing movement in organic forms hitherto deemed inanimate, microscopic films were therefore praised as opening up a new realm of the visible that, in its turn, engendered a generative perception of nature as 'in a turmoil of life'.[39]

This idea gained a new lease of life from 1911. This was down to the appearance of 'speed magnification' as devised by Percy Smith, who took charge of nature films at Urban (and later its subsidiary Kineto) after Duncan left in 1908 for reasons unknown. In the UK, Smith is often remembered for the *Secrets of Nature* (1922–1933) film series he made in conjunction with his wife and editor Mary Field, first for British Instructional Films and later British-Gaumont Instruction. Like Duncan, Smith began his career as a gentlemanly naturalist and member of London's Quekett Microscopical Club. There he served as secretary from 1904 to 1909 and met Urban, who convinced him to abandon his job as a clerk for the Board of Education and embrace filmmaking as a profession. Although, as Smith once recalled, he 'had been taking "dead" nature photos for several years', it was only when he 'met Mr. Urban and he induced me to take up film work' that he began to produce *living* pictures.[40]

An ingenious inventor and tinkerer, Smith developed and perfected what we today call the time-lapse technique by attaching bits and pieces to a camera so that it would automatically take snapshots at repeated intervals. Although time-lapse had been used in professional and amateur experiments before, his films were the first to be 'shown widely to general audiences' and far predated those of botanical life developed by Dr. Jean Comandon in France.[41] By compressing time, Smith's films revealed the organic processes, or 'life-history', of animals and plants in a couple of minutes. One such film, *From Egg to Chick* (1911), traced 'the development of the embryo in a hen's egg through all the stages of the hatching of the chicken', according to its catalogue description.[42] A promotional article drove the point home: here was

38 Ibid., 4.

39 'Pictures of the Unseen', Reprints of Press Articles, 4.

40 F. Percy Smith, 'Experiences of Mr Percy Smith', *The Kinematograph & Lantern Weekly* (30 March 1911), Charles Urban Papers, URB-8-1.

41 Gaycken, *Cabinets of Curiosity*, 67. See also Oliver Gaycken, 'The Secret Life of Plants: Visualizing Vegetative Movement, 1880–1903', *Early Popular Visual Culture* 10:1 (2012): 51–69; Hannah Landecker, 'Microcinematography and the History of Science and Film', *Isis* 97:1 (2006): 121–132. On Jean Comandon's films, see Paula Amad, *Counter-Archive: Film, the Everyday, and Albert Kahn's Archives de la Planète* (New York: Columbia University Press, 2010), 226–260.

42 'From Egg to Chick', New Kineto Film Catalogue (1911); 'From Egg to Chick', promotional description (no date or source), both URB-8-1.

'a process never witnessed by human eyes'.[43] Other films – many using Urban's newly launched Kinemacolor, a two-grading colour system – focused on the previously unseen motion of flowers and plants, which, Smith reiterated, were 'shown upon the screen as *living objects*'.[44] The description of *The Germination of Plants* (1911) in the catalogue is emphatic: it notes that 'our partial human vision is often deceived and fails to recognise phenomena not clearly manifest' and concludes that the 'Kinematograph reveals to our astonished eyes the wondrous happenings' of seeds, 'open[ing] up a new world to our gaze'.[45]

Judging from the press response, this astonishing new world was perceived as similar in kind to the one made visible by microcinematography both on aesthetic and ontological levels: the former because the wondrous visual experience was never before seen, the latter because the botanical forms on display were revealed as endowed with agency and volition, literally brought to life. One reviewer wrote: by 'means of speed magnification, the actual movements of bud and foliage takes place as the bloom bursts from the bud, and thus transforms them, as it were, into sentient beings'.[46] This idea was reiterated by Leonard Donaldson in his 1912 book *The Cinematograph and Natural Science*, which championed cinema as an aid for scientific research:

> The modern student of plant life no longer rewards the objects of his study as so many things which merely demand classification and arrangement, and whose history is exhausted as soon as a couple of Latin or Greek names have been appended to each specimen [...]
> To the cinematographer the plant is no longer a kind of half inanimate being, but stands revealed in so many feet of film as an organism exhibiting sensitiveness, often showing likes and dislikes, possessing its own way of life.[47]

Donaldson's reference to the traditional, and in his view now exhausted, tenets of natural history (classification, arrangement and morphological description), as well as his mentioning of a 'modern botany' concerned with sentient and organic life, expose the crossroads at which the study of nature

43 '"From Egg to Chick"', *The Bioscope* 253 (17 August 1911), 345.

44 'Experiences of Mr Percy Smith', *The Kinematograph & Lantern Weekly* (March 30 1911), 74, URB-8-1, emphasis mine.

45 'The Germination of Flowers', New Kineto Film Catalogue, URB-8-1.

46 'Botany by Bioscope: Birth of a Flower Depicted on the Screen', *Evening Standard* (11 March 1911), URB-8-1.

47 Leonard Donaldson, 'Sidelights on Scientific Cinematography: 1. The Motion Picture in Botany', *The Bioscope* 291 (9 May 1912), 407–408; 407.

found itself around the turn of the twentieth century, and how cinema was recruited to the project of modernising natural history in tune with the emergence of new biological sciences. He ends by noting that, thanks to the cinema, 'the experimentalist may be led to still deeper questions of the philosophy of organic nature', which is 'the fundamental problem which underlies all biology – the origin and development of the varied forms of life that people our globe'.[48]

The films by Duncan and Smith are often situated against this new field of experimentalism brought about by biology as a science of life. I hope to have shown that many of their films were in fact praised in terms of their contribution to a developmental vitalism in conformity with the aesthetic and processual sensibilities of the new biology, and that herein resided their perceived aesthetics of wonder. As Hannah Landecker observes, turn-of-the century uses of cinema for scientific research 'presented the haunting possibility of capturing over time phenomena that had escaped static means of representation such as histology, photography, or drawing', which underlined 'the importance of moving from a static to a dynamic medium in the study of life'.[49] Cinematic techniques such as microcinematography and time-lapse were, in this context, regarded as assets from which biological studies could profit. While Duncan and Smith pertained to amateur natural history rather than the professional sciences, their films were also conceptualised on the basis of the dynamism they brought to an understanding of nature and its vital processes against static modes of representations such as 'dead' pictures, fossils, dioramas and taxidermy. In making visible the hitherto imperceptible and unnoticeable movement of macro and microorganisms, their films injected a new *vitality* into natural history.

Never Before Seen

Both Duncan and Smith neatly fit into the category of the naturalist as defined by Pratt: 'a benign, often homely figure, whose transformative powers do their work in the domestic contexts of the garden or the collection room'.[50] From this perspective, it might seem incongruous to place them next to early expedition and wildlife filmmakers such as Cherry Kearton and Captain Robert Falcon Scott, both of whom are instead closer

48 Donaldson, 'Sidelights on Scientific Cinematography', 408.

49 Landecker, 'Microcinematography and the History of Science', 122, 125.

50 Pratt, *Imperial Eyes*, 33.

to what Pratt characterises as the imperialist figure of the 'navigator or the conquistador'.[51] From the perspective of natural history's aesthetics of wonder, however, both modes of filmmaking, I wish to argue, vigorously complemented each other. For, if Duncan and Smith deployed techniques to show invisible and microscopic organic phenomena on a local level, Kearton and others promised to reveal the unseen as incarnated in big game in natural surroundings on a global stage. In particular, the discourse of wonder surrounding wildlife films in terms of a direct access to unperturbed life and untouched nature is one that persists to this day, thus demanding that we unpack its conceptual ramifications and affiliations before we move on to examine contemporary nature programming.

That wildlife and popular science films were seen as two sides of the same coin is confirmed by their critical reception. One 1911 article, for example, noted that 'the stay-at-home camera expert also has trials of patience and endurance, his victories and disappointments, that may be quite as interesting in their way as those [...] engaged in the search for pictures in wild and uncivilized quarters of the globe'.[52] Leaving aside for a moment the author's troubling if customary conflation of wildness and non-civilisation, later in the article Smith himself is quoted saying:

> Natural history work of a 'homely' type might be thought rather tame, but is not by any means the case. It sounds great when one hears of the hours of excited waiting indulged in by the photographers of big game and rare birds; but I can recommend plenty of little back garden studies which would drive the average man crazy.[53]

Patience, endurance and waiting were indeed the virtues most highlighted in the promotional discourses surrounding wildlife cinema's optical revelations, and Smith's allusion to big game and rare birds leaves no doubt that he had the photographer and filmmaker Cherry Kearton in mind.

Born in 1871 in Swaledale, Kearton moved to London in 1887 to work as a commercial clerk at Cassell publishers. This was the workplace of his older brother Richard, whose first natural history book, *Birds' Nests, Eggs, and Egg Collecting*, appeared in 1890. In 1895, Richard published his second book, *British Birds' Nests: How, Where and When to Find and Identify Them*, which was illustrated with Cherry's photographs and comprised, as

51 Ibid., 33
52 'Experiences of Mr Percy Smith', 74.
53 Ibid.

John Bevis notes, 'the first systematic *photographic* catalogue of a natural taxonomy'.[54] Cherry embarked on a career as a professional photographer, and the collaboration between the brothers continued in Richard's *Wild Nature and a Camera* (1897), *Wild Nature's Ways* (1903) and *Wild Life at Home* (1907), to cite just a few.

The Keartons' developing branch of nature photography coincided with the rise of conservation movements in the UK, with the Society for the Protection of Birds, for example, founded in 1889.[55] Theirs was a project predicated on the idea that it respected the wildness of the natural world via pictures of living animals taken 'direct from nature' (the subtitle of *Wild Nature's Ways*), in contrast with those of animals in captivity and of taxidermic specimens. Just as importantly, 'direct from nature' in the brothers' conception referred to the fact that these photographs purported to show animals, from a close distance and without the help of telephoto lens, unaware of human presence. For the Keartons this was paramount since they believed that humans triggered unnatural reactions on the part of animals when these became aware of being watched.

Perils and frustrations thus had to be endured. Not only was it necessary to find the right spot from which to survey and capture undisturbed life, a number of artefacts had to be concocted to help the brothers 'to observe whilst remaining unobserved, to hear without being heard'.[56] These included bespoke camouflage suits of reversible material, as well as a number of makeshift hideaways and props: an artificial tree trunk made out of bamboo struts and galvanised mesh and fabric (figure 2.3); an artificial rock made of stretched canvas on deal frames; an artificial rubbish heap; a tent covered with heather; and, most bizarrely, a stuffed ox (figure 2.4) and a stuffed sheep, designed by a taxidermist, from whose empty insides Cherry could shoot his photographs through a peep hole.[57]

However, if these resourceful devices were meant to secure the invisibility of the photographer in the natural world, through a reversal gesture they were made visible for the human observer as a way of boosting revelationism. Among the more than 200 illustrations of 'birds and wild beasts' in *Wild Nature's Ways*, for example, several depict the hides mentioned above, as well as Richard and Cherry in positions and situations intimating physical

54 Bevis, *The Keartons*, 38, my emphasis. My biographical account of the Keartons is indebted to Bevis.

55 Ibid., 148.

56 Richard Kearton, *Wild Nature's Ways* (London, Paris, New York and Melbourne: Cassell and Company, 1903), 1.

57 Bevis, *The Keartons*, 86; see esp. chapter 4, 85–105, for a detailed discussion of these hides.

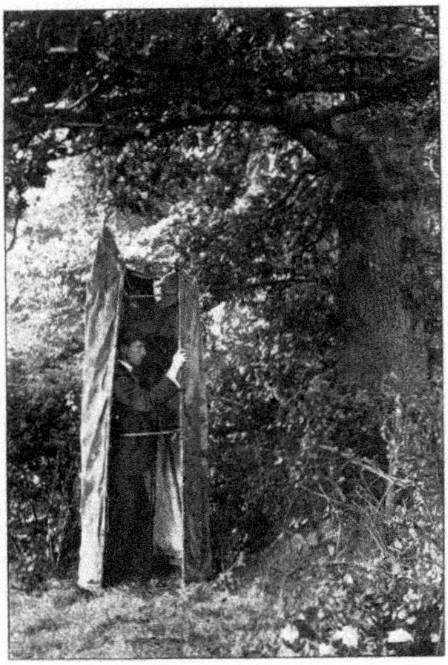

Figure 2.3 Cherry Kearton photographing from an artificial tree trunk. Image from *Wild Life at Home* (1898).

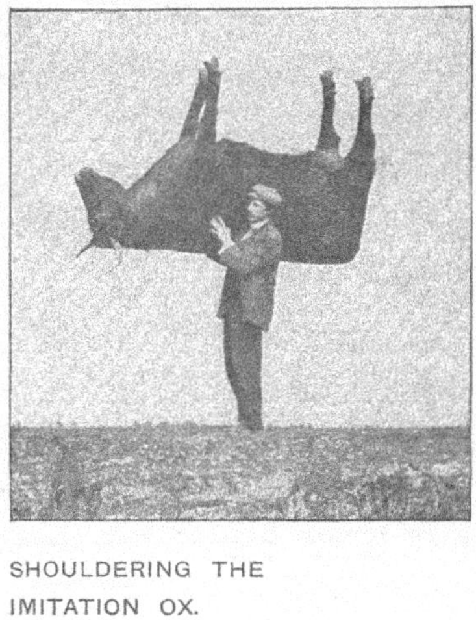

Figure 2.4 Cherry Kearton carrying a stuffed ox used as a hideaway to photograph animals. Image from a 1909 edition of *Wild Nature's Ways*.

tribulations and operational difficulties.[58] Behind-the-camera proceedings then made their way into their first films for Charles Urban, *Sea-Bird Colonies* (1907) and *Wild Birds at Home* (1908), both of which, one review noted, were filmed with an 'absolutely noiseless moving picture camera' so as not to disturb the birds.[59] Whereas in the former film, according to its catalogue description, Cherry is seen 'hiding underneath a dummy sheep', in the latter his 'hazardous descent' and 'perilous work' is recorded as he is roped down a cliff to take pictures of sea birds.[60] The catalogue concludes: '[n]ever before have animated pictures of mother birds feeding, caring for, tenderly brooding their young, been secured'.[61]

Hazardous and perilous shoots, in fact, quickly became Cherry Kearton's trademark, especially from 1910, when he embarked on a solo filmmaking career. Refashioning his identity as a global expedition filmmaker of big game, Kearton's first, critically flopped attempt was *Roosevelt in Africa*

58 Kearton, *Wild Nature's Ways*, 4–5, 32.

59 Thomas Bedding, 'Photographing Outdoor Subjects', *The Bioscope* 158 (21 October 1909), 29.

60 'Urbanora: The World's Educator, Catalogue: Scientific and Educational Subjects' (August 1909), catalogue, URB-10–16, 25, 27.

61 Ibid., 25, 27.

(1910). The result of a premeditated encounter with the former US president and amateur naturalist while he was on a hunting safari, it was followed by *Lion Spearing* and *Lassoing Wild Animals in Africa*, both released in 1910.[62] By 1913, Kearton was already known as *The Man Who Dared*, to cite his 1913 film, promoted as his encounter with 'wild and ferocious denizens', which had 'never been done before and can never be done again' (figure 2.5).[63] Kearton trailed on the colonial networks of the British Empire and reaped its benefits with the mission to document *Wild Life across the World*: the title of his 1913 photography book, which, as it turned out, included a foreword by Roosevelt himself reiterating the authenticity of its animal pictures.[64]

Kearton's films were often deemed to convey not only 'real life in place of guide books' by showing animals going about their business, but also to offer virtual global expeditions, the 'equivalent to a three years' journey through India, Borneo, America and Africa', to cite one commentator at the time.[65] Many of his films set on the African continent during the 1910s and 1920s subscribe to Cynthia Chris's contention that 'ethnographic and wildlife filmmaking would often conflate, representationally and discursively, observations of animal behavior and assumptions about the behaviors of nonwhite people'.[66] Thus, as late as 1929, a publicity programme for Kearton's *Tembi: A Story of the Jungle* boasted its disclosure of 'well-known and little-known animals [...] unmindful of, or indifferent to, the camera eye', while 'revealing something of the life and habits of a coloured race in the Dark Continent'.[67] But Kearton's films were problematic less for their stereotypical representations of the 'savage' than for what Nwachukwu Frank Ukadike describes as early cinema's disproportionate 'portrayal of Africa as a reservoir of wild animals and impenetrable jungle', often to the detriment of 'human life'.[68] To that end, the newest technology had to be procured and promoted, such as the Aeroscope, used in his *Across Central Africa from East to West* (1914) and described in one publicity booklet as 'a

62 On Kearton's critically flopped *Roosevelt in Africa*, see Gregg Mitman, *Reel Nature: America's Romance with Wildlife on Film* (Seattle and London: University of Washington Press, 2009), 5–16.

63 '*The Man Who Dared, or a Naturalist's Wanderings with Cinema Camera in Many Lands*', supplement to *The Bioscope* 341 (24 April 1913), xxxiv.

64 Mitman, *Reel Nature*, 13.

65 The Compleat [*sic*] Traveller, 'Round the World by Film', *The Bioscope* 340 (17 April 1913), 171.

66 Cynthia Chris, *Watching Wildlife* (Minneapolis: University of Minnesota Press, 2006), 2.

67 '*Tembi: A Story of the Jungle*, Polytechnic Cinema Theatre', cinema publicity programme (London), week commencing 4 November 1929, Bill Douglas Cinema Museum (BDCM), University of Exeter, UK. Item number: 19966.

68 Nwachukwu Frank Uradike, *Black African Cinema* (Berkeley, Los Angeles and London: University of California Press, 1994), 40; as also noted by Chris, *Watching Wildlife*, 5.

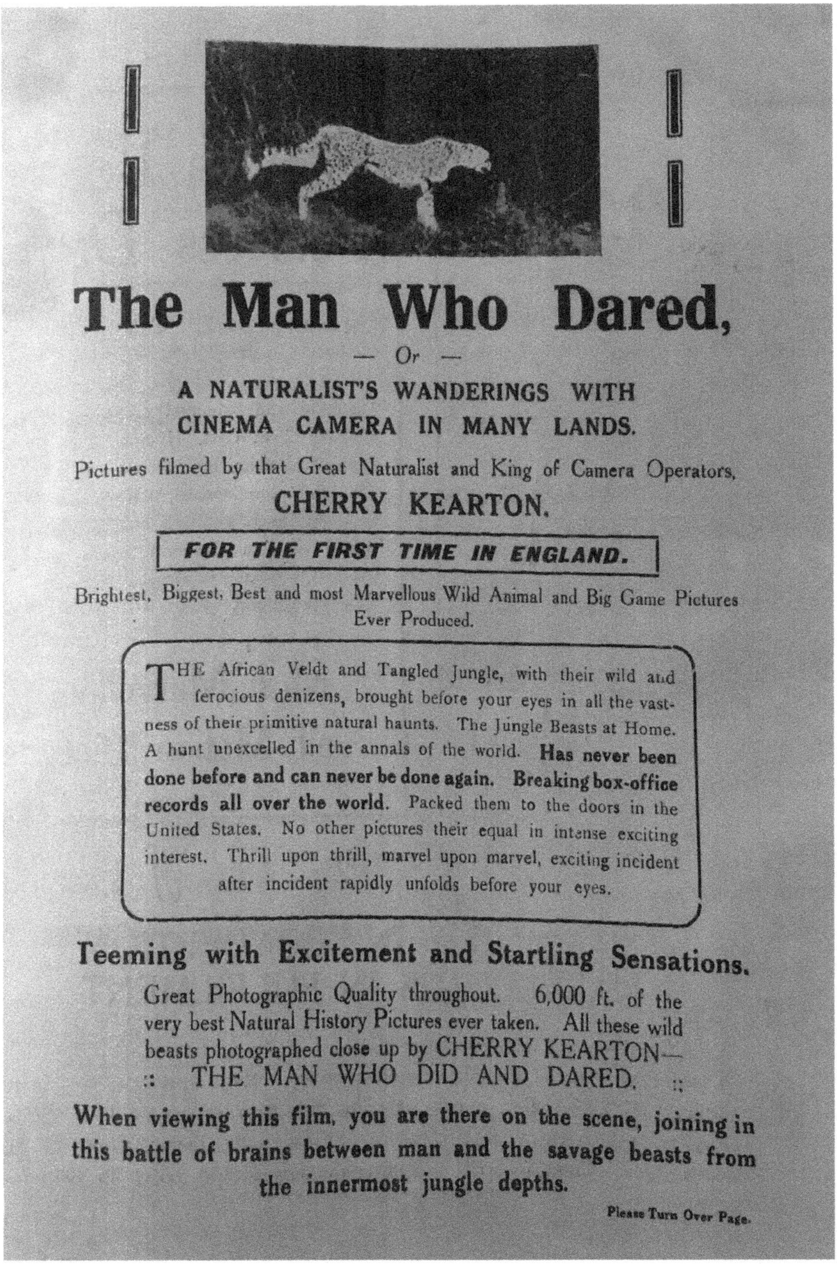

Figure 2.5 'Has never been done before and can never be done again': advertisement for Kearton's *The Man Who Dared* (1913). Source held at the British Library.

new air-driven machine of a silent type' that allowed 'photographs of the most suspicious animals to be secured'.[69]

In his essay 'Why Look at Animals?' (1977), John Berger contends that the turning of animals into visual spectacle in modernity is a response to their disappearance from urban centers: 'Zoos, realistic animal toys and the widespread commercial diffusion of animal imagery, all began as animals started to be withdrawn from daily life'.[70] Looking at animal photography books, Berger comments:

> Technically the devices used to obtain ever more arresting images – hidden cameras, telescopic lenses, flashlights, remote controls and so on – combine to produce pictures which carry with them numerous indications of their normal *invisibility*. The images exist thanks only to the existence of a technical clairvoyance.[71]

Kearton was one of the pioneers of such a mode of seeing: from sheep decoys through to air-driven equipment, his images were said to elicit wonder precisely because they made visible this 'normal invisibility' of animals. As such, they participated in a contemporaneous discourse according to which, to cite Gregg Mitman, 'nature's wilderness was the therapeutic balm that healed the debilitating effects of modern urban life'.[72]

Yet disappearance in Kearton's work also emerged in tandem with growing concerns over animal extinction, especially in Africa. As Heise notes: 'From the 1920s on, conservationists of various stripes [...] focused in a colonial manner on the preservation of wildlife in Africa for big-game hunting'.[73] Although Kearton's first expedition films showed graphic images of hunting and killing, and while he was adept at hunting during his early travels, he somehow managed to build for himself the image of someone benevolent towards and trusted by animals: 'the first man in the world to shoot with a camera rather than a rifle', as his wife Ada recalled in the 1950s.[74] Take for instance this 1913 cartoon (figure 2.6), published in *The*

69 '*Across Central Africa from East to West*: The Thrilling Story of a Naturalist's Journey with a Cinema-Camera through the Heart of the Dark Continent', publicity press book, BDCM. Item number: 18668.

70 John Berger, *About Looking* (London: Bloomsbury, 2009), 26.

71 Ibid., 16, emphasis in original.

72 Mitman, *Reel Nature*, 13.

73 Heise, *Imagining Extinction*, 20.

74 Bevis, *The Keartons*, 147–167; Ada Cherry Kearton, *On Safari* (London: Robert Hale, 1956), 49.

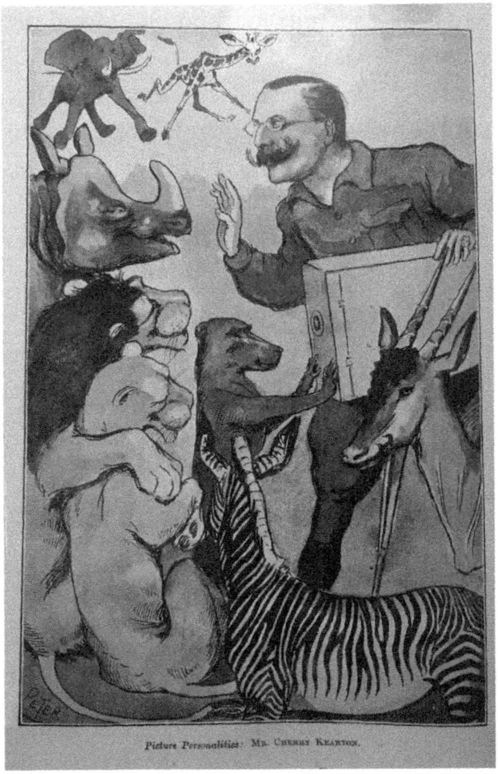

Figure 2.6 An illustration in the trade magazine *The Bioscope* (1913) pictures Kearton as a trusted friend of wild (and friendly) animals. Source held at the British Library.

Bioscope, where Kearton is surrounded by docile animals, including a lion, a lioness, a rhinoceros and a zebra, all devotedly posing and smiling for the camera. By 1923, upon the publication of *Photographing Wild Life across the World* (an updated version of his 1913 book), Kearton was expressing his dismay that 'animal life is rapidly disappearing' in Africa thanks to 'the thoughtless selfishness' of settlers and big game hunters.[75] While, as Bevis notes, Kearton's defence of untamed wildness is marred with contradictions given his career is replete with examples that prove otherwise, his work lent credence to a perception of nature and animal life as increasingly threatened by humankind; hence his belief, widely reinforced by natural history systems of representation, that 'the territory of "wild nature" vanishes once humans are detected within its frontiers'.[76]

75 Cherry Kearton, *Photographing Wild Life across the World* (London: J.W. Arrowsmith, 1923), 14.

76 Bevis, *The Keartons*, 147.

In some ways, the wildness of this 'wild nature' was discursively pitched against the 'melancholy, apathetic' behaviour of 'animals [filmed] in a zoological garden'.[77] The perceived values of on-spot wildlife cinema when compared with pictures taken in the zoo, negatively conceptualised in terms of its 'artificial collection of animals' and 'unnatural surroundings', was a dominant perception at the time.[78] As one reviewer put it: 'The cinema enables us to transplant African forests, Australian bushes, and American plains into any town or village, so that by its means we can study at leisure and with safety every species of creation'.[79] In this light, cinema was seen as both a better alternative than zoological gardens and an aid for natural history museums: indeed, Kearton himself was of the opinion that a collection of 'cinematograph pictures dealing with animal life should be placed in every important museum'.[80]

Kearton's allusion to museum collections takes us back to his roots in amateur natural history. Jean-Baptiste Gouyon even goes on to argue that '[i]n the same way that Kearton's practice of filming amounts to that of collecting specimens, each of the footages he displayed can be considered as a *naturalium*, the name given to the natural objects exhibited in early modern cabinets of curiosities'.[81] This comparison makes sense in a number of ways, yet, as with Duncan and Smith, I would add that wildlife filmmaking was perceived as qualitatively reenergising the wonders of natural history displays by way of cinematic motion. This was highlighted, first, in terms of a footage selection that discarded uneventful moments in favour of 'only the notable and the unique, relieving you of the thousands of tedious commonplaces which you could not avoid if you made the journey alone'.[82] Second, the novel sight of animal life in the field was for some equal to the microscopic revelation of the 'minute marvels of nature' in that it brought 'home even to the expert a great many points which are overlooked in the immense out of doors'.[83] The reviewer concludes: 'For example, who was

77 The Editor, 'Education by Cinematograph', 7.

78 Man About Town, 'New Wonders of Darkest Africa', *Pathé Cinema Journal* 1:7 (15 November 1913), 29; Joseph Best, 'Is the Zoo Cruel? Real Nature Study by the Cinema', *Pathé Cinema Journal* 2:27 (4 April 1914): 43–44; 43.

79 Best, 'Is the Zoo Cruel?', 43.

80 Cherry Kearton, quoted in *The Times*, reprinted in 'Cherry Kearton: *The Man Who Dared*', *The Bioscope* 343 (8 May 1913), 448–449.

81 Jean-Baptiste Gouyon, 'From Kearton to Attenborough: Fashioning the Telenaturalist's Identity', *History of Science* 49:1 (2011): 25–60; 37.

82 The Compleat Traveller, 'Round the World by Film', 171.

83 Percival Westell, 'What the Cinema Can Tell Us of Animal Life', *Pathé Cinema Journal* 2:31 (2 May 1914), 52.

aware [...] that the hedgehog was such a good swimmer; that the lizard was so fond of drinking; that such a ferocious beast as the polecat could be made such a pet; that the squirrel was so very ratlike [*sic*]'?[84] Like a global macroscope of outdoor nature, expedition wildlife film revealed an unseen world by widening the scope of the world into the whole world.

Yet, as contemporaneous evidence also shows, the lasting impact of these cinematic wonders was already up for debate. Take this 1913 article by Frederick A. Talbot:

> The modern picture palace-goer is a difficult person to please. He has become so accustomed to the cinema, has had so many remarkable scenes unfolded before him therewith, that he has come to consider nothing impossible. No effort is spared by enterprising firms to fulfil this demand, and extreme risks are incurred to appease the public appetite for the extraordinary and novel.[85]

For Talbot, who had authored the book *Moving Pictures: How They Are Made and Worked* (1912), freshly captured images of the Antarctic from the Shackleton and Scott expeditions, as well as many films being concurrently filmed at the North Pole, attended to this public demand to be astonished by the 'extraordinary and novel'. Worthy of note here is that filmed expeditions were regarded both as exploration films and 'natural history' pictures. While Duncan had acted as an 'advisory photographic expert' to Shackleton, the ill-fated Scott expedition, filmed by cinematographer Herbert G. Ponting, was touted as showing 'the most marvellous revelations of the animal kingdom ever recorded by means of the camera'.[86] Indeed, these films even received subtitles such as *Animal and Bird Life in the South Pole* and *Animal Life in the Antarctic* in subsequent versions of Ponting's footage.

Joining terrestrial animals in jungles and at the poles were sea creatures, thanks to advancements in underwater filmmaking. An example is the 1912 film *An Otter Hunt*, described in *The Bioscope* as 'the revelation of the unknown' and as 'not a "fake" picture, made with a glass box; in this case the photographer was in the tank, not the fishes', filming 'under absolutely

84 Westell, 'What the Cinema Can Tell Us', 52.

85 Frederick A. Talbot, 'Risks and Trials Incurred by the Cinema Camera Man', *Pathé Cinema Journal* 1:7 (15 November 1913), 17.

86 'The Bioscope in the Antarctic', *The Bioscope* 142 (1 July 1909), 9; 'With Captain Scott in the Antarctic, and Animal and Bird Life in the South Polar Regions' (n/d), promotional cinema programme, BDCM. Item number: 18663.

novel conditions, in a manner never before attempted'.[87] Another is the *The Williamson Expedition Submarine Motion Pictures* (1914), by the brothers Ernest and George Williamson, as described in a promotional booklet:

> When the two poles had been discovered, it seemed as if the whole earth was now an open book; that the limits of geographical conquest had been reached; and having no further regions to explore, man could rest on his laurels.
> But all who thought thus obviously forgot that there still remained vast regions under the oceans and seas almost entirely unknown.[88]

As with the promotion of Kearton's work, the booklet continues by noting how the pictures were obtained at great, ingenious effort via the construction of a 'deep-sea tube, which has made possible the exhaustive examination of submarine life'.[89] And it concludes, immodestly: 'This, then, is the story of how the sea has given up its secrets, and at last bowed to the restless genius of man'.[90]

Denis Cosgrove has noted that the turn-of-the-century rush to 'conquer' the poles, 'global destinations of ultimate inaccessibility', offered 'individuals and nations a competitive sense of global mastery comparable only to circumnavigation by sea or air or the ascent of high mountains'.[91] The deployment of film to record these voyages, which took place at the same time as Kearton's colonial adventures, reveals how cinema itself was recruited as a tool of conquest, while 'the earth is being ransacked for pictures', as one commentator put it.[92] Such ransacking was made particularly visible in a series of advertisements for camera equipment by the British company Walturdaw, published in the pages of *The Bioscope* during 1912–1913. In one of these, a camera sits on a support at the South Pole while admired by 'locals' pushing carts with ice (figure 2.7). In another, a camera morphed with a human body and replacing its upper part is mounting an enormous ostrich and, in another ad, a horse jumping, here meant to emphasise the

87 'Pictures Under Water: *An Otter Study*, by Kineto, Limited', *The Bioscope* n. 288 (18 April 1912), 205.

88 'The Story of the Williamson Expedition Submarine Motion Pictures', 3. Publicity booklet, BDCM. Item number: 18272.

89 Ibid., 5.

90 Ibid., 12.

91 Denis Cosgrove, *Apollo's Eye: A Cartographic Genealogy of the Earth in the Western Imagination* (Baltimore, MD, and London: The Johns Hopkins University Press, 2001), 217.

92 Man About Town, 'New Wonders of Darkest Africa', 29.

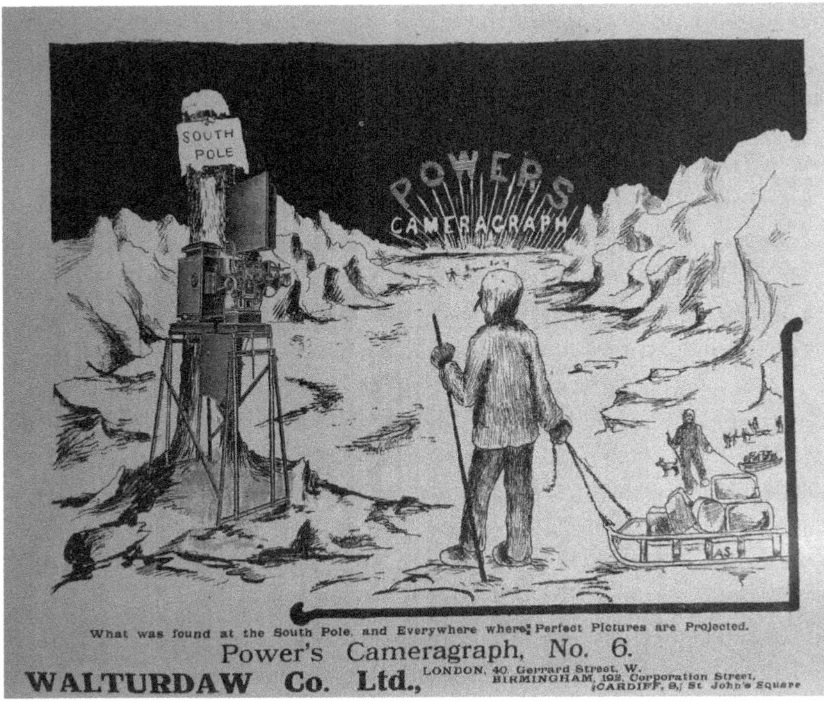

Figure 2.7 Advertisement for Power's Cameragraph (1912) nods to contemporaneous films shot at the South Pole. Source held at the British Library.

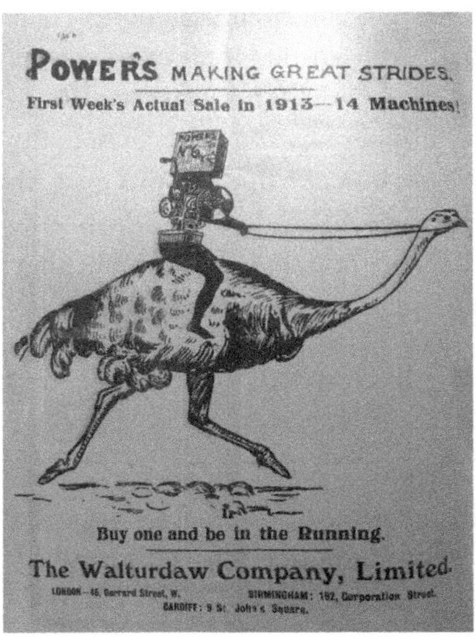

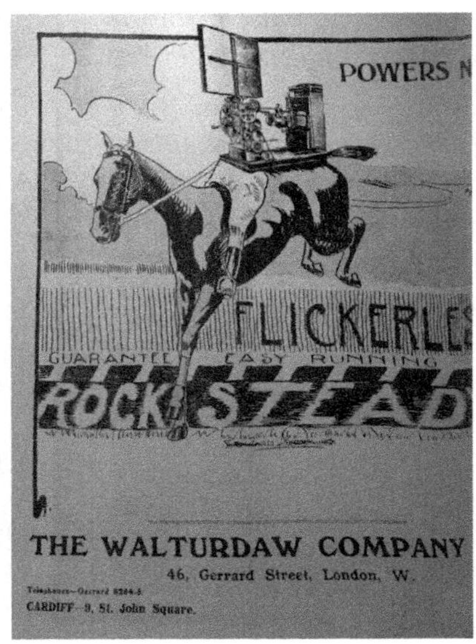

Figures 2.8 & 2.9 Advertisements for Power's Cameragraph (1912–13) make recourse to animals so as to laud the camera's 'flickerless' agility. Source held at the British Library.

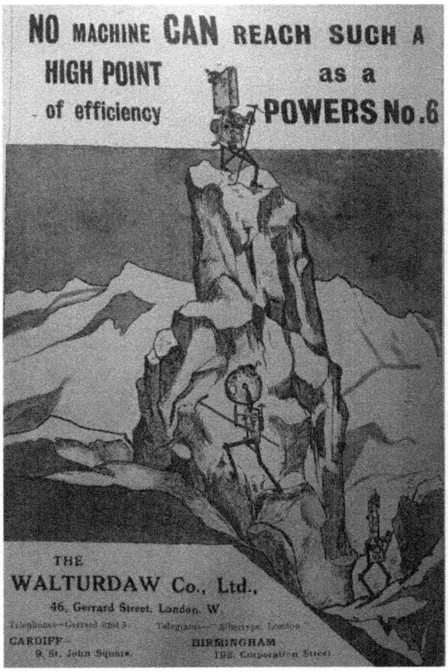

Figure 2.10 Advertisement for Walturdaw (1913) lauds the camera's flexibility by referencing expedition films. Source held at the British Library.

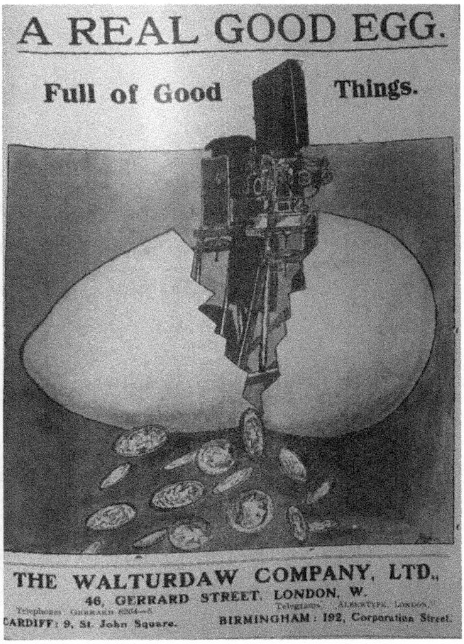

Figure 2.11 Advertisement for Walturdaw (1913) makes a reference to 'hatching egg' films. Source held at the British Library.

camera's 'flickerless' quality (figures 2.8 and 2.9). Another ad shows a variation of this rudimentary human-camera figure standing atop the peak of a mountain in the position of a conqueror (figure 2.10), and yet another depicts a camera hatching from 'a real good egg full of good things', with coins tumbling down (figure 2.11).

These images are striking not only because they make visible cinema's visual conquest of global space, but also because they directly allude to early cinema's natural history genres, be it specific popular science films, as in Smith's hatching egg; via recourse to the figures of animals to emphasise the agility and flexibility of cameras; or through references to expedition films in relation to the South Pole and mountain peaks. Alongside the promotional articles, publicity programmes and critical writings examined above, they evince the centrality of nature filmmaking in early cinema in the formulation of a hyperbolic discourse of the never-before-seen that aimed to augment the wondrous experience of visual surprise promised by the films themselves. They therefore provide us with a much-needed historical lens with which to examine the re-newing of this discourse in contemporary nature programming.

Never Before Seen (Again)

Shortly after Captain Scott's death in the Antarctic, in 1912, the letter and diaries he wrote during the expedition were made public. One sentence, advertised in the press as his 'last wish' to Lady Scott, read: 'Make the boy interested in natural history if you can; it is better than games. They encourage it in some schools'.[93] In 1955, Peter Scott, dutifully acting upon his father's advice, debuted on British television as the presenter of the BBC natural history series *Look*, two years before the formalisation of the BBC Natural History Unit (NHU) in Bristol.[94] Produced by Desmond Hawkins and broadcast live from Scott's Wildfowl Trust (located in his own house), *Look* consisted of interviews with naturalists who brought amateur footage of animals in the field.[95] Over 50 years later, the first episode of the 2011 BBC seven-part series *Frozen Planet*, 'The Ends of the Earth', recalled Captain Scott's fated expedition. In a helicopter flying over boundless icy expanses, Attenborough tells the viewer that we are 'following the route of Scott'; later, now standing on the polar ice cap next to the flags of Great Britain, the US, Norway, France and other countries, he remarks that it is exactly 100 years since the first human beings 'stood right here: [Norwegian explorer Roald] Amundsen, followed by Scott'. He continues:

> In those days, reaching the poles was regarded as the ultimate human endeavour and endurance, and a source of great national pride. Today, the polar regions have a rather different significance, because now we've come to understand that what happens here and in the north affects every one of us, no matter where we live on this planet.

Drawing the episode to a close, these words set the tone for the series as a whole. On the one hand, *Frozen Planet* is both a homage to, and a perfecting of, earlier explorations via new technology. Thus, in episode 6, Attenborough states that 'Shackleton's men [...] had no idea of the extraordinary spectacle that lay beneath their feet' when climbing Mount Erebus, which provides the cue for images of the 'network of caves' with 'ice formations that occur nowhere else on Earth'. On the other hand, the justification for this exploratory impetus is to be found not in national pride but in a planetary ecology. Channelling an anthropocentric story of global conquest into a holistic view

93 J. Best, 'Scientific and Educational', *Pathé Cinema Journal* 1:7 (15 November 1913): 41-42; 41.
94 Jean-Baptiste Gouyon, 'The BBC Natural History Unit', 429.
95 Ibid.

of the Earth, this opening crystallises, as we shall see, the conflicting forces BBC wildlife series often have to negotiate, with varying degrees of success.

Gouyon has shown that, after WW2, 'the British practice of natural history film-making did not resume as part of the cinema industry but became part of the development of television'.[96] Although a history of British wildlife television falls outside the scope of this chapter, some observations are in order.[97] As Gouyon points out, 'in the early 1950s two new approaches to presenting wildlife on TV emerged, each informed by a different cultural repertoire'. He goes on:

> One originated in Bristol. Involving celebrity naturalist Peter Scott, it was rooted in the culture of amateur natural history. The other, elaborated in London, featured celebrity wildlife film-makers from the 1930s, Armand and Michaela Denis, and extolled the culture of imperial big game hunting [especially their *Filming Wild Animals*, 1954]. These two cultures were two aspects of the Victorian culture of natural history, itself enmeshed in British imperialism and a nascent culture of mass consumption.[98]

These two strands moreover contrasted with the first wildlife programme ever broadcast on British television in the late 1930s: David Seth-Smith's *Friends from the Zoo* (initially *The Zoo Today*), which consisted of showing animals being sent from London Zoo to a studio in Alexandra Palace (or sometimes outside in the case of big animals). Although transmission was halted in 1939 due to the outbreak of the war, this genre was briefly revived in the early 1950s with George Cansdale's *Looking at Animals* (1951–1954).[99]

Yet there was another wildlife genre emerging in the 1950s, one identified with a young David Attenborough, who by the end of that decade had already become 'the face of natural history on the BBC, the equal to a Peter Scott'.[100] Entitled *Zoo Quest* (1954–1963), Attenborough's inaugural programme on British television showed the recording of animals in their natural habit and their subsequent capture as part of scientific expeditions, and then the live presentation of the captured animals in a studio, with

96 Gouyon, 'From Kearton to Attenborough', 40.

97 My thanks to Helen Wheatley for her precious advice on this section of the chapter.

98 Jean-Baptiste Gouyon, *BBC Wildlife Documentaries in the Age of Attenborough* (London: Palgrave, 2019), 19. On Armand and Michaela Denis's 'safari' documentaries, see also Helen Wheatley, *Spectacular Television: Exploring Televisual Pleasure* (London and New York: I.B. Tauris, 2016), 87–97.

99 Gouyon, *BBC Wildlife Documentaries*, 18–19.

100 Gouyon, 'From Kearton to Attenborough', 48.

Attenborough and specialists from the London Zoo. In some ways tapping into the zoological inflection of early television wildlife broadcasting, *Zoo Quest* simultaneously perpetuated the exploration genre still in vogue at the time and whose origins harked back to Kearton's colonial filmmaking. As Gouyon has shown, from an early stage Attenborough fashioned his identity as a trusted 'telenaturalist' in line with Kearton's image of 'intimacy with natural world' and 'the ability to film nature and animals undisturbed'.[101] But just as importantly, in its partnership with specialists from a learned society, the Zoological Society of London (ZSL), *Zoo Quest* also signalled the scientific bent that was to become identified with BBC wildlife programming and Attenborough himself as 'trustworthy sources of knowledge about nature'.[102] This was cemented when Attenborough became the head of BBC 2 in 1965 and started conceiving and producing natural history series in alliance with the scientific community and based on the understanding that the footage produced by the BBC supported and generated knowledge of animal behaviour.

Although Attenborough stepped down as head of BBC 2 in 1972, this scientifically inflected conception of wildlife programming proved enduring, even if, as with the films by Duncan and Smith, the boundaries separating science from a natural history paradigm were never completely erected so much as they became complexly entwined. This was complicated by the impact of the so-called 'blue-chip' format developed in Disney's *True Life Adventure Series* (1948–1960), the main tenets of which, according to Derek Bousé, included the 'depiction of mega-fauna', 'visual splendor', 'dramatic storyline', 'absence of historical reference points' and 'absence of people'.[103] As Morgan Richards notes, 'When the BBC finally embraced the blue chip format in the late 1960s, it cultivated a more rigorously scientific approach that served to differentiate its programmes from those of [its British rival] Survival Anglia', while simultaneously crafting a 'unique niche in the international television market'.[104] This hybrid conception received a new boost in terms of production values, global scope and aesthetic embellishment with the first multipart, or landmark, Attenborough series: *Life on Earth* (1979). Produced in conjunction with Warner Bros, *Life on Earth*'s constitutive aesthetic and institutional features – colossal budget, international co-production with

101 Ibid., 47.

102 Gouyon, *BBC Wildlife Documentaries*, 3.

103 Bousé, *Wildlife Films*, 14–15.

104 Morgan Richards, 'Greening Wildlife Documentary' in Libby Lester and Brett Hutchins (eds.) *Environmental Conflict and the Media* (New York: Peter Lang, 2013), 171–185; 176.

a view to global sales and distribution, the combination of the scientific and the spectacular – are still to this date deemed synonymous with BBC natural history programming.

The genealogies sketched above reveal the conflicting forces shaping wildlife television, including the scientific imperatives of the BBC as a public broadcasting service, the hybrid and competing traditions co-existing in television in its first decades, and the transnational influence of the blue-chip aesthetic. Without denying the importance of these lineages and their entanglements, in the remainder of this chapter I want to pursue a different route so as to explore the way in which contemporary BBC wildlife programming recycles an early-cinema aesthetic of natural wonder through the revelatory wonders of technology. In this respect, I want to suggest that the Victorian culture of natural history is still an enduring paradigm underpinning such programming in spite of its more scientific aspirations, and that this in fact may explain these series' much-debated ecological ambivalences.

The project of natural history, as we have seen, is indissociable from ocular revelationism, and as illustrated by titles such as *Look* and *Looking at Animals*, this has also been part of its configuration through television history. That said, as I hope to show, the discourse of the 'never-before-seen' has received a new boost over the last two decades owing to two interrelated phenomena. On the one hand, the appearance of high-definition technologies has translated into a new, more ferocious chapter in wildlife programming's self-promotion as wondrous revelationism through 'an aesthetics of astonishment' reminiscent of early cinema's flaunting of technology. On the other hand, the growing awareness of the global environmental crisis has imparted a new significance to unseen natural phenomena by intimating that such phenomena – landscapes, oceans, biodiversity – may never be seen again.

While *Life on Earth* is often credited with cementing the features we today associate with BBC wildlife programming, the release of *The Blue Planet* in 2001 also constituted an important milestone. As Richards notes, produced on an £8 million budget and designed to 'trial HD filming techniques', the show tellingly featured Attenborough as a voiceover narrator rather than on-camera presenter in order to heighten 'the spectacular appeal of HD wildlife footage' itself.[105] This was followed by *Planet Earth* (2006), the first BBC natural history series almost entirely recorded in

105 Morgan Richards, 'Global Nature, Global Brand: BBC Earth and David Attenborough's Landmark Wildlife Series', *Media International Australia* 146:1 (2013): 143–154; 152.

high definition and whose release coincided with changing domestic viewing habits occasioned by the appearance of rectangular and flat plasma screens. As Helen Wheatley has argued, while the natural history genre certainly 'preexisted the advent of HD broadcasting', 'it was clear that certain spectacular forms and genres were being brought to the fore at a moment when an expensive new technology needed to be sold to the masses'.[106] Eleanor Louson has similarly noted that 'even though *Planet Earth* employed many familiar elements from earlier wildlife series, it left an indelible mark on the genre thanks to its scope and visual spectacle', both achieved through 'innovative film technologies' and 'new possibilities for home viewership'.[107] Since then, a discourse of technological novelty and superlative spectacle has been progressively intensified each time a new BBC landmark series is released.

It is hard not to be impressed by the figures surrounding such series. To give one of the most recent examples: before airing, *Blue Planet II* had already been sold to over 30 countries.[108] When it premiered in 2017, it drew 14 million viewers, making it the UK's most-watched natural history series in 15 years and the third most-watched show ever in five years.[109] Shot over a four-year span in 39 countries, the show involved 125 expeditions, 6,000 hours of filming and 1,000 hours of filming in submersibles.[110] *Blue Planet II* was the crowning achievement to come out of the NHU, with a history of successes since the early 2000s, including the aforementioned *Planet Earth*, produced on a £16 million budget, and *Frozen Planet* (2011) and *Planet Earth II* (2016); while the budgets for these last two have not been released (nor that for *Blue Planet II*), it is estimated costs were 'in the vicinity of £16 million'.[111] All of these series were produced in partnerships between the NHU and international companies and TV channels (BBC America, NHK, Tencent, WDR, France Télevisions and CCTV9), and since 2009, they became part

106 Helen Wheatley, 'Beautiful images in Spectacular Clarity: Spectacular Television, Landscape Programming and the Question of (Tele)Visual Pleasure', *Screen* 52:2 (Summer 2011): 233–248; 234.

107 Louson, 'Taking Spectacle Seriously', 16–17.

108 'BBC Worldwide Makes Waves with Blue Planet II Series', *BBC* (9 October 2017), https://www.bbc.co.uk/mediacentre/worldwide/2017/blue-planet-2-presales (accessed 15 February 2019).

109 Stewart Clarke, 'BBC's *Blue Planet* Makes Ratings Splash', *Variety* (6 November 2017), https://variety.com/2017/tv/news/bbc-blue-planet-ratings-splash-david-attenborough-1202607834/ (accessed 15 February 2019).

110 Steven McIntosh, '*Blue Planet II*: 22 Things to Know about the New Series', *BBC* (29 October 2017), https://www.bbc.com/news/entertainment-arts-41692370 (accessed 18 February 2019).

111 Richards, 'Global Nature, Global Brand', 154.

of the 'BBC Earth' brand to ensure that 'sales were firmly linked with the BBC's broader brand image'.[112]

With each new series came new technologies: *Planet Earth* was heavily promoted on the basis of its aerial views achieved through a 'heligimbal' stabiliser mounted on helicopters that permitted both smooth flyover imagery and close-up shots of animals unaware of being filmed from high above. To show seasonal change in the Arctic and the Antarctic in glorious time-lapse, *Frozen Planet* relied on a purposefully designed motion-control rig, with marine life captured on ultra-high-speed cameras. *Planet Earth II*, filmed in ultra-high-definition (4K), anchored its claims for unprecedented visuals in the application of heat-sensitive and Go-Pro-like cameras alongside stabilising rigs and drones; the resulting footage, according to one critic, made you 'feel you are among the animals'.[113] And to get at unexplored sea depths, *Blue Planet II* devised a mechanism that allowed the camera to be automatically deployed off the submarine in deep water, while a 'megadome' housing camera piece enabled the simultaneous recording of under and above water for the first time.[114]

Not unlike an early-cinema nature film, these series splice together short-duration moments all of which – to cite Gunning's canonical 'Aesthetics of Astonishment' – 'offer the viewer a moment of revelation' via a display of movement, whether by making recourse to the techniques pioneered by Duncan and Smith, or else, à la Kearton, by cutting out the boring bits where nothing happens in order to show undisturbed nature in action.[115] Indeed, if the appearance of early nature films was conceptualised in relation to the idea that 'rest, repose, stillness, seem absolutely opposed to the principles of Nature', to cite again a review of the time, this statement could be readily applied to contemporary BBC series.[116]

Take the first episode of *Blue Planet II*, which shows, in the following order: a cuttlefish that changes its colour so as to hypnotise crabs and turn

112 Ibid., 150.

113 David Pierce, 'The Crazy New Camera Tech That Made *Planet Earth 2* Possible', *Wired* (26 March 2017) https://www.wired.com/2017/03/crazy-new-camera-tech-made-planet-earth-2-possible/ (accessed 18 February 2019).

114 Alessandra Potenza, 'Squid Wars and Methane Explosions: How *Blue Planet II* Captured Never-Before-Seen Views of the Ocean', *The Verge* (19 January 2018), https://www.theverge.com/2018/1/19/16907172/bbc-america-blue-planet-2-orla-doherty (18 accessed February 2019). See also McIntosh, 'Blue Planet II'.

115 Gunning, 'An Aesthetic of Astonishment', 122. This point has also been made by Thomas Austin in his *Watching the World: Screen Documentaries and Audiences* (Manchester and New York: Manchester University Press, 2007), 137.

116 'Pictures of the Unseen', Reprints of Press Articles, 4.

them into meals; an octopus and a grouper unexpectedly joining forces to capture prey; time-lapses of opening coral polyps in microcinematography; a sea turtle, with a camera attached on its back, taking us on a ride through the sea; dolphins playing games with rocks; aerial views of manta rays synchronically creating a vortex on the sea surface; a lion fish being attacked by a bobbit worm. Granted, most of these standalone vignettes are narrativised according to generic categories (comedy, horror, suspense), each of which entails a careful arrangement of formal elements such as montage, soundtrack and voiceover. Yet all of them aim to invoke wonder both at the fact that these events occur in the natural world *and* at the fact a camera was able to record them at the right time or over a period of time. Recalling an early-cinema lecturer, Attenborough's inimitable voiceover maximises visual impact by repeatedly and excitedly drawing attention to the fact these events were rarely or never captured on camera before.

And if all of this was not enough, a making-of segment closing each instalment of *Blue Planet II* – or for that matter, any BBC Earth series – ensures that viewers can marvel at the equipment and devices utilised to film these happenings, as well as the hardships and trials faced by crews in perilous shoots. Whether we consider behind-the-scenes footage of a submersible going down into depths 'never explored' before, the waiting endured by the camera team so that they could film the 'rarely witnessed' boiling sea phenomenon in high waters, or the remote underwater cameras utilised to film sea lions in a cove, these episodes, to cite Adrian Ivakhiv, 'celebrate the expansion of the human colonization of the world through technology'.[117] Tellingly, the format and image quality of these segments contrast with the preceding visual spectacle: they resemble more traditional documentaries, including interviews with the crew and scientists, while the image, captured by cheaper cameras, does not exude the glossy sheen of the actual series. By including these segments at the end of each episode, BBC natural history series thus kill three birds with one stone: the visual splendour of the series is heightened by comparison; the making-of footage serves as proof that such visual splendour has been found in the natural world (and not forged at post-production); and the images convey the idea that it has taken ingenuity, bravery and endurance to record that which was never before seen. The Keartons would have been proud.

By foregrounding continuities with early cinema, it is not my wish to dismiss the way contemporary images may be perceived as novel. I am in no

117 Adrian J. Ivakhiv, *Ecologies of the Moving Image: Cinema, Affect, Nature* (Ontario: Wilfrid Laurier University Press, 2013), 212.

position to deny that the image of an octopus building a makeshift disguise made out of assorted rocks and shells down in the ocean has 'never been recorded before' (*Blue Planet II*). Or that the breaking up of an ice shelf in the Antarctic seen from above is 'an enormous event that has never been filmed' (*Frozen Planet*). One may be equally astounded by microscopic shots of marine and land creatures as by the sweeping aerial views of large expanses and the translucent colours of nature rendered through HD technology and time-lapse. Whether or not these images astonish all viewers is ultimately impossible to ascertain, though judging from the furore they cause in public forums and the audience ratings of these series, it seems that they do indeed. Whatever the case, one thing is certain: these are images aggressively packaged in a mode of address the aim of which is spectatorial wonder. When viewed within a longer history of natural history's wondrous lineage and its revitalisation by way of cinema in its first decades, we must therefore ask ourselves why this discourse has returned with such force in our time.

It seems to me that Gunning provides some answers to this question in an essay that revisits the discourse and aesthetics of astonishment. He notes that 'the two ends of the Twentieth Century hail each other like long lost twins. Both periods generate inventions revolving around reproduction and communication and, perhaps even more clearly, both mine these new technologies for theoretical and aesthetic implications'.[118] For Gunning, both are periods of 'cultural wonder', eras of 'technological acceleration and transformation of the environment' that are ferociously branded as such.[119] He goes on: 'neither astonishment nor habit derive simply from individual cognition of single objects, but are triggered by changing relations to the world, guided or distracted by language, practice, representation and aesthetics'.[120] In this light, the arsenal of publicity surrounding both early nature films and contemporary nature programming serve the discursive function of maximising optical newness and aesthetic astonishment.

And yet, from a different angle, it could be argued that the technologies advertised in these recent series are, ontologically speaking, not that new. Certainly, HD equipment, stabilisers, lipstick and high-speed cameras, state-of-the-art time-lapse rigs and so on are all innovations in the recording and visualising of the world, as are new television sets and plasma screens, all of which bring new sensual, aural and visual qualities to the phenomenology of the viewing experience. However, these novelties are changes in degree, not

118 Gunning, 'Re-Newing Old Technologies', 51.

119 Ibid., 42.

120 Ibid., 42, 46.

in kind. Though they introduce new modes of mediating and experiencing the natural world, it would be unwise to put them on an equal footing with the appearance of moving-image recording technologies and the experience of wonder they initially provoked.

We must therefore also consider the 'changing relations to the world' which, for Gunning, make us perceive technologies and their abilities to astonish in a new light. This is paramount considering that wildlife programming has had to reckon, since the early 2000s, with the growing awareness of climate change and related environmental catastrophes around the world. Whether or not we agree with Anthropocene discourses and/or terminology (discussed in the Introduction), the natural world of natural history, a nature historically seen at a remove and separate from humans for the purposes of scopic pleasure and aesthetic enjoyment, has been troubled in recent times thanks to global anthropogenic change. In this context, there is a general perception that BBC series have been quite coy about the state of the world so as not to taint their scintillating displays of nature with uncomfortable truths. In the last section of this chapter I want to complicate this view by showing that these series have in fact often relied on the idea of a disappearing world in order to maximise its wondrous aesthetics from within an elegiac register.

Never to Be Seen Again

One of the most virulent critics of BBC nature programming, the journalist and environmentalist George Monbiot, noted in 2018:

> For many years, wildlife film-making has presented a pristine living world. It has created an impression of security and abundance, even in places afflicted by cascading ecological collapse. The cameras reassure us that there are vast tracts of wilderness in which wildlife continues to thrive. They cultivate complacency, not action.[121]

Implicit here is the widespread idea that BBC Earth series have turned a blind eye to global warming and related ecological disasters. By way of concluding this chapter, I hope to show that this idea does not entirely

121 George Monbiot, 'David Attenborough Has Betrayed the Living World He Loves', *The Guardian*, (7 November 2018), https://www.theguardian.com/commentisfree/2018/nov/07/david-attenborough-world-environment-bbc-films (accessed 25 February 2019).

stand up to scrutiny when we look both at the epistemological centrality of disappearance in such series *and* notable textual changes discernible across BBC nature programming over the last two decades.

The idea that the natural world is a disappearing world has gradually gained momentum in BBC landmark series. In the eight-part *The Blue Planet*, for example, the first series to be partly shot in HD in 2001, as we saw earlier, no mention or allusion to ecological issues is ever made. Since *Planet Earth*, however, BBC documentaries are often structured around the verbalised idea that the viewer will be taken to 'the last wildernesses' of our planet: 'places barely touched by humanity', as Attenborough announces in the first episode, 'From Pole to Pole'.[122] *Frozen Planet*, released five years later, was more emphatic. It opens with Attenborough noting that the polar regions are the 'planet's last true wilderness and one that's changing just as we begin to understand it'; in the series, he goes on, viewers will thus have the opportunity 'to witness its wonders, perhaps for the last time'.[123] A revealing promotional article on the BBC website presses the point home: 'the series uses the latest in filming technology to portray the Arctic and Antarctic as *they have never been seen before – and may never be seen again*'.[124] For its part, *Blue Planet II* begins with Attenborough's recognition that we must face an 'uncomfortable fact: the health of our oceans is under threat. They are changing at a faster rate than ever before in human history', which provides the justification for the show's underwater exploration: 'never has there been a more crucial time to reveal what is going on beneath the surface of the sea'.

More than simply punch lines, these statements reveal a carefully constructed premise on which these series' mode of address relies. Rather than dismissing the idea that the environment is in danger, they in fact openly engage with 'the end of nature' by coupling never-before-seen imagery with the trope of the never-to-be-seen-again. In so doing, these documentaries lend visual form to what Déborah Danowski and Eduardo Viveiros de Castro have termed the 'mythic theme' of the 'Edenic world', which

> persists nowadays in the idea of *wilderness*, those ever more restricted spaces of a pure nature not corrupted by human presence, *horti conclusi* (enclosed gardens) that bear witness to a past that is supposed to have

122 See Richard Beck, 'Costing *Planet Earth*', *Film Quarterly* 63:3 (2010): 63–66.

123 As also noted by Wheatley, *Spectacular Television*, 121.

124 'Frozen Planet: The Last Great Wilderness', *BBC* (26 October 2011), https://www.bbc.co.uk/news/in-pictures-15452539 (accessed 25 February 2019), my emphasis.

> managed to survive 'untouched' from the dawn of time until now – but which would today be under threat of disappearing as a result of Western civilization's blindly predatory action.[125]

This 'strain of environmentalism', they continue, 'considers the existence of human beings as an essentially *denaturing* force'.[126] Although no mention is made of 'Western civilization's blindly predatory action', BBC wildlife programming subscribes to this belief by asking the viewer to imagine a not-so-distant future where such places and species are no more, hence the importance of seeing these images for one more time – perhaps for the last time.

Granted, in some ways this ecological inclination rehashes concerns already on display in early wildlife photography and film. As we saw in relation to Cherry Kearton, his bird photographs and big game films, with their emphasis on untouched and undisturbed nature, must be placed alongside the rise of the conservation movement in Britain and a growing awareness of the extinction of megafauna on the African continent. More broadly, the popularisation of natural history cannot be dissociated from the onslaught of modernity and the threat it was deemed to pose to pristine nature, which in turn acquired a renewed visual significance in view of its possible decline. However, there is a crucial difference related to the fact that the environmental problems we face today – ocean acidification, sea-level rise and land erosion, ice melting, biodiversity loss, to cite a few – are not only relatively new but also far more disastrous in their *global* implications. In other words, it is no longer the case that a few localised species or sites are endangered but that human action is now widely deemed responsible for changing the climate to the extent that the health of the entire planet and its interdependent ecosystems are now at stake, something I will return to at the end of this chapter.

Let us note for now that it would be unfair to say that the BBC has merely employed disappearance as an ornamental boost for its aesthetics of wonder. There have been efforts to integrate ecological issues into its programming over the years, both within and outside landmark series. As Richards notes, *Life on Earth* 'began the tradition, continued in subsequent landmark series, of addressing human impacts and broader environmental issues in [Attenborough's] final to-camera statements'.[127] For Richards, the

125 Déborah Danowski and Eduardo Viveiros de Castro, *The Ends of the World*, trans. Rodrigo Nunes (Cambridge and Malden, MA: Polity Press, 2017), 23, emphasis in original.
126 Danowski and Viveiros de Castro, *The Ends of the World*, 24, emphasis in original.
127 Richards, 'Greening Wildlife Documentary', 179–180.

turning point of what he calls the emergence of 'green chip programming' was the release of the three-part documentary *The State of the Planet* (2000), followed by the two-part *The Truth About Climate Change* (2006) and the three-part *Planet Earth: The Future* (2006). On the other hand, as Richard Beck notes, *Planet Earth: The Future* was telling insofar as it was launched as a companion, entirely separate series to *Planet Earth* in order 'to satisfy some environmentalists who had criticized the series for underplaying or even ignoring the precarious state of the world's wilderness and wildlife'.[128] A conventional documentary featuring interviews with scientists and environmentalists, *Planet Earth: The Future* evinced that ecological issues, when tackled head-on in their causes, consequences and complexities, had to be treated as an appendix to BBC nature shows.

This splitting has also been verifiable in the structure of landmark series. An example is *Frozen Planet*, which reserves its last two episodes for the theme of climate change in a more expositional mode.[129] In fact, so textually distinct was the last episode ('On Thin Ice') from the rest of the series that not only was it dubbed 'the global warming episode' in the press as it also became involved in a telling controversy when the BBC was accused of offering the series without the episode on the international market in order to boost sales.[130] *Blue Planet II*'s final episode, 'Our Blue Planet', equally tackles challenges related to global heating, plastic pollution and ocean acidification more openly.

Blue Planet II further signalled a shift in BBC nature programming by containing references to ecological problems *within* all of its episodes. These ranged from images of coral reefs bleached by warming seas to plastic in the oceans and its impact on the lives of whales, which have been deemed to trigger a '*Blue Planet* effect' in the UK by raising awareness of these issues and prompting environmental action.[131] More recently, intratextual attention to environmental problems has been cemented with the release

128 Beck, 'Costing *Planet Earth*', 63.

129 The last episode of *Planet Earth II*, 'Cities', also stands out by focusing on humanmade environments.

130 Discovery was equally rumoured to have considered dropping the episode in the US owing to global warming denial in the country (the episode was eventually aired). See Andy Bloxham, 'BBC Drops *Frozen Planet*'s Climate Change Episode to Sell Show Better Abroad', *The Daily Telegraph* (15 November 2011) https://www.telegraph.co.uk/news/earth/earthnews/8889541/BBC-drops-Frozen-Planets-climate-change-episode-to-sell-show-better-abroad.html (accessed 23 June 2020).

131 Fiona Gell, 'The *Blue Planet* Effect: The Plastics Revolution is Just the Start', *The Guardian* (25 March 2019) https://www.theguardian.com/commentisfree/2019/mar/25/plastics-revolution-marine-life (accessed 26 June 2020).

of *Our Planet* (2019), Netflix's foray into nature programming in partnership with the production team behind the *Planet Earth* and *Blue Planet* series, featuring Attenborough as narrator. Tellingly, *Our Planet* is described in the Netflix catalogue not in terms of the planet's 'last wildernesses' that 'may never be seen again', but rather as a series that 'will celebrate the natural wonders that remain and reveal what we must preserve to ensure people *and* nature thrive'.

The shift from that which is to disappear to that which remains and must be preserved is subtle but nonetheless striking. Attenborough's warnings about ecological disturbances are no longer restricted to the opening or closing of episodes, but felt more heavily throughout and in more concrete and localised descriptions that highlight the planetary interdependence of ecosystems. For some critics this shift is proof that the BBC shows' reluctance to grapple with ecological issues was down to the NHU, even though the BBC's own *Seven Worlds, One Planet*, released in 2019, follows in the footsteps of *Our Planet* by treating environmental issues far more regularly than its predecessors and in all its episodes.[132]

No doubt this gradual effort at addressing environmental concerns responds to mounting public pressure on the perceived responsibility of these series to educate viewers, even if the actual results may still be considered timid and in some cases dubious, including an exaggerated belief in geoengineering as the anthropocentric solution to set the world right. For example, in *Frozen Planet*'s 'The Last Frontier', Attenborough draws attention to the 'most high-tech scientific research station' ever built, in 2006, at the South Pole, while situating this triumph within a lineage of polar explorations and celebrating the way 'humanity has achieved the extraordinary and opened up the last frontier'.[133] With the exception of a few episodes, by and large humans are still nowhere to be seen in these shows, and though there has been an increase in the number of shots of damaged landscapes over the years, the bulk of the imagery remains unspoiled, lush nature.[134]

132 Lucy Mangan, '*Our Planet* Review – Attenborough's First Act as an Eco-warrior', *The Guardian* (5 April 2019), https://www.theguardian.com/tv-and-radio/2019/apr/05/our-planet-review-david-attenborough-netflix-eco-warrior-activist-bbc (accessed 26 June 2020).

133 On Anthropocene geoengineering, see Christophe Bonneuil and Jean-Baptiste Fressoz, *The Shock of the Anthropocene: The Earth, History and Us*, trans. David Fernbach (London and New York: Verso, 2017), 65–96.

134 In addition to these environmentalist episodes, *Planet Earth II* devoted one episode to animals living in cities.

In highlighting these shortcomings, my intention is less to dismiss such attempts than to demonstrate the veritable struggle on the part of BBC shows to effectively confront ecological imbalance. While this struggle has been linked with these shows' debt to the blue-chip model, I believe it is equally if not more revealing of their firm anchoring in the aesthetic paradigm of natural history, which, as Michael Jeffries observes, postulates nature as 'balanced and harmonious' in opposition to principles of 'change, disorder and turmoil'.[135] In his analysis of *Planet Earth*, Beck notes that 'all nature films, either explicitly or implicitly, answer the question "What is nature good for?"' For him, 'the answer given by *Planet Earth*, over the course of its eleven episodes, is: "Nature is good for looking at"'.[136] This is certainly the case, yet in some ways this is also a self-evident fact when one considers BBC wildlife programming's deep roots in natural history, which, since its popularisation in the nineteenth century, has defined itself as an optical awakening of nature. As Louson notes, to accuse contemporary wildlife documentary of using spectacular imagery to 'impress viewers and elicit wonder' to the detriment of educational and informative strategies risks missing 'its complexity as a cultural product and resonance with historical forms of nature on display', as in nineteenth-century museums of natural history.[137]

However, to simply recognise this historical lineage is in my view also insufficient. Rather, I believe we must also assess the implications of this mode of visualisation for our environmentally precarious times by shifting the terms of the debate in relation to the fixity of natural history itself. For, indeed, as Jean-Marc Drouin and Bernadette Bensaude-Vincent remark: 'In spite of the conspicuous changes which natural history has undergone, there is a remarkable continuity in the rationales given for its popularization, among which the simple desire to open the eyes of the public to the wonders of the living world remains foremost'.[138] Nowhere is this better exemplified than in the recent commingling of the never-before-seen with the never-to-be-seen-again, which indicated that concessions may be made to the aesthetic predicates of natural history insofar as these continue to 'open the eyes of the public to the wonders

135 Jeffries, 'BBC Natural History', 528–529. Richards has attributed the environmental reluctance in BBC wildlife series to the blue-chip model; see 'Greening Wildlife Documentary', 172–173. To be fair, Richards himself cites Jeffries, but his argument is focused on the incompatibility of the blue-chip format with ecological issues.

136 Beck, 'Costing *Planet Earth*', 63.

137 Louson, 'Taking Spectacle Seriously', 30; 34.

138 Drouin and Bensaude-Vincent, 'Nature for the People', 424.

of the living world', but now in elegiac mode. The real issue to confront, then, is not why BBC nature programming privileges nature in terms of its visual pleasures, but why natural history struggles to make concessions to its scopic postulates when faced with versions of nature that pose a challenge to its foundational belief that nature ends as soon as humans are spotted within its domain.

In his now landmark essay 'The Climate of History', Dipesh Chakrabarty notes that global warming has spelled 'the collapse of the age-old humanist distinction between natural history and human history' in light of the scientific consensus that humans have irremediably changed the climate and are now deemed themselves 'a force of nature in the geological sense'.[139] Chakrabarty does not refer to natural history in terms of the popular field and practice discussed in this chapter, but more widely in terms of a longstanding perception of nature as the unchanging background for the discipline of history understood as the arena of human action and experience. Yet his thesis that the bifurcation of history into distinct human and natural realms has been thrown into disarray by the Anthropocene finds an even more literal resonance when applied to the history of 'natural history'. For, as we have seen, the nature of natural history is one that strictly comes into fruition through mediation: microscopes, magnifying glasses, glass cases, displays, hidden cameras and plasma screens that screen nature from humans and ensure a scopic when not voyeuristic regime. It is a nature, to recall Berger again, where animals and the natural world at large are always the observed and never the observers.

Taking this line of thinking further, it could be said that nature is now looking back at us through unpredictable acts of its own accord and volition as a response to anthropogenic global heating, which dissipates once and for all the idea that nature was ever really separate from humans. The growing public awareness of this state of affairs has occasioned important changes in contemporary BBC nature shows, even if it remains to be seen whether they will manage to fully break out of their glass case. Yet one thing seems certain: the environment will keep on pressing until the glass breaks.

139 Dipesh Chakrabarty, 'The Climate of History: Four Theses', *Critical Inquiry* 35: 2 (2009): 197–222; 201, 207.

3 The Universal Equality of Things

Abstract
Cinema's ability to record and catalogue anything and everything is the focus of this chapter. Adopting Kracauer's reflections on photography and cinema's rapport with 'things' as a methodological guide, the chapter first explores around-the-world stereocard boxes as well as the early-cinema catalogues of UK-based entrepreneur Charles Urban. The second section turns to the 'world symphony' genre, looking specifically at Vertov's *A Sixth Part of the World* (1926) and Ruttmann's *Melody of the World* (1929). The last section turns to contemporary web-based and -sourced projects dealing with the avalanche of images flooding the Internet, including Perry Bard's *The Global Remake: The Man with a Movie Camera* (2007–2014), Natalie Bookchin's *Mass Ornament* (2009), the YouTube documentary *Life in a Day* (2011) and the parodic global symphony *In Praise of Nothing* (2017).

Keywords: things, catalogue, contingency, database, Kracauer, world symphony

Never before has an age been so informed about itself, if being informed means having an image of objects that resembles them in a photographic sense.
Siegfried Kracauer

Our planet has never been as recorded as it is today. From surveillance systems and satellite imaging through to personal smartphones and web cameras, virtually anything, anywhere can now be photographed and filmed. Images also circulate as never before. They are the default mode through which we engage with other people and with the world. We automatically take photos of sights that catch our attention. We record videos of unexpected as well as quotidian happenings. And if we so wish we may make such images immediately available to a global audience at the click of a button.

de Luca, T. *Planetary Cinema. Film, Media and the Earth.* Amsterdam: Amsterdam University Press, 2022
DOI: 10.5117/9789463729628_CH03

Although, as Joanna Zylinska warns us, we should be wary of the 'rhetorical strategies of the mathematical sublime' accompanying this global phenomenon, it is hard not to be impressed by it.[1] As Nicholas Mirzoeff notes:

> As early as 1930, an estimated 1 billion photographs were being taken every year worldwide. Fifty years later, it was about 25 billion a year, still taken on film. By 2012, we were taking 380 billion photographs a year, nearly all digital. One trillion photographs were taken in 2014. There were some 3.5 trillion photographs in existence in 2011, so the global photography archive increased by some 25 percent or so in 2014. In that same year, 2011, there were 1 trillion visits to YouTube. Like it or not, the emerging global society is visual.[2]

Many people don't like it. Mirzoeff himself has elsewhere expressed concerns over what he terms our current regime of 'post-panoptic visuality', where 'the entire planet is now taken to be the potential site for insurgency and must be visualized as such'.[3] Before him, Paul Virilio had informed us that '[t]he Earth, that phantom limb, no longer extends *as far as the eye can see*; it presents all aspects of itself for inspection in the strange little window [of computers]', heralding a 'virtual vision which is supplanting the vision of the real world around us'.[4]

Concerns over the proliferation of images are, however, not exclusive to our time. Already in 1928, in his famous essay on photography, the German cultural critic Siegfried Kracauer expressed his unease at the 'flood of images' circulating by way of illustrated magazines in Weimar Germany, with their cumulative and all-encompassing logic of global coverage. Prefiguring Virilio's disquiet, Kracauer regarded the 'assault of this mass of images' as a threat that could sweep 'away the dams of memory' and 'destroy the potentially existing awareness of crucial traits' of the world.[5] He wrote: 'In the illustrated magazines, people see the very world that the illustrated magazines prevent

1 Joanna Zylinska, *Nonhuman Photography* (Cambridge, MA, and London: MIT Press, 2017), 19.

2 Nicholas Mirzoeff, *How to See the World: An Introduction to Images, from Self-Portraits to Selfies, Maps to Movies, and More* (New York: BasicBooks, 2016), 13.

3 Nicholas Mirzoeff, *The Right to Look: a Counterhistory of Visuality* (Durham, NC, and London: Duke University Press, 2011), 19.

4 Paul Virilio, *The Information Bomb*, trans. Chris Turner (London and New York: Verso, 2005), 18, 15, emphasis in original.

5 Siegfried Kracauer, *The Mass Ornament: Weimar Essays*, ed. and trans. Thomas Y. Levin (Cambridge, MA, and London: Harvard University Press, 1995), 58.

them from perceiving'.[6] In contrast with 'memory images', which 'retain what is given only insofar as it has significance', the 'blizzard of photographs', on the contrary, 'betrays an indifference toward what the things mean'.[7]

For Kracauer, however, the obliteration of meaning that arose with photography, understood as 'the spatial configuration of a moment', also carried within it a revolutionary potential, constituting for him, in fact, 'the *go-for-broke game* of history'.[8] This was because, he maintained, 'for the first time, the inert world presents itself in its independence from human beings', which held the promise of breaking 'every habitual relationship among the elements of nature'.[9] It followed that photography could liberate human consciousness from the given order and allow a confrontation with the world as it appeared newly disintegrated, divested of preconceived meanings. By levelling the world out and reordering it into 'unusual combinations which distance them from human proximity', photographs evinced 'that the valid organization of things remains unknown', that is: the possibility of change.[10]

As Thomas Y. Levin has shown, if Kracauer's early writings on modernity testify to a nostalgia for more stable eras, through the 1920s his position switches to a dialectical materialism that speaks to the need to confront modernity's sensory-perceptual reorganisation of lived reality *through* the mechanical media that both reflect and effect such reorganisation.[11] In this context, one of the changes heralded by photography resided in its potential to make the human confront the nonhuman by granting all things in the world – from 'the chignon and the corset as well as the high-Renaissance chair with its turned spindles' (objects that appear in the picture of a grandmother, possibly Kracauer's) – a strange, new peculiarity of their own, 'like an ocean-dwelling octopus'.[12] Accordingly, as Miriam Hansen observes, cinema became for Kracauer a tool to 'recover, transform, and reanimate the world of things', offering 'a mode of perceptual experience that blurs analytic distinctions between subject and object and allows things to appear in their otherness'.[13]

6 Kracauer, *The Mass Ornament*, 58.

7 Ibid., 50, 58.

8 Ibid., 56, 61; emphasis in original.

9 Ibid., 62.

10 Ibid., 62, 63.

11 Thomas Y. Levin, 'Introduction' in *The Mass Ornament*, 13–30.

12 Ibid., 56, 62.

13 Miriam Bratu Hansen, *Cinema and Experience: Siegfried Kracauer, Walter Benjamin, and Theodor W. Adorno* (Berkeley, Los Angeles and London: University of California Press, 2012), 16–17.

Over the last two decades, a surge of interest in *What It's Like to Be a Thing* – as captured in the title of Ian Bogost's book (2012) – has been discernible in theories such as speculative realism, new materialisms and object-oriented ontology, which are bound together through a concern with debunking the idea that the world can only be accessed via human subjectivity.[14] In many ways, as Bill Brown notes, this thingly turn may be interpreted not only as a reaction to the fact that 'our most precious object, the earth, seems to be dying', but also in relation to digital culture's '"dematerialization" of the world', which supplants material objects with their images.[15] Yet this turn to things has also not gone unchallenged. While some, such as Andrew Cole, have deemed it the philosophical iteration of commodity fetishism, others, such as Mirzoeff, have taken issue with its racial insensitivity.[16] In tune with these ideas, Elizabeth Ezra's incisive *The Cinema of Things: Globalization and the Posthuman Object* (2017) explores popular cinema as 'a medium that allows us to chart the dehumanization of people triggered by hyperconsumption, which begins as the supplementation of people by objects and results in the supplementation of objects by people'.[17]

Without denying the importance of these debates, this chapter will veer in a different direction when it comes to a cinema of things. Rather than drawing on object-oriented philosophies, I will instead build on Kracauer's dialectical thinking as both a source of inspiration and a methodological helping hand. This is not only because his ruminations prefigure by many decades current discussions on things and a dematerialised world of images, but also because for him mechanical images simultaneously and paradoxically held the promise of a liberatory non-anthropocentrism.[18]

14 Ian Bogost, *Alien Phenomenology, or What It's Like to Be a Thing* (Minneapolis and London: University of Minnesota Press, 2012). See also Steven Shaviro, *The Universe of Things: On Speculative Realism* (Minneapolis: University of Minnesota Press, 2014); Graham Harman, *Object-Oriented Ontology: A New Theory of Everything* (London: Penguin, 2018); Jane Bennett, *Vibrant Matter: A Political Ecology of Things* (Durham, NC: Duke University Press, 2010).

15 Bill Brown, *Other Things* (Chicago and London: University of Chicago Press, 2015), 12–13.

16 Andrew Cole, 'Those Obscure Objects of Desire: Andrew Cole on the Uses and Abuses of Object-Oriented Ontology and Speculative Realism', *Artforum* (2015), 318–323; Nicholas Mirzoeff, 'It's Not the Anthropocene, It's the White Supremacy Scene' in Richard Grusin (ed.) *After Extinction* (Minneapolis and London: University of Minnesota Press, 2018), 123–149.

17 Elizabeth Ezra, *The Cinema of Things: Globalization and the Posthuman Object* (London, Oxford, New York, New Delhi and Sydney: Bloomsbury, 2017), 21.

18 Kracauer, *The Mass Ornament*, 62. In this, Kracauer prefigures both André Bazin's and Stanley Cavell's reflections on the nonhuman ontology of photography and cinema. See chapter 6. For an excellent account of Kracauer in relation to speculative realism, see Chelsea Birks, 'Objectivity, Speculative Realism, and the Cinematic Apparatus', *Cinema Journal* 57:4 (2018): 3–24.

In this sense, this chapter aims to show another way through which nonhuman forms, and specifically material objects, gained currency in early recording media, although my take on 'things' must be primarily understood in terms of the concept's semantic capaciousness, that is, its ability to refer to 'every-thing'.

Mary Ann Doane has shown that part of the fascination of photography and cinema at their dawn resulted from their allegiance to contingency, the idea that, by virtue of their indexical properties, *anything* was potentially *a thing* worth recording. She notes: 'Indexicality has acted historically not solely as the assurance of realism but as the guarantee that anything and everything – any moment whatever – is representable, cinematographic'.[19] For Doane, the rapid appearance of narrative structures in early cinema was a response to this ontological promiscuity, which posed epistemological anxieties concerning the dissipation of meaning of which Kracauer was all too aware. Yet there is another way through which cinema has sought to avert the meaninglessness of contingency, especially when it comes to nonfiction cinema. As Lev Manovich argues, historically competing with narrative as an organisational mode to 'make meaning out of the world' is what he calls a 'database complex'.[20] In this chapter I turn to the catalogue as a structuring principle that prefigures the database impulse and underpins a range of nonfiction modes and genres in cinema from its inception to this date. More specifically, my aim is to explore how this cataloguing thrust often necessitates globality as an ordering category in order to modulate and regulate a world of things.

As a system, the catalogue aims to encompass an abundance of things while simultaneously promising to contain such abundance via semantic, linguistic and/or geographical parameters, often through recourse to whole-world figures and tropes. Here I will focus on three modes, which I will conceptualise as: the *encyclopedic* stereocard box set and early cinema catalogue; the *integrated whole* of the 1920s world symphony; and the contemporary *collaborative database* documentary. Whereas these categories are not historically self-contained and in some cases overlap with one another, it is my hope that they can help us discern three distinctive models of structuring what Walter Benjamin, recalling Kracauer, once termed 'the universal equality of things' ushered in by the mechanical reproduction

19 Mary Ann Doane, *The Emergence of Cinematic Time: Modernity, Contingency, the Archive* (Cambridge, MA, and London: University of Harvard Press, 2002), 31.
20 Lev Manovich, *The Language of New Media* (Cambridge, MA, and London: MIT Press, 2002), 233–234.

of images.[21] My contention is that these models testify to three historical moments where the proliferation of images acquires particularly delineated and even celebratory contours through a conceptual and geographical widening of the remit of the filmable.

The chapter is accordingly divided into three sections. The first section explores the way in which stereography and cinema invoked global figures and around-the-world narratives in order to contain and organise a plurality of objects and subjects. Here, I focus on stereocard box sets, as well as the catalogues and series produced by entrepreneur Charles Urban in the first two decades of the twentieth century. The second section travels to Weimar Germany and the Soviet Union in the late 1920s in order to explore Dziga Vertov's *A Sixth Part of the World* (*Shestaya chast mira*, 1926) and Walter Ruttmann's *Melody of the World* (*Melodie der Welt*, 1929), both of which testify to a broadening of the city symphony film into a global stage, although from different ideological standpoints. The inclusion of these two films is necessary for my argument insofar as they constitute fascinating attempts to update an encyclopedic impulse via a montage design that aims to create a more integrated sense of the whole, thus providing a productive bridge between the expositional thrust of early cinema and the database aesthetics of contemporary audiovisual media. Indeed, as the last section investigates, it is impossible not to consider these earlier 'world symphony' films when exploring online-sourced projects such as *Man with a Movie Camera: The Global Remake* (2007–2014), *Mass Ornament* (2009), *Life in a Day* (2011) and *In Praise of Nothing* (2018), all of which attempt to organise the 'flood of images' in the age of the Internet, now that virtually anyone anywhere can record anything anytime.

The Encyclopedia, or 'The Sun is No Respecter of Persons or of Things'

This section will consider how around-the-world stereographic boxes and early cinema's world-encompassing catalogues conflated global travel narratives and the figure of the encyclopedia as a way to assuage the potential loss of meaning identified with the contingent. As Brooke Belisle has noted: 'The sense that anything could be photographed, and the idea that any photograph could be infinitely copied positioned photography as a kind

21 Benjamin, *Illuminations*, ed. Hannah Arendt and trans. Harry Zohn (New York: Schocken Books, 2007), 223.

of universal currency, a medium into which anything could be translated, circulated, stored, and exchanged'.[22] One of the first to expound on this phenomenon was the American writer and physician Oliver Wendell Holmes. The inventor of an 1861 handheld stereoscope, Holmes had reflected on the device in an 1859 article titled 'The Stereoscope and the Stereograph', marvelling at the instrument's ability to make 'surfaces look solid' (solid being the Greek meaning for *stereo*) through the collusion of near-identical images placed next to each other.[23] The result, he proclaimed, was that 'the stereoscopic figure spares us nothing': 'all must be there, every stick, straw, scratch, as faithfully as the dome of St Peter's, or the summit of Mont Blanc, or the ever-moving stillness of Niagara'. His conclusion: 'The sun is no respecter of persons or of things'.[24]

For Holmes, photography's radical, unselecting vision was therefore a reason to celebrate. Yet its simultaneously all-levelling and all-encompassing optics, compounded by the illusion of depth in the stereograph, instilled anxiety in view of what he perceived as 'a frightful amount of detail' in the image.[25] To appease a potential dissolution of anthropocentric parameters, Holmes thus suggested that, in stereocards, 'a human figure adds greatly to the interest of all architectural views, by giving us a standard of size'.[26] Later, when commenting that '[e]very conceivable object of Nature and Art will soon scale off its surface for us' as it becomes an object of attention for the camera, he concluded that the 'consequence of this will soon be such an enormous collection of forms that they will have to be classified and arranged in vast libraries'.[27]

If, as noted, the digital dematerialisation of things is currently a cause of concern for some, for Holmes photography was to be commended precisely because it divorced the appearance of things from their material bases. However, as he hastened to add, systems of classification and arrangement were going to be required for the purposes of manageability. In Holmes's account, as Belisle summarises, 'the world is replete with images, as if the very density and depth of the visible world was a proliferation of photographs

22 Brooke Belisle, 'Total Archive: Picturing History from the Stereographic Library to the Digital Database', *Mediascape* (2013), 3, http://www.tft.ucla.edu/mediascape/pdfs/Winter2013/PicturingHistory.pdf (accessed 30 July 2020).

23 Oliver Wendell Homes, *Soundings from the Atlantic* (Boston: James R. Osgood and Company, 1872), 140.

24 Holmes, *Soundings from the Atlantic*, 148.

25 Ibid.

26 Ibid., 160.

27 Ibid., 162.

waiting to be developed', which betrayed an instrumentalist and even colonialist logic of appropriation.[28] In this context, Holmes's parametric and classificatory stipulations aimed to tame anxieties concerning the legibility of photography not only because of its seemingly indiscriminate visuality but also because of its potentially ceaseless reproduction.

Holmes's comments proved prescient. For stereography did quickly conform to visual codes designed to standardise its promiscuous imaging of the world. For example, the format was soon understood as suitable for pictures of outdoor landscapes, historical monuments, street scenes and public buildings – often framed within a colonial imaginary and sometimes, as suggested by Holmes, with human figures posing in the foreground and taken in such a way as to emphasise different planes in the image. Likewise, stereographic cards were sold as part of thematic sets that imparted order and narrative to otherwise unrelated images. Themes included education, religion and, most notably, travel and around-the-world travel, especially at the turn of the century, when stereography underwent a revival owing to the appearance of more portable cameras. Often sold in a box simulating a book or a double-book which made visible their encyclopedic ambitions (figure 3.1), these travel-themed sets contained numbered lists with captions, as well as maps and routes depicting the places from which the pictures were taken. Booklets listed topographical and historical details while stressing the need to follow the geographical order of the pictures, arranged in 'such a way as to really make it a continuous journey' (as noted in *Traveling in the Holy Land through the Stereoscope*, Underwood and Underwood, 1900).[29]

For their part, individual cards had on their backs summaries with instructions and directions to ensure the image's legibility. Take for instance the first card in the Underwood and Underwood box-set *Around the World through the Stereoscope*. Its caption reads: 'From Empire Building (N.) past Trinity Church Steeple, up Broadway, New York, U.S.A.', and the image shows, from an elevated perspective, a long avenue flanked by sky-scrapers receding into the distance.[30] The description on the card's back starts by noting that it is 'a 20-story building on which we are standing, more than 200 feet above the street level. The Battery, at the S. end of Manhattan island, is half a

28 Belisle, 'Total Archive', 4.

29 Jesse Lyman Hurbult, *Traveling in the Holy Land through the Stereoscope* (New York, London, Ottawa and Toronto: Underwood and Underwood, 1900), 6. Booklet held at the National Science and Media Museum (NSMM), Bradford, UK. Object number: 1990-5036/7753.

30 *Around the World through the Stereoscope*, volumes 1 and 2 (New York, London Ottawa and Toronto: Underwood and Underwood, n/d). Double book-type storage box, NSMM. Object number: 1990-5036/7763.

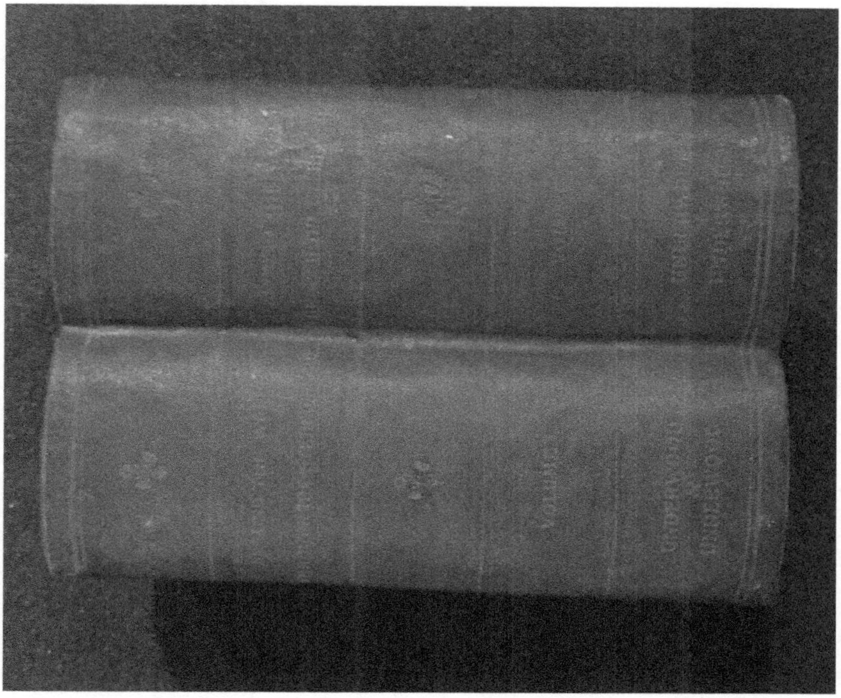

Figure 3.1 Underwood & Underwood around-the-world stereocard box set reveals its encyclopedic ambitions. Box set held at the National Science and Media Museum, Bradford, UK.

mile behind us'. It then provides historical commentaries on the buildings and objects, noting: 'Trinity church-spire, at our left, once dominated the street' and the 'early colonists of New York shrewdly foresaw commercial advantages in this location'. And it concludes with the observation that the need to grow 'sky-ward' was solved by 'the use of steel skeletons, merely faced with stone'.[31]

Such a detailed explanatory text foreshadows what Roland Barthes has termed 'the linguistic message' that often accompanies photographs as captions in mass media, serving as a '*repressive*' anchorage that dictates '*the correct level of perception*' and 'directs the reader [...] towards a meaning chosen in advance'.[32] In its positioning of the viewer 'in the scene', both in terms of where exactly the picture was taken ('200 feet above the street level') and its geographical surroundings, the description delineates the position of the observer and in so doing assuages the unusual – disembodied

31 Ibid., my emphasis.

32 Roland Barthes, *Image, Music, Text*, trans. Stephen Heath (London: Flamingo, 1984), 39–40, emphasis in original.

and elevated – perspective of the camera. The 'meaning chosen in advance' here is also unmistakable, with a wealth of references designed to laud New York as the cultural symbol of American modernity. All in all, the description ensures that the picture is not simply a 'spatial configuration of a moment', as Kracauer would have it decades later, but that it is endowed with geographical, contextual, historical and ideological significance.[33]

Nevertheless, the combinatorial system of stereographic boxes sometimes left room for unusual correspondences. Though touristic sites and outdoor sceneries prevailed, a set such as *Around the World through the Stereoscope*, for example, also contained images of an empty, fully furnished imperial room in Berlin, a banana plantation in the Hawaiian Islands and the picture of two parallel lines of hanging pork carcasses in a packing house in Chicago. Placed alongside the White House, the Taj Mahal and the Grand Canal, these less predictable images metonymically conveyed their geographical location much like those monuments, all of which were made equivalent and interchangeable as a result. Granted, the global travel narrative built into *Around the World through the Stereoscope* was not tied to a closely followed itinerary, as in *Traveling in the Holy Land through the Stereoscope*. Yet, by promising to take the viewer 'around the world in 60 minutes', to cite a dominant tagline in advertisements for the stereoscope (figure 3.2), these boxes did provide an ordering and thematic idea around which different pictures could be rudimentarily inventoried and made sense of. In other words, by harnessing the idea of global tourism as a structuring principle, stereography managed to bring together and contain within a box otherwise unrelated images, thereby imparting organisation to a miscellaneous assortment of things in the world.

Like stereography, cinema immediately latched on to the global as a paradigmatic category around which it organised its heterogeneous imagery. As Tom Gunning has shown, early cinema is better understood not so much in terms of individual films, but in relation to the unit of the programme, one that 'could, and frequently did, draw on the global as a readily understood means of uniting a variety of interrelated attractions'.[34] Titles such as 'Round the World in Two Hours', 'The World in Motion', 'The World on Wings' and 'Globe Express Excursions', as noted in the introduction to this book, were

33 Kracauer, *The Mass Ornament*, 56.

34 Tom Gunning, 'Early Cinema as Global Cinema: the Encyclopedic Ambition' in Richard Abel, Giorgio Bertellini and Rob King (eds.) *Early Cinema and the 'National'* (Bloomington: Indiana University Press, 2008), 11–16; 13.

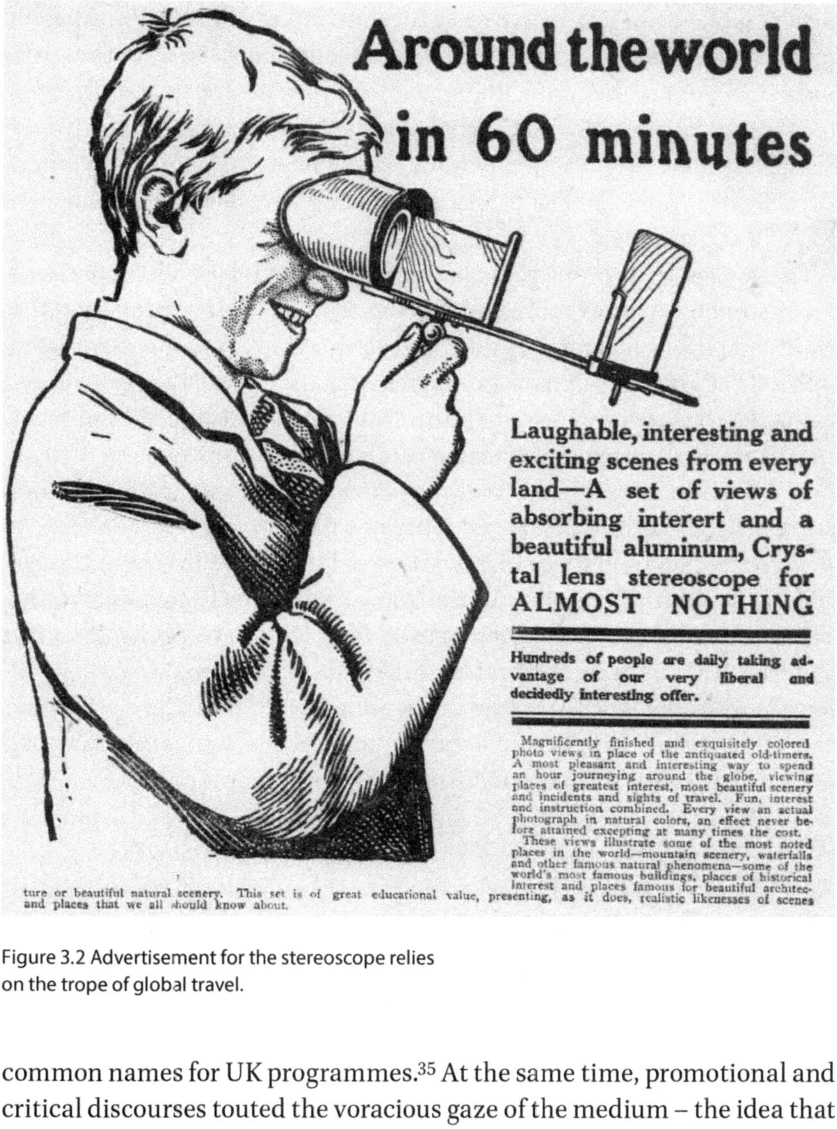

Figure 3.2 Advertisement for the stereoscope relies on the trope of global travel.

common names for UK programmes.[35] At the same time, promotional and critical discourses touted the voracious gaze of the medium – the idea that 'almost anything can be portrayed' (as per one article of the time) – while conflating that gaze with cinema's travelling properties.[36] To cite a common

35 These titles appear in the following articles and programmes: 'A "Travel Feature"', *The Bioscope* 347 (5 June 1913); '"The World in Motion" at Wigan', *The Bioscope* 169 (18 November 1909), 39; 'The World in Motion at Loughborough', *The Bioscope* 181 (3 March 1910), 37; 'The World on Wings', *The Bioscope* 217 (8 December 1910), 15; 'Globe Express Excursions by the Bigograph', programme no. 18 for the Scenorama and Globe Choir Federation, n/d, held at the Bill Douglas Cinema Museum (BDCM), University of Exeter, UK. Item number: 18536.
36 Arthur S. Newman, 'The Exhibition and Progress', *The Bioscope* 336 (20 March 1913), 877.

perception: 'one can get a marvelously accurate idea of what foreign countries are like, and, in addition, of many interesting incidents, both amusing and pathetic'.[37] For another critic, the spectator 'drops into one of these shows and sees, pictorially presented, perhaps Macbeth, motor skating, the Victoria Falls, glass blowing, a Passion Play, the latest aviation meeting in France, a Texas melodrama, and King Edward opening a bazaar'.[38] Anything was possible at the cinema.

Catalogues, as Gunning goes on, similarly 'offered the world in the form of consumable images', with the 'global sensibility' often providing the 'all-encompassing metaphor'.[39] Whereas such a sensibility is often associated with the Pathé brothers' production company and corresponding catalogues, in the remainder of this section I turn instead to a less studied figure but one whose fascinating early cinema catalogues and projects testify to an even more obsessive attempt to transmute a proliferating world of things into the figure of the whole world: Charles Urban.

An American from Cincinnati who moved to London in 1897 at the age of 30, Urban initially worked for the Warwick Trading Company until he founded his own production company in 1903, the Charles Urban Trading Company (CUTC). In the second half of the 1900s, nonfiction cinema enjoyed considerable popularity in Britain, accounting for 50% of all films produced, half of which were CUTC releases encompassing a staggering range of subjects and topics.[40] As Urban historian, Luke McKernan, notes:

> Pathé came closest to Urban in offering a sophisticated programme of non-fiction material that spanned the travel film to the science film. But Urban anticipated their diversity of non-fiction material by some years, and no producer of the time was to match him for his special advocacy of the genre, and for his determined search for what amounted to an alternative mode of cinema.[41]

Combining science and natural history films (as explored in chapter 2), current events and affairs, travelogues, exploration films, humorous sketches, and many other subjects, the idea was that the CUTC 'Put the World Before You', as its main logo suggested, often through recourse to global images and

37 'The World Before You', *The Bioscope* 215 (24 November 1910), 17.

38 'The Bioscope a Necessity', *The Bioscope* 179 (17 March 1910), 43.

39 Gunning, 'Early Cinema as Global Cinema', 13.

40 Luke McKernan, *Charles Urban: Pioneering the Non-Fiction Film in Britain and American, 1897–1925* (Exeter: University of Exeter Press, 2013), 66.

41 Ibid., 36.

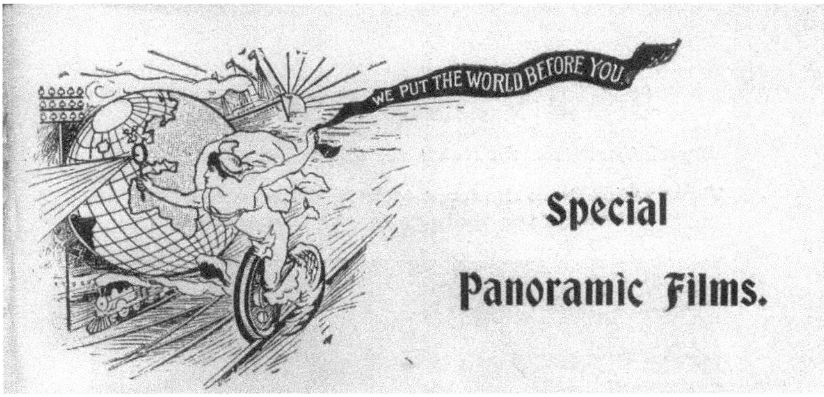

Figure 3.3 An Urban logo pictures the globe as a thing to be inspected with a magnifying glass. Courtesy of Luke McKernan.

figures that served as unitary symbols circumscribing the overwhelming wealth of topics and themes.

On the front cover of the 1903 catalogue, for example, banners displaying the logo encircle a globe around which we can see four human figures, some stereotyped according to perceptions of the time: in clockwise order from the top left, an Englishwoman holding a croquet mallet, an African holding a shield, a Native American wearing a headdress and what appears to be an Indian woman, her forehead jewelled, with her hands resting on her knees.[42] Inside the catalogue, an even more suggestive image shows a winged-Mercury on fast-moving wheels that recall film reels (figure 3.3). Mercury holds a magnifying glass in one hand and a banner with the CUTC logo in the other. Behind him a globe slides on a railway track, surrounded by three visual motifs: a transatlantic cruise in the background, a locomotive in full speed underneath, and an electric tower on the left-hand side. By positioning a racing human of divine qualities in the foreground overtaking the globe, the message is hardly unambiguous: electricity, transportation and communication had changed the face of the world, which, thanks to the cinema, could now be put before you much like an object – a thing – to be unveiled, inspected and dissected as though it was under a magnifying glass.

An equally suggestive image graced the front cover of the 1909 catalogue (figure 3.4).[43] Here, a turning globe is internally perforated by a tripod, with

42 'We Put the World Before You', Urban catalogue available at https://archive.org/details/weputworldbeforeoounse/page/n6 (last accessed 31 December 2020)

43 *Catalogue, Scientific and Educational Subjects – Urbanora: The World's Educator*, August 1909, Charles Urban Papers (URB), NSMM, URB 10-6.

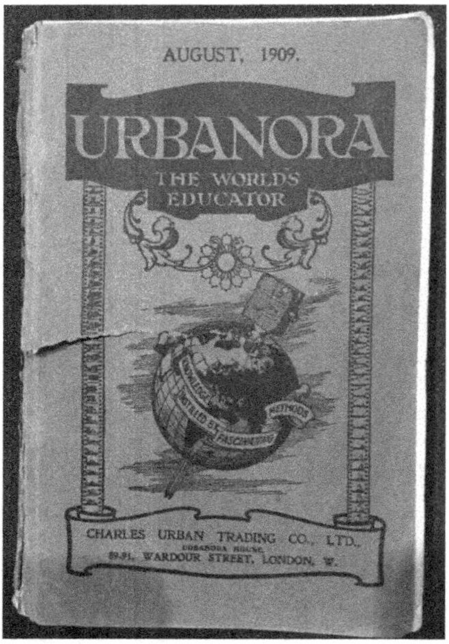

Figure 3.4 An Urbanora catalogue depicts cinema and the world as one and the same *thing*. Courtesy of National Science and Media Museum, Bradford, UK.

the camera sitting on the north polar cap, thus conveying the idea that cinema and the world have indissolubly become one and the same thing. Like the 1903 cover, with its stereotypical rendition of othered subjects, this image suggests the imbalanced power relations of a cinema that belonged to a certain part of the world while seeing as its mission the appropriation, via recording, of the whole world. With the camera literally standing atop the world, the cover recalls the tradition of putting up flags to signal conquest of territory, only in this case it is the planet itself that has become an object conquered by the recording device, and not before an act of perforation. Yet, as was often the case in the early cinema discourses that attempted to legitimise the cultural cachet of film, the recording of the world is equated here with instruction and education, with banners around the globe bearing the inscription 'Knowledge Instilled by Fascinating Methods' under the logo 'Urbanora: The World's Educator'.

For Urban, nonfiction cinema held an educational value that set it apart from cheap entertainment. In 1907, he published a manifesto titled *The Cinematograph in Science, Education and Matters of State*, where he listed the benefits cinema could bring to such fields and areas. If for Holmes photography threatened to overwhelm the subject with a 'frightful amount

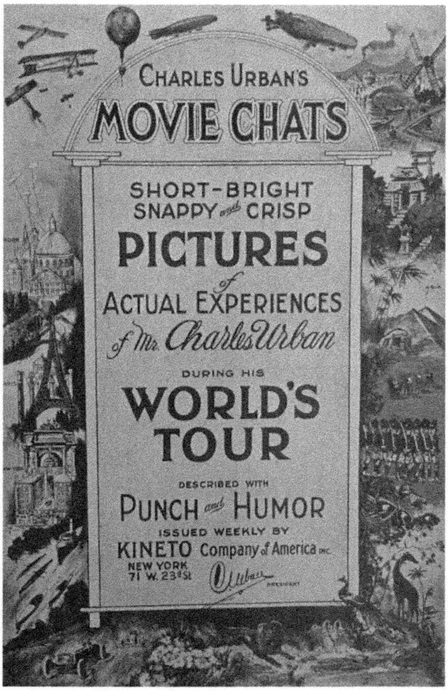

Figure 3.5 A poster for Urban's *Movie Chats* illustrates its quest to show 'every sight and experience accessible to man'. Courtesy of National Science and Media Museum, Bradford, UK.

of detail', Urban marvelled at cinema's ability to render such details in movement: 'every detail *in motion* of the subject under consideration'.[44] As Urban saw it, not unlike the philanthropist Albert Kahn discussed in chapter 6, 'the accurate and truthful eye of the camera' pointed up the archival importance of film, since 'animated pictures of almost daily happenings, which possess no more than a passing interest now, will rank as matters of national importance to future students'.[45]

Urban's archival conception of cinema gains in significance when examined next to his own self-promotional and cumulative footage archive. Overflowing with images acquired over two decades, this personal archive acted as the foundation for two globalist cinemagazines he produced back in the US after the war: *Charles Urban's Movie Chats* (from 1919) and *Kineto*

44 Charles Urban, *The Cinematograph in Science, Education and Matters of State* (1907), booklet reprinted in *Charles Urban, Motion Picture Pioneer*, http://www.charlesurban.com/documents_cinematograph.html (last accessed 3 May 2019), my emphasis.
45 Ibid.

Review (from 1921). Thus, the inaugural *Movie Chats* catalogue emphasises Urban's 'vast library of instructive subjects from all parts of the world, containing over 1,300,000 feet of negatives'. Defining itself as a weekly one-reel series featuring 'Extensive Travel, Scenics, Natural History, the Arts and Sciences of all countries,––Native Costumes and Customs, Architecture, Chemistry, Electricity, Physics, Sports and Training', the catalogue proudly concludes that 'every sight and experience accessible to man' is included.[46] And an enclosed poster visualises this idea, with airplanes, balloons, historical monuments, animals, natural sceneries, automobiles and airplanes encircling a notice board with the title of the cinemagazine, misleadingly advertised as 'Pictures of Actual Experiences of Mr. Charles Urban during his World's Tour' (figure 3.5).

The first catalogue for *Kineto Review* was similarly ambitious. It includes an illustration of an owl perched on a globe, with the inscription 'The Living Book of Knowledge', while lauding Urban's '2,000,000 feet of Educational pictures', accumulated 'during a period of twenty years' (figure 3.6). It then notes that 'Mr. Urban has undertaken the gigantic and exacting task of assembling this Encyclopedia of the Sights, Historic Episodes and Wonders of the World'. However, unlike *Movie Chats*, which assorted disparate subjects within its weekly episodes, the *Kineto Review* organised Urban's images according to specific themes: the aim was 'to publish 1000 reels, dealing with every conceivable subject'.[47]

Their hyperbolic rhetoric and lofty ambitions notwithstanding, Urban's globalist cinemagazines appear as rather insignificant when placed next to his previous achievements in the UK and contemporaneous aspirations in the US, as McKernan notes.[48] Also worth noting is the fact that Urban had resourcefully reused footage for a plethora of films since the beginnings of his career at CUTC.[49] Yet *Movie Chats* and the *Kineto Review* evinced a more radical expression of this practice as connected with an increased awareness of the vast archive of things that cinema had been generating for more than two decades.[50] If, to cite Gunning, 'early film companies and at least certain viewers [...] conceived of cinema as encyclopedic, an

46 *Charles Urban's Movie Chats: The Wonders of the World–Catalog of Subjects*, 1920, URB 10-12, my emphasis.

47 *Kineto Review: The Living Book of Knowledge–Catalog of Subjects*, 1920, URB 10-11.

48 McKernan, *Charles Urban*, 178.

49 Ibid., 202.

50 On cinema's globally expanding archive, see Michael Cowan, 'Cutting Through the Archive: *Querschnitt* Montage and Images of the World in Weimar Visual Culture', *New German Critique* 120, 40:3 (Fall 2013), 1–40; 6.

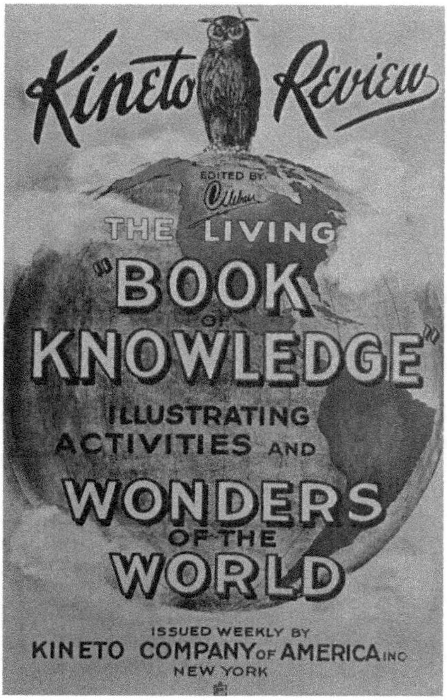

Figure 3.6 'A Living Book of Knowledge': cinema is lauded as a global visual encyclopedia in a poster for *Kineto Review*. Courtesy of National Science and Media Museum, Bradford, UK.

ever renewable catalogue made up of new editions through the addition of new films', then Urban's 'vast library' and 'living book', while wholly embodying this idea, equally suggested an encyclopedia bursting with images, hence the replacement of addition with recombination, accumulation with reshuffling.[51]

On an aesthetic level, however, Urban's repurposing efforts ultimately did little more than subscribe to the modular editing formula of the travelogue, which, as Jennifer Lynn Peterson has shown, 'creates a sense that the shots form a collection rather than a unified whole'.[52] On reading the sequence-by-sequence description of the *Kineto Review*'s 'Village Life' episode, for example, which includes headings such as 'Away Up in the Mountains' and 'The People on the Farms', one concludes that they could have been arranged in any given order, with no sense of integration or sequentiality aimed for.

51 Gunning, 'Early Cinema as Global Cinema', 15.

52 Jennifer Lynn Peterson, *Education in the School of Dreams: Travelogues and Early Nonfiction Film* (Durham, NC, and London: Duke University Press, 2013), 18.

As Peterson notes: 'The stand-alone quality of the shots' and a seemingly 'arbitrary' editing pattern are two of the travel genre's staples, these being films that to a large extent preserve 'the contingency of the real world [...] Travelogues certainly do not lack a system, but theirs is a system that makes order out of contingency'.[53]

Around the turn of the previous century, both stereography and cinema attempted to make order out of contingency through recourse to narratives of world travel and cataloguing models of organisation, often by combining and conflating the two. The recurrence of the tropes of the book, the library and the encyclopedia in this context in both critical and promotional discourses – whether we think of Holmes's 'vast library' of stereographs, stereocard boxes literally packaged as encyclopedias, or Urban's 'Encyclopedia of Sights' and 'Living Book of Knowledge' – is therefore more than a coincidence. For what these tropes reveal is a quest to tame and organise the contingency unveiled by recording media by promoting it as education, instruction and knowledge. Largely expositional and enumerative, these early cataloguing efforts were to be made outmoded by new trends in nonfiction cinema in the 1920s. These trends, many of which were aligned with avant-garde sensibilities, likewise endeavoured to combine fragments of the world, but they now shunned modular systems in favour of a montage aesthetics that attempted to create a more unified sense of the whole that in some cases was figured as the whole world.

The Integrated Whole, or 'An Instantaneous Survey of the World'

In an essay that returns to the discursive origins of the documentary mode, Bill Nichols compellingly demonstrates that the 1920s avant-gardes were often sidelined in such discourses due to their modernist aesthetics. Citing as examples films such as *Rien que les heures* (Alberto Cavalcanti, 1926), *Berlin: Symphony of a Metropolis* (*Berlin: die Symfonie der Grosstadt*, Walter Ruttmann, 1927), and *Man with a Movie Camera* (*Chelovek s kinoapparatom*, Dziga Vertov, 1929), Nichols notes that it was 'precisely the power of the *combination* of the indexical representations of the documentary image and the radical juxtapositions of time and space allowed by montage that drew the attention of many avant-garde artists to film'.[54]

53 Ibid., 18.

54 Bill Nichols, 'Documentary Film and the Modernist Avant-Garde', *Critical Enquiry* 27:4 (Summer 2001): 580–610; 595, emphasis in original.

Nichols's pantheon often goes by the city symphony moniker, a genre that enjoyed remarkable popularity through the 1920s and 1930s, with over 80 films produced in many countries.[55] He reminds us of the documentary basis of these films, which 'imaginatively reconstructed the look of the world with images, or shots, taken of this world'.[56] In this section I briefly revisit Ruttmann's and Vertov's landmark city films as a point of entry to explore how their creative take on the catalogue as a formal principle is geographically stretched out in what can be considered their 'world symphonies': *A Sixth Part of the World* (*Shestaya chast mira*, 1926) and *Melody of the World* (*Melodie der Welt*, 1929). Although these films display differences in form, content and ideology, I argue that they nonetheless testify to a similar impulse to celebrate and organise, via montage, a world of things in the context of a rapidly enlarging global archive of images.

If early cinema catalogues and programmes strung together an assortment of images, films and topics, the emergence of the city symphony in the 1920s showed that diversity and capaciousness could be achieved within the parameters of a single film. For Kracauer, both *Berlin: Symphony of a Metropolis* and *Man with a Movie Camera* testified to a renewed excitement at recording the world: whereas the former resulted from 'the voracious appetite of a man [cameraman Karl Freund] starved for reality' and who wanted 'to show everything', *Man with a Movie Camera* was 'infatuated with every expression of real life'.[57] This did not mean, however, that the city film shunned structures and parameters. On the contrary, in order to manage its everything-ness it recruited on the one hand a temporal framework spanning a whole day from dawn to dusk and on the other a spatial demarcation as derived from the city's geographical confines (though in some cases, as in *Man with a Movie Camera*, different cities – Moscow, Kiev and Odessa – morphed into one imaginary location).

For Manovich, *Man with a Movie Camera* is 'perhaps the most important example of a database imagination': a 'mechanical catalog of subjects that one could expect to find in the city of the 1920s––running trams, city beach, movie theaters, factories'.[58] It is important to note, however, that this was

55 For an excellent and exhaustive account of the city film see Steven Jacobs, Eva Hielscher and Anthony Kinik, 'Introduction: The City Symphony Phenomenon' in Steven Jacobs, Eva Hielscher and Anthony Kinik (eds.) *The City Symphony Phenomenon: Cinema, Art, and Urban Modernity Between the Wars* (New York: Routledge, 2018), 3–42.

56 Nichols, 'Documentary Film', 596.

57 Siegfried Kracauer, *From Caligari to Hitler: A Psychological History of the German Film* (Princeton, NJ: Princeton University Press, 1966), 183, 185.

58 Manovich, *The Language of New Media*, 239.

a new way of cataloguing things cinematically. First, this was because the city film fostered visual experimentalism, often achieved through skewed compositions and unusual perspectives, themselves the result of a more playful and adventurous application of the camera as a recording device. *Man with a Movie Camera* is exemplary in this regard, featuring as it does the titular camera operator (Vertov's brother Mikhail Kaufman) going to great lengths to show the world literally from new angles, even digging holes underneath a rail track and climbing factory chimneys.

Second, in contrast with the modular editing of nonfiction early cinema, the individual shot in the city film was both chronologically motivated and modulated on the basis of its ability to connect with other images in terms of convergences, parallels, resonances and dissonances.[59] As László Moholy-Nagy, who had himself drafted a film script titled *Dynamic of a Metropolis* in 1921–1922, summarised in 1927:

> We have – through a hundred years of photography and two decades of film – been enormously enriched [...] We may say that we see the world with entirely different eyes. Nevertheless, the total result to date amounts to little more than a visual encyclopaedic achievement. This is not enough. We wish to produce systematically, since it is important for life that we create *new relationships*.[60]

No longer an enumerative visual encyclopedia, cinema was accordingly hailed as a medium whose power was attributed to the relational configurations it engendered: not only did it show the things of the world, it amassed and combined them into a systematic whole.

Yet, for Kracauer, these wholes did not turn out the same. Whereas the whole of *Man with a Movie Camera* was 'quivering with revolutionary energies that penetrate its every element' and 'has a significant shape of its own', that constructed by *Berlin: Symphony of a Metropolis* was instead 'a shapeless reality, one that seems to be abandoned by all vital energies'.[61] Crucial in Kracauer's appraisal was the idea that the Ruttmann film lavished excessive attention on formal metrics to the detriment of content, when the Berlin

59 Although shots of sunsets were not uncommon in early travelogues, these hardly adhered to sequentiality, with sunsets instead meant to convey 'an emblematic sense of "something picturesque"', as Peterson notes in her *Education in the School of Dreams*, 19.

60 Lazlo Moholy-Nagy, *Painting, Photography, Film*, trans. Janet Seligman (London: Lund Humphries, 1969), 29.

61 Kracauer, *From Caligari to Hitler*, 185–186.

of 1927 'cried out for criticism, for interpretation'.[62] The film's method, he concluded, was 'tantamount to a "surface approach", inasmuch as it relies on the formal qualities of the objects rather than on their meanings'.[63] Although Kracauer's indictment may appear surprising in light of his belief that modernity's surface phenomena had to be confronted rather than averted, this indictment coincides with a shift in his writings in the late 1920s, when, in the context of a turbulent Weimar Republic, he deemed it the duty of the critic to 'expose the social conceptions and ideologies hidden in average films' – a project he would develop fully in *From Caligari to Hitler: A Psychological History of the German Film* (1947).[64]

Kracauer's analysis provides a useful point of entry for a comparison between Vertov's *A Sixth Part of the World* and Ruttmann's *Melody of the World*. Previously relegated to a second place in the scholarship on these directors, these two films have received renewed critical attention more recently, sometimes in discussions that, like Kracauer's, pit the political Vertov against the formalist Ruttmann.[65] Without denying the usefulness of these discussions, in the remainder of this section I hope to demonstrate that, from a different angle, both films can be productively examined in terms of correspondences and as celebratory expressions of the medium's evidentiary and combinatory possibilities for a globalist project. This is the case, I argue, when we consider how they not only enact a literal or implied global widening of the boundaries of the city film but also relativise the human in relation to the things of the world, even if from within distinct aesthetic systems and ideological traditions.

Commissioned by the film syndicate Tobis Klangfilm and the shipping line HAPAG as an advertising film for its global cruises, *Melody of the World* was released two years after the international success of *Berlin: Symphony*

62 Ibid., 187.

63 Ibid. 184.

64 Kracauer, quoted in Levin, 'Introduction', *Mass Ornament*, 23–24.

65 See Malte Hagener, *Moving Forward, Looking Back: The European Avant-Garde and the Invention of Film Culture, 1919–1939* (Amsterdam: Amsterdam University Press, 2007), chapter 6, 205–234; Michael Cowan, *Walter Ruttmann and the Cinema of Multiplicity: Avant-Garde–Advertising–Modernity* (Amsterdam: Amsterdam University Press, 2015), chapter 2, 55–98; Laurent Guido, 'Toward an Archaeology of Global Rhythms: *Melodie der Welt* and Its Reception in France' in Elena Gorfinkel and Tami Williams (eds.) *Global Cinema Networks* (New Brunswick, NJ: Rutgers University Press, 2018), 97–116; Oksana Sarkisova, 'Across One Sixth of the World: Dziga Vertov, Travel Cinema, and Soviet Patriotism', *October* 121 (2007), 19–40; Lilya Kaganovsky, '"The Threshold of the Visible World:" Dziga Vertov's *A Sixth Part of the World* (1926)' in Lilya Kaganovsky, Scott MacKenzie and Anna Westerstahl Stenport (eds.) *Arctic Cinemas and the Documentary Ethos* (Bloomington: Indiana University Press, 2019), 46–67.

of a Metropolis. As recent scholarship on Ruttmann has emphasised, the commissioned nature of the film was in fact the norm in his oeuvre and should not make us see it as a minor work.[66] Essentially a montage film, Ruttmann was given the task of selecting and editing 16,000 metres of film shot by Heinrich Mutzenbacher and the HAPAG film crew during a world tour aboard the ship Resolute, with the finished film reduced to 1,000 metres.[67] Ruttmann's non-involvement in the shoot might go some way towards explaining why *Melody of the World* looks less formalist than *Berlin: Symphony of a Metropolis*, with footage exuding a more unfinished, ethnographic quality. It may also explain why the film does away with chronology, given the likely difficulty in arranging the images longitudinally across the span of a day.

Whatever the case, it is striking that Ruttmann did *not* opt for a travelogue or episodic mode. Granted, the film does feature a fictional sailor (Iwan Kowal-Samborski) and his wife (Renée Stobrawa). These appear at the film's opening and on a few occasions throughout, providing the spectator with the Western eyes with which he or she is meant to view the world on parade through the film. Yet *Melody of the World* largely shuns a geographical conception of the planet in favour of a tripartite cataloguing structure organised around categories, listed in intertitles before each act. Whereas Act 1 includes 'buildings', 'streets', 'religions' and 'war' and Act 2, 'children', 'sport' and 'races', Act 3 closes the film with 'languages of the world', 'meals', 'dance and music', 'work' and 'spectacles', among others. This three-part structure thus evinces the film's ties with an encyclopedic aesthetic, even if, as we shall see, the editing system articulated here goes beyond mere listing by combining images into creative configurations that rearrange the boundaries separating the human from the nonhuman.

To be sure, *Melody of the World*'s self-declared aim is 'to increase understanding for the diverse forms of human life and make visible the commonalities between people', as per its introductory title card. The film therefore participated in a transnational discourse of 'international understanding' in the interwar period that in some ways built upon and updated the idea of film as a universal language, discussed in chapter 5.[68] The novelty of

66 See Cowan, *Walter Ruttmann*, 12–13. As Jacobs et al. further note, *Berlin: Symphony of a Metropolis* 'can be understood on some level as a commissioned work, as it was produced as part of a quota program ("Kontingentfilm") for Fox Europe'. See 'Introduction: the City Symphony Phenomenon, 1920–40', 7.

67 Cowan, 'Cutting through the Archive', 30.

68 Cowan, *Walter Ruttmann*, 82. At the end of Act I, a war sequence reiterates the film's ties to this internationalist discourse.

Melody of the World within this discourse related to the fact it was a sound film – indeed, the film was often billed, erroneously, as the first German sound film. As it turned out, however, it featured for the most part a musical score by Wolfgang Zeller, with the exception of the inclusion of machinic sounds throughout and human speech in one sequence.[69] This sequence featuring human languages is, moreover, telling insofar as it starts with the sound of barking dogs over the image of two men talking. It then proceeds by superimposing images of crowds in the visual track and languages being spoken in the aural track, thus suggesting linguistic confusion rather than harmony. One way of seeing the film's articulation of universality, then, might be to explore the way it parades diverse forms of dance as an 'ideal of the rhythmic synchronization of peoples' rather than emphasising human language, as Laurent Guido has argued in his persuasive analysis.[70] Yet here I want to take a different route in order to highlight that, in spite of its avowed aims, the universalism of the film is ultimately more-than-human rather than human.

Take for instance the film's sparse but nonetheless exhibitionistic use of sound. At the film's beginning, the sound of machine wheels inside the ship provides the cue for Zeller's score, which propels both the ship and the film into an exploration of the world. In another sequence, a rattling train becomes the sonic cue for a percussive beat (played by a man facing the camera), which is then made to interact with tram bells, car horns and fire-brigade sirens. Towards the end of the film, this percussion reappears, alongside instruments such as a violin, a drum, a piano and a sax, all played by human hands in close-up shots as part of a rhythmic montage. Taken together, these sequences fuse machinic and instrumental sounds into a melodic form where the noises of modernity both compete and resonate with musical instruments. It was this concretist aural quality that enticed the French filmmaker Abel Gance to declare that *Melody of the World* was far superior to the 'talkies' and presaged a future cinema in which dialogue is 'harmonized with all sounds of nature and life'.[71]

As for its visual track, the film is equally interested in widening its range of content into the nonhuman world. Undoubtedly, *Melody of the World* devotes the majority of its running time to cataloguing humans engaging in universally

69 Cowan, 'Cutting through the Archive', 32.

70 Guido, 'Toward an Archaeology of Global Rhythms', 107.

71 Abel Gance, 'Images of Yesterday, Voices of Tomorrow' in Richard Abel (ed.) *French Film Theory and Criticism: A History/Anthology, 1907–1939*, volume 2 (Princeton, NJ: Princeton University Press, 1988) 41–42; 41. Incidentally, Ruttmann travelled to France after *Melody of the World* to collaborate with Gance in his ill-fated *La fin du monde* (1930).

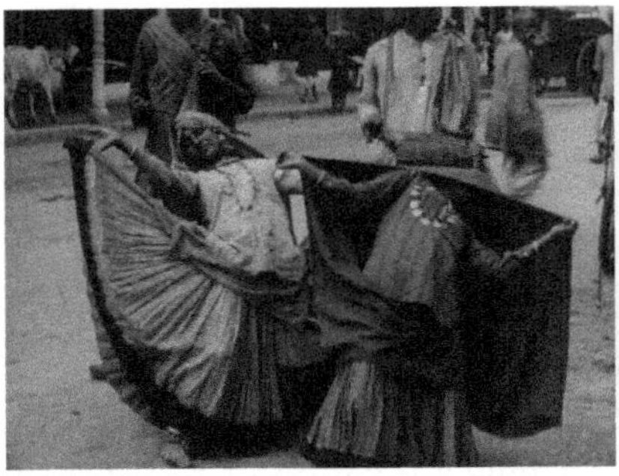
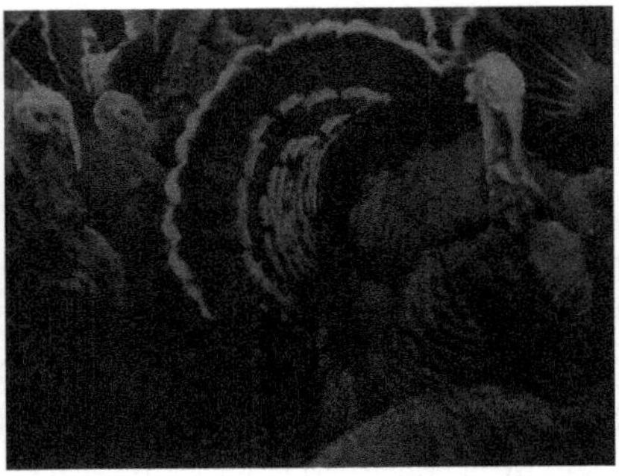

Figures 3.7 & 3.8 A woman's skirt finds a surprising correspondence with a turkey in *Melody of the World* (1929).

recognisable activities, including dancing, eating, waking up, working and so on. Yet the film also builds two comparative systems, each informed by montage sequences subscribing to an associative and/or metaphorical logic and whose aim is to broaden its human horizons. On the one hand, animals are put on the same footing with objects and machines according to an anthropocentric logic of serviceability. For example, one sequence intercuts shots of automobiles and buses in urban centers with images of camels and elephants transporting people in rural areas, whereas another shows in succession two-wheel ox-carts and a tractor mowing the land. A similar logic is in place in the sequence that intercuts dog- and horse-racing with races featuring motorised vehicles. In these cases, animals are depicted as the counterpart and/or precursors of modernity's mechanised objects of speed and efficiency, as things whose use-value is measured for and by humans.

On the other hand, *Melody of the World* complicates this utilitarian logic by frequently positing animals as analogous to humans. Thus, one sequence depicting mother-and-children relationships around the world includes images of a cat with suckling kittens and an elephant with its calf, while sumo fighters and sword players are related to bulls and tigers engaged in fighting. Other sequences postulate human-animal analogies through similarities of shape and form. At the film's beginning, for example, the image of the fictional sailor climbing up a mast is replaced with that of a man climbing a tree in a tropical country, which is then followed by that of an ape mounting a similar tree. Further examples include visual echoes between a woman's skirt and a turkey (figures 3.7 and 3.8), running children and a flock of pigeons, and seagulls and airplanes. In these cases, an instrumentalist logic gives way to a celebratory thrust to rejoice in the film medium's creative ability to disrupt anthropocentric hierarchies via unexpected formal analogies.

Now, given the fact that for a long period in human history nonwhite peoples were considered non-persons, or things, *Melody of the World* may understandably raise suspicions, especially when one considers that some of the subjects depicted in terms of analogies with nonhuman entities and animals are themselves nonwhite. In the film's defense, however, humans of all ethnicities are shown, and while the film may not escape the accusation that it turns a blind eye to the world's asymmetries of power (more of which shortly), it does not subscribe to a colonialist perspective. Indeed, here I would side with Michael Cowan when he notes that 'the lack of colonial critique in Ruttmann's montage does not necessarily make *Melodie der Welt* a work of colonialist propaganda', with the film displaying none of the 'nostalgia for Germany's lost colonial possessions that informed colonialist educational films of the 1920s', and constructing 'global space as a space of universal correspondences, in which we see little if any hierarchical distinction between the "European" and the "exotic", the "primitive" and the "modern", or makers and receivers of culture'.[72]

Melody of the World must be instead placed alongside interwar artistic movements and practices that engaged with a rapidly expanding photographic archive and which, to different extents, relied on collage and montage techniques, including Surrealism, Cubism, Dadaism and Constructivism. Closer to home, the film warrants comparisons with art historian Aby Warburg's monumental *Mnemosyne Atlas* (1928–1929), which comprised more than sixty panels each of which assembled photographic reproductions of recurring tropes and motifs through the history of Western Art.[73] Yet, as Cowan has convincingly shown, it is the *Querschnitt*, or cross-section concept, pervading 1920s Weimar culture that offers the most compelling framework when it comes to *Melody of the World*'s quest to make sense of a flood of images.

As Cowan points out, the 'proliferation of indexical representations of the world in the 1920s formed the backdrop for the emergence of *Querschnitt*-style montage'.[74] Looking at the photomontages in the journal *Der Querschnitt*, many of which concocted pairings via thematic and compositional analogies, Cowan observes how 'they sought to strike a balance between the schematic and the contingent'.[75] That is, if on the one hand the cross-sectional principle

72 Cowan, Walter Ruttmann, 89.

73 Guido also notes the film's resonances with Élie Faure's *Historie de l'art: L'Esprit des formes* (1928), yet another contemporaneous art-historical project concerned with charting similarities of form across time, 'Toward an Archaeology of Global Rhythms', 103–104.

74 Cowan, 'Cutting through the Archive', 20.

75 Ibid., 21.

sought to impart order and regularity to otherwise unrelated peoples, places and things, it privileged on the other surprising or unexpected connections, with examples including, in *Der Querschnitt*, couplings such as a tiger's open mouth and an opening orchid, and a dancer leaning on his elbow and a sculpture. Something similar happens in *Melody of the World*, which, at the same time as it offers itself as a programmatic visual catalogue listing worldly peoples and things under particular categories, gestures towards the contingent via startling, playful juxtapositions.

The popularity of the cross-section as a formal method, as Cowan shows, was made possible by photography, with 'its ability to bring objects, images and people closer together in time and space and reduce them to a common scale'.[76] In turn, Cowan's words recall Benjamin's contention that the loss of aura in modernity resulted from 'the desire of contemporary masses to bring things "closer" spatially and humanly' by way of the mechanical reproduction of such things.[77] Benjamin writes:

> To pry an object from its shell, to destroy its aura, is the mark of a perception whose 'sense of the universal equality of things' has increased to such a degree that it extracts it even from a unique object by means of reproduction. Thus is manifested in the field of perception what in the theoretical sphere is noticeable in the increasing importance of statistics.[78]

Locating homogenising impulses in photography's levelling of the world, Benjamin was cautious about the universality of image reproduction, and *Melody of the World* may prove to some extent that his concerns were not unfounded. By evading considerations of historical processes and power relations in order to highlight the sameness of the world chiefly from a Western perspective, the film has been seen both when it was released and more recently as a shallow and superficial exercise in world-making.[79] Malte Hagener, for example, argues that *Melody of the World* shows a 'disinterestedness towards the material from different cultures' it combines, remaining 'very much on the surface of things'.[80] This statement leads us back to Kracauer who, in line with his criticism of *Berlin: Symphony of a Metropolis*, deemed Ruttman's 'world melody' as 'void of content, because

76 Ibid.

77 Benjamin, *Illuminations*, 223.

78 Ibid.

79 On the negative responses the film generated, see Cowan, 'Cutting through the Archive', 31–32.

80 Hagener, *Moving Forward, Looking Back*, 220.

his concern with the whole of the world leads him to disregard the specific content of each of the assembled melodies'.[81] Such an appraisal in turn resonates with Kracauer's aforementioned unease with photography's logic of global coverage seen in illustrated magazines' 'complete reproduction of the world', whose temporal equivalent he moreover related to a hypothetical 'giant film depicting the temporally interconnected events from every vantage point'.[82]

However, it is important to emphasise that both Benjamin and Kracauer tackled mass visual culture dialectically and assessed some of its perceptual rearrangements as welcoming impacts on the collective human sensorium. Benjamin distinguished positive and democratic features arising from the erosion of aura given its elitist roots, while praising the 'shock effect' of montage in its aesthetic modulation of modernity's bombardment of chance impressions. As Doane explains, 'Montage functions for Benjamin not so much to confer order or meaning but to rapidly accumulate and juxtapose contingencies. In this, the film form mimics and displays for the spectator the excesses of a technologically saturated modern life'.[83] Kracauer likewise ascribed a revolutionary value to similar photographic techniques, including 'unusual combinations which distance them from human proximity' and a 'capacity to stir up the elements of nature [...] to create strange constructs'.[84]

In this context, as much as we should approach *Melody of the World* with caution and as, in some measure, a statistical quest to reproduce the surface of the world, I would like to suggest we may also choose to adopt the dialectical thinking championed by Benjamin and Kracauer as a way to discern in the film moments that do justice to their praise of cinema's ability to shake the world into new configurations and 'strange constructs', some of which may jolt the viewer to this date.[85] From this perspective, *Melody of the World* can be seen as a celebration of cinema's imaginative possibilities, in line with Kracauer's belief that film could undercut 'a larger anthropocentric worldview' and in the process, as Hansen notes, defy 'traditional distinctions between the human and the nonhuman, the living and the mechanical,

81 Kracauer, *From Caligari to Hitler*, 209.

82 Kracauer, *The Mass Ornament*, 50.

83 Doane, *The Emergence of Cinematic Time*, 15.

84 Kracauer, *The Mass Ornament*, 62.

85 This impression was confirmed in a public screening organised by me as part of a film season related to this book. For more information on this season, see Tiago de Luca 'Envisioning the World: How Film Shapes the Earth, film programme, 8–22 June 2019, Close-Up Film Centre', *Transnational Screens* 11:1 (2020): 78–81, DOI: 10.1080/25785273.2019.1692604.

the unique (integrated, inner-directed) individual and the mass subject'.[86] Whether we are looking at planes turning into seagulls, masts into trees or children into pigeons, *Melody of the World* sometimes enacts a creative ordering of the world that shows 'that the valid organization of things remains unknown'.[87]

Ruttmann's montage aesthetic, while certainly indebted to the Soviet school, is often deemed antithetical to Dziga Vertov's filmic praxis. Laurent Guido, for example, notes: 'In Vertov's work [*A Sixth Part of the World*], juxtapositions are explicitly discursive, analytical, and political. For Ruttmann [in *Melody of the World*], they refer, in a more engrossed and lyrical fashion, to the great aesthetic and cultural paradigms of an era fascinated by the possibilities of the cinematic apparatus'.[88] And yet I want to suggest in the remainder of this section that, from a different angle, *both* world symphonies, in spite of conceptual and political differences, can be considered products typical of 'an era fascinated by the possibilities of the cinematic apparatus'. This is the case not only when we consider the equal (if unfulfilled) importance Vertov ascribed to an expanding global archive of images, but also in terms of the epistemological centrality that a world of 'things' comparably acquired in a Soviet context. Unlike *Melody of the World*, the subject matter of *A Sixth Part of the World* was not the world itself but, as per its title, a large part of it as embodied in the Soviet Union. Yet, as its title also indicates, the film is concerned with demarcating and implicating that nation within a global space, and in so doing, as I hope to show, the film prefigures Vertov's dreams of a cinema made by and for the whole world in line with a Marxist, utopian globalism.

A filmmaker who actively theorised his own practice, Vertov's writings and manifestos consistently extolled cinema's ability to capture the world in its multifaceted dimensions. For Vertov, the superior eye of the camera, or 'Kino-Eye', was both a mutating vision that could morph into the perspective of other things – moving 'apace with the muzzle of a galloping horse' and 'ascending with an airplane' – and a seemingly indiscriminate gaze that spared nothing, capturing a 'whirlpool of colliding visible phenomena, where everything is real, where people, tramways, motorcycles, and trains

86 Miriam Bratu Hansen, 'Introduction', in Siegfried Kracauer, *Theory of Film: The Redemption of Physical Reality* (Princeton, NJ: Princeton University Press, 1997), vii–xlv; xvii.

87 Kracauer, *The Mass Ornament*, 62.

88 Guido, 'Toward an Archaeology of Global Rhythms', 113. See also Cowan, *Walter Ruttmann*, 88–89.

meet and part'.[89] Yet, according to Vertov, it was also imperative that these fragmented, kaleidoscopic impressions of human and nonhuman worlds be made to cohere, via montage, into a 'rhythmic and aesthetic whole'.[90] The Kino-Eye in this sense could potentially engender not only 'the conquest of time (the visual linkage of phenomena separated in time)', but also 'the conquest of space, the visual linkage of people throughout the entire world'.[91]

Vertov's third feature-length film – released after *Kino-Eye* (*Kino-Glaz*, 1925) and *Forward, Soviet!* (*Shagai, Soviet!*, 1926) and before *Man with a Movie Camera* – *A Sixth Part of the World* is often regarded as a transitional work, combining the newsreel form of his *Kino-Pravda* series (particularly its travelogue issues) and the more daring style of his later films.[92] Filmed in a number of expeditions across the Soviet territory, the ethnographic footage depicts rural lifestyles and manual labour activities, with locals not infrequently looking at the camera. After a prologue devoted to the exploitative 'countries of capital' (more of which later), the film begins its survey of ethnic plurality as embodied in Tatars, Buryats, Uzbeks, Kalmyks and so on. Then it moves on to chart the geographical expanse of the country – 'From the Kremlin to the Chinese Border' – followed by an inventory of its nonhuman animals, from 'the goats at Ulu-Usen' to 'the camels on the river steps of Kyrgyzstan'. The focus then shifts to the production and trade of goods (many of which derived from animals) crisscrossing the land and finally to the modernisation sweeping the country, from irrigation and electricity to the explosion of factory plants.

A large part of *A Sixth Part of the World* is therefore concerned with enumerating the peoples, animals and things making up the nation. Yet enumeration is not the only logic in place here. For these ethnographic, zoological and topographical inventories are not only combined into an interconnected and geographical whole, they also subscribe to a teleological idea of progress, or to cite Jeremy Hicks, 'a judiciously conceived, over-arching expositional frame' that aims to chart the country's technological

89 Dziga Vertov, *Kino-Eye: The Writings of Dziga Vertov*, trans. Kevin O'Brien and ed. Annette Michelson (Berkeley, CA, and London: University of California Press, 1984), 17, 167.

90 Dziga Vertov, 'Artistic Calling Card (1917--1947)' in Österreichischen Filmmuseum, Thomas Tode and Barbara Wurm (eds.) *Dziga Vertov: The Vertov Collection at the Austrian Film Museum* (Vienna: SYNEMA, 2006), 79–158; 83.

91 Vertov, *Kino-Eye*, 87–88.

92 Jeremy Hicks, *Dziga Vertov: Defining Documentary Film* (London and New York: I.B. Tauris, 2007), 45; Kaganovsky, '"The Threshold of the Visible World"', 52.

development, one in which every-thing has a fundamental and irreplaceable role to play.[93]

As it happens, *A Sixth Part of the World* was commissioned by Gostorg, the state organisation then in charge of imports and exports in the USSR. This is an important detail given the extended time the film devotes to such matters, especially in its second half, which turns to how the exports of goods (cattle, fur, grains) to capitalist countries are necessary for the purchase of 'machines that will produce machines' and thus ensure the country's self-sufficiency. At this point in the film, its networked figuration of the nation acquires more visibly delineated contours, with title-cards emphasising that 'along every path [...] roll goods for export' and images showing the range of means by which goods are transported across the national territory, often via working animals. This idea comes to a head in an uncharacteristically long sequence 'in the snow banks of the endless Tundra', whose placement in the film serves the function of highlighting that even the remotest geographical regions in the country play a part in its socialist model of production, as the collected fur is to be traded with capitalist countries for machines.[94]

In the world of *A Sixth Part of the World*, then, every human and nonhuman entity has a role to play within an interlinked system that propels the country forwards. If analogy provides the connective tissue for the whole of *Melody of the World*, here the dominant organisational principle is interdependence. Rural labourers cultivate the land and breed animals in order to produce goods traded in turn for machines that modernise the nation. In conformity with the artistic doctrine of Constructivism, according to which art had to fulfil a collective and utilitarian function, *A Sixth Part of the World* endeavours to visualise a model of production as a 'giant system of collaboration between humanity and the spontaneous forces of nature', to cite contemporaneous art critic Boris Arvatov.[95]

Of course, this productive model was overly anthropocentric. For it suggested a hierarchy in which nature and animals were regarded as things to be used and exploited, with the natural world in the film conceptualised in terms of the resources it generates in a chain of relations whose ultimate goal is technological modernisation. In this respect, as much as *A Sixth*

93 Hicks, *Dziga Vertov*, 48.

94 Kaganovsky, '"The Threshold of the Visible World"', 55–64.

95 Boris Arvatov, 'Everyday Life and the Culture of the Thing (Toward the Formulation of the Question)', trans. Christina Kiaer, *October* 81 (1997): 119–128; 127. On productivism's ideological promiscuity and its hold on Soviet culture, see Malcolm Turvey, *The Filming of Modern Life* (Cambridge, MA, and London: MIT Press), 138–143.

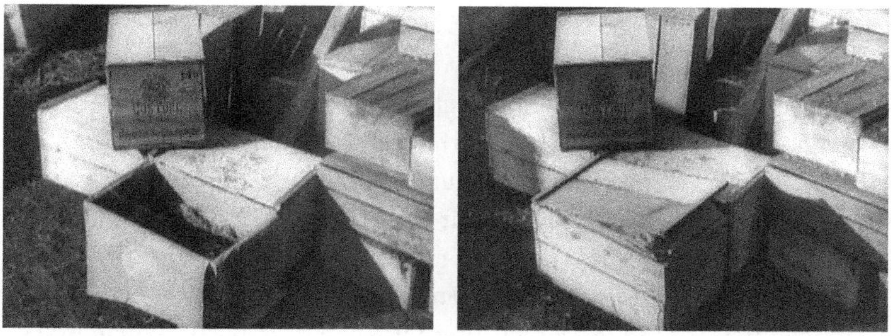

Figures 3.9 & 3.10 A fruit box becomes an animate thing in *A Sixth Part of the World* (1926).

Part of the World may be diverse and expansive in its nonhuman focus, it is inflected, like *Melody of the World*, with the human-centric logic of its time. That said, questions concerning what constituted a 'thing' and what role things were to play in a socialist economy were central, widely debated issues at the time. An example is Arvatov's 1925 essay 'Everyday Life and the Culture of Thing', in which the thing is mobilised 'as a force for social labor, as an instrument and as a co-worker', in contradistinction with 'the Thing as something completed, fixed, static, and, consequently, dead' prevailing in capitalist societies.[96]

In this context, as Christina Kiaer has shown, the thing in Soviet artistic practices often appeared as 'an active, almost animate participant in social life' rather than a fetishised commodity.[97] This idea is not only made visible in the film through its emphasis on the movement of goods along interconnected pathways, but also made literal in a sequence that shows individual fruits magically packing themselves into a box, which in turn closes itself and leaps, as if on its own volition, on to the top of a pile of stacked boxes (figures 3.9 and 3.10). Achieved through stop-motion, a technique that supresses the sight of human intervention so as to endow inanimate things with mobility, this sequence crystallises a conception of material things as if they were indeed animate and alive.

Annette Michelson has argued that *Man with a Movie Camera* is a 'synthetic articulation of the Marxist project, concretized in every detail of an unprecedented complexity of cinematic design'.[98] For her, the film constructs, through editing, 'a community of productive forces in an "all-around

96 Arvatov, 'Everyday Life and the Culture of the Thing', 122, 124.

97 Christina Kiaer, *Imagine No Possessions: The Socialist Objects of Russian Constructivism* (Cambridge, MA: MIT Press, 2005), 1.

98 Annette Michelson, 'Introduction', *Kino-Eye*, xv-lxi; xxxvii.

dependence" [Marx's term]'.[99] Though lacking the rhythmic density of *Man with a Movie Camera*, *A Sixth Part of the World* can be similarly understood as an attempt to manufacture this social model of 'all-around dependence' through a montage style that unites the country as one interlinked collective force. Yet, as its title suggests, equally important when it comes to the film's Marxist project is the fact that it gestures towards the utopian global imagination underpinning Marx's communist ideals.

Here it is worth noting that *A Sixth Part of the World* starts with a sequence in the 'countries of capital', which alternates, via an oppositional montage, images of the wealthy in a jolly mood laughing and dancing with images of workers in factories and presumably slaves in India and Africa. This opening in turn resonates with the film's ending, in which the building of a socialist society is exalted both as a 'hub for the workers of the West' and as a model 'for the people of the East who stand up to fight' – title-card statements that are illustrated with images of work rallies, manifestations and military training in different countries. The film concludes, now over images of rural plantations and a freight train carrying goods and workers, with the assertion that 'oppressed countries [are] gradually leaving the world of capital. They will pour forth into the stream of the united socialist economy'.

A Sixth Part of the World's nod to a united, socialist world materialises what Martin Jay, in his masterful *Marxism and Totality*, defines as Marxism's optimistic and 'genuinely "longitudinal" use of the concept of the totality', according to which 'societies could be understood only in relation to each other, that is, as part of a coherent whole'.[100] Already in *The German Ideology* (1845–46), and as a response to the encroachment of global capitalism, Marx and Engels noted that 'the liberation of each single individual will be accomplished in the measure in which history becomes transformed into world history':

> Only then will the separate individuals be liberated from the various national and local barriers, be brought into practical connection with the material and intellectual production of the whole world and be put in a position to acquire the capacity to enjoy this all-sided production of the whole earth (the creations of man). *All-round* dependence, this natural form of the *world-historical* co-operation of individuals, will be

99 Ibid.

100 Martin Jay, *Marxism and Totality: The Adventures of a Concept from Lukács to Habermas* (Berkeley: University of California Press, 1984), 31, 34.

> transformed by this communist revolution into the control and conscious mastery of these powers.[101]

A few years later, while noting capitalism's 'need of a constantly expanding market [...] over the whole surface of the globe' in *The Communist Manifesto* (1848), Marx and Engels welcomed one side-effect of this phenomenon, namely a 'world literature' that surpasses 'national one-sidedness and narrow-mindedness'.[102]

In many ways, Vertov attempted to realise for the medium of film the globalism Marx and Engels positively identified with a world literature. As Vertov saw it, the 'Kino-eye' was to fulfil the 'goal of establishing a visual bond between the workers of the whole world'.[103] In his published notes and diaries, one finds numerous ideas and plans for a 'world [seen] with the eyes of millions', captured by an 'army of kinoks' – or film scouts – who would regularly generate images and comprise a global 'mass authorship'.[104] One of these projects was a film he provisionally titled *A Day Throughout the World*, which would 'attend to the need to produce newsreels "on the life of the peoples of all lands" [Lenin's phrase]'.[105] Conceiving of the film as a 'footage film' whose 'script in this instance can be composed after the selection of material', Vertov notes that a 'selection of footage will be made in foreign film archives', which 'should enable us to arrange shots taken by capitalist firms in a combination to be directed against capitalism'.[106] In another project titled *A Minute of the World*, Vertov dreams up a film about 'the world from the perspective of a single minute on a certain day in a certain year of a certain era':

> We use this minute to show how many joys, fears, dramas, births, deaths, acts of heroism at work and at war, how many events, big and small, have taken place in these brief sixty seconds. The little minute turns into the Minute of the World, into an instantaneous survey of the world, into

101 Karl Marx and Friedrich Engels, *The German Ideology: Part One*, ed. C.J. Arthur (London: Lawrence & Wishart, 1970), 55, emphasis in original.
102 Karl Marx and Friedrich Engels, *The Communist Manifesto* (New York: International Publishers, 1948), 12–13.
103 Vertov, *Kino-Eye*, 52.
104 Vertov, 'Artistic Calling Card', 96, 86.
105 Vertov, *Kino-Eye*, 299.
106 Ibid., 297–298.

> a mighty document meant not only for contemporaries, but for future generations as well.[107]

In this light, *A Sixth Part of the World* can be regarded as a prototype of Vertov's utopian, Marxist world cinema by amassing footage shot on the Soviet nation while implicating the whole world of which that nation was a part.

Whereas *Melody of the World* cannot be disentangled from an interwar internationalist discourse of commonality, *A Sixth Part of the World* instead builds upon the collectivist principles of a Marxist globalism in order to extol the power of the film medium to figure and connect a larger whole. And yet, although some miles away from Ruttmann's film, Vertov's realised and unrealised projects must also be seen as celebratory expressions of a historical moment in which cinema was envisaged both as a machine capable of capturing the world from new, surprising angles and as a growing global archive of images that, in its flattening out of material phenomena and boundless combinatory possibilities, promised not only an aesthetic reorganisation of the world but also the revolution.

The Database, or 'YouTube is the World Stage'

In his *The Language of New Media* (2001), Manovich conceptualises the computerised database, understood as a 'collection of items on which the user can perform various operations – view, navigate, search', as the major logic underpinning new media objects and the Internet at large.[108] As we saw earlier, however, for Manovich the database is not exclusive to contemporary storage media, with *Man with a Movie Camera* hailed by him as 'perhaps the most important example of a database imagination'.[109] Moreover, as a cultural form premised upon collection rather than narrative, or a paradigmatic rather than syntagmatic relation between images, the database finds even older precursors in recording media such as photography and stereography, which have since their inception privileged 'catalogs, taxonomies, and lists'.[110]

107 Ibid., 315.
108 Manovich, *The Language of New Media*, 219.
109 Ibid., 239.
110 Ibid., 234

In this last section, however, my use of the concept of the database will specifically attend to its contemporary affiliations in terms of how it often (though not exclusively) implies the operations and/or inputs of a user on the Internet. I choose to retain the contemporaneity of the term first because I believe its use as an all-encompassing category risks losing a historically informed grasp of the diverse ways in which the catalogue can be cinematically brought to life – especially when it comes to non-fiction cinema, where, as we have seen, the integrated montage efforts of the global symphony differ quite significantly from the expositional collection mode of early-cinema programmes. Second, in maintaining the connection between the database and the Internet, my aim is to highlight the collective and/or participatory forms that such a connection has ushered in as an updated version of documentary modes premised on the catalogue. Indeed, as I hope to show, it is not a coincidence that the world symphony has received a boost in our time, when the current ease of recording and uploading images has exponentially expanded both the realm of the filmable and the global archive of images.

Of crucial importance in this discussion is the storage-video website YouTube, which, as James Leo Cahill has noted, 'stokes a utopian fantasy of total repository: nothing escapes record, no moment is lost, everything is shareable'.[111] If, to cite Doane, one of the consequences of the emergence of recording media is that 'time appears to be [...] reducible to no system or hierarchy (any moment can be the subject of a photograph; any event can be filmed)', then we can say that the appearance of YouTube has radicalised this phenomenon in both qualitative and quantitative terms.[112] As a seemingly endless archive that allegedly expands by 100 hours of video per minute, YouTube not only raises questions about the meaning and power of images in today's mediatised and virtualised world, it also generates anxiety in relation to how to navigate, look at and make sense of a world of images.[113] In order to explore these questions, I turn below to online-sourced projects – including *Man with a Movie Camera: The Global Remake* (2007–2014), *Mass Ornament* (2009) and *Life in a Day* (2011) – which can be seen as both reflections on this expanding image-world *and* symptoms of a quest to structure unbounded contingency in times of hyperconnectivity. I conclude the chapter by treating the collaborative documentary *In Praise of Nothing*

111 James Leo Cahill, 'A YouTube Bestiary: Twenty-six Theses on a Post-Cinema of Animal Attractions' in Paul Flaig and Katherine Groo (eds.) *New Silent Cinema* (New York: Routledge, 2016), 263–293; 268.

112 Doane, *The Emergence of Cinematic Time*, 10.

113 Cahill, 'A YouTube Bestiary', 268, 278.

(2017) as a parodic world symphony that attempts to subvert the genre from within by embracing rather than resisting the erosion of meaning associated with the contingency of the world.

Let us start with *Man with a Movie Camera: The Global Remake* (hereafter *The Global Remake*), a participatory online experiment conceived by the Canadian video-artist Perry Bard. Launched in 2007 and terminated in 2014, *The Global Remake* asked people from all over the world to upload 'remade' frames from *Man with a Movie Camera* on to a specially designed website. Each image, archived next to the film's original frames, was accompanied by information such as author, location and uploading date, and a software-automated film comprising the uploaded frames was produced daily. According to Bard, the project took inspiration both from Manovich's analysis of *Man with a Movie Camera* as a database form and from a desire to accomplish 'Vertov's vision of a future in which an army of "kinoks" (cameramen) will update world news every four hours'. She concludes: 'Today we have that army of kinoks, and YouTube is the world stage'.[114]

Vertov's utopian global cinema has, indeed, acquired a renewed significance in our time. It is hard to think of his unrealised *A Day Throughout the World* and *A Minute of the World* and not be reminded of strikingly similar titles such as *Life in a Day* and *One Day on Earth*, both of which were released in 2011 featuring footage collected across the planet shot over the span of a single day. Sponsored by YouTube and directed by Kevin Macdonald, *Life in a Day*'s 90-minute running time was edited out of 4,500 hours of received footage shot on 24 July 2010. On the other hand, it might seem perverse to align *Life in a Day* with Vertov's communist goals, especially given the fact that the film has been accused of typifying, to cite Jon Dovey, the 'kinds of creative expression afforded by the digital as a form of free labour exploited and appropriated for capital accumulation by corporations, brands and advertising', in this case YouTube itself.[115] One may equally suspect that Vertov would not have approved of the film's sentimental humanism and lack of political energies.

In fact, *Life in a Day*'s aim of portraying the Internet as an all-inclusive global forum was thwarted from its inception. Its makers admitted that they had to send 'hundreds of digital cameras to far-flung places across

114 Perry Bard, 'When Film and Database Collide' in Geert Lovink and Rachel Somer Miles (eds.) *Video Vortex Reader II: Moving Images Beyond YouTube* (Amsterdam: Institute of Network Cultures, 2011), 322–329; 323.

115 Jon Dovey, 'Documentary Ecosystems: Collaboration and Exploitation' in Kate Nash, Craig Hight and Catherine Summerhayes (eds.) *New Documentary Ecologies: Emerging Platforms, Practices and Discourses* (Hampshire, UK: Palgrave Macmillan, 2014), 11–32; 13.

the globe [...] that don't necessarily have easy access to the YouTube, or the Internet'.[116] The curious effect of this 'commissioned' footage, which often exudes a voyeuristic quality with reluctant subjects on camera, is that it makes visible, by virtue of its contrast with the otherwise spontaneously filmed materials, the film's effort to paper over the inequalities of the digital world in the name of a 'we-are-the-world' message. Indeed, as Bard herself realised after the launch of *The Global Remake*, its planetary aspirations soon became revealing of a digital divide materialised both through 'lo-res and high-def uploads' and 'the absence of uploads from parts of the world that are unrepresented' – a gap she chose to leave visible so that the project would retain a critical eye on the unevenness of global digital access.[117]

Notwithstanding their many differences, however, both *The Global Remake* and *Life in a Day* are contemporary efforts to structure the web's avalanche of contingent images. As Macdonald once remarked in relation to the received footage: it 'was all serendipity, it was all chance'.[118] The challenge for the editing team thus consisted in linking up the images meaningfully, a task that the film attempts to accomplish by borrowing both the city symphony's daily temporal span and its dominant montage principles of similarity and simultaneity. Granted, structuring tendencies were already in place at the film's commissioning phase, since users were asked to send recorded footage by answering on-camera questions chosen in advance ('What do you have in your pocket?'; 'What are you are scared of?', etc.). But, on a broader scale, the editing performed by Macdonald and his team could be seen as the updated, digital version of Ruttmann's footage selection for his *Melody of the World* in the age of YouTube – though, it must be noted, without Ruttmann's flair in creating surprising and striking juxtapositions.

For its part, *The Global Remake* rehabilitates *Man with a Movie Camera* by using it as a model against which images can be compared, made equivalent and, in the process, made less contingent (figure 3.11). This is especially visible on the project's website, which provides an alphabetical listing of the shots' visual content, ranging from 'aerial', 'ambulance' and 'airplane' in section A through to 'water', 'worker' and 'woman sleeping' in the last section. This inventory lays bare the pan-visuality of the filmic original, while at the same time offering itself as a scheme to be modelled on. That said, the sense of humour arising out of the combinations derives from the fact that the analogies envisaged by each contributor do not necessarily

116 Joe Walker, 'Kevin and Joe Promos', extras in *Life in a Day* (2011), DVD, Entertainment One.
117 Bard, 'When Film and Database Collide', 324, 328.
118 Kevin Macdonald, 'Kevin and Joe Promos'.

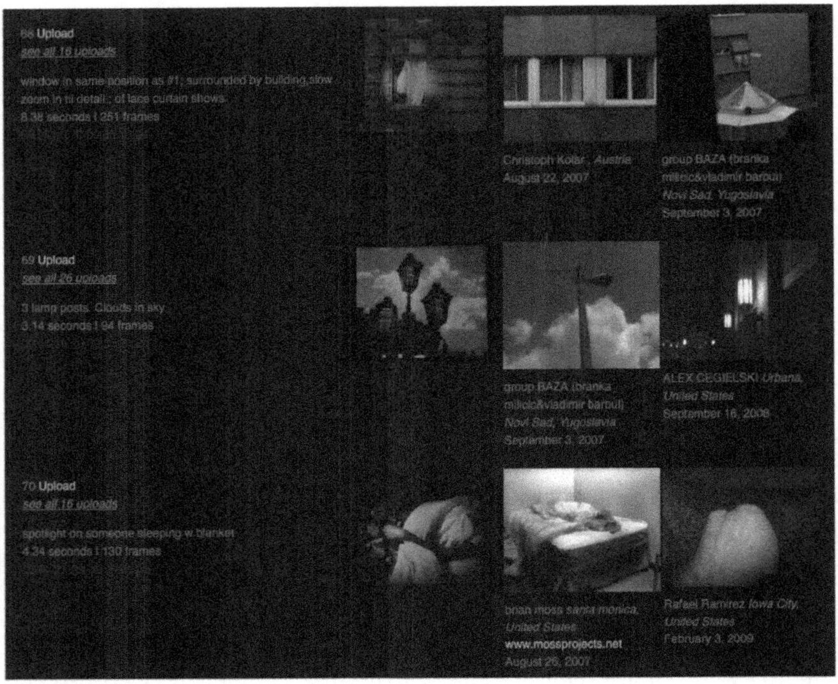

Figure 3.11 Pictures uploaded to *Man with a Movie Camera: The Global Remake* often bring an element of the unexpected when compared with the original shots. Courtesy of Perry Bard.

conform to compositional or graphic equivalence. To give one example: under the tag 'Bottles', the film's original shot of a crate with symmetrically lined up bottles appears alongside images that range from cardboard beer boxes stacked up in a garage to a litter overflowing with empty bottles.[119] While the tagging logic promises a rudimentary comparative system, the threat of disruption of that system remains as the uploaded pictures introduce unprogrammed and unexpected elements.

A balancing act between structure and contingency is also evident in Natalie Bookchin's multiple-channel installations, many of which are available as videos on the Internet. By sifting through online videos that become, in her words, 'a vast, largely untapped stream of constantly updated source material', Bookchin's aim is to find correlations between images and make them cohere into a bigger picture.[120] In the four-channel video installation *Round the World* (2007), images from online security webcams,

119 See http://dziga.perrybard.net/tags/show/bottles (last accessed 18 June 2019).

120 Natalie Bookchin and Blake Stimson, 'Out in Public: Natalie Bookchin in Conversation with Blake Stimson', in *Video Vortex Reader II*, 306–317; 307.

many showing roads, seas and means of transport, are alternated with the verbal transcription of Thomas Edison's imagined 1888 around-the-world trip recorded on his phonograph, while the recording itself plays on the aural track. In the single-channel, 64-minute *trip* (2008), YouTube images of travels from 70 different countries filmed from window cars are stitched together with the aim of creating a global journey recorded through the eyes of missionaries, truckers, soldiers and migrants. In *Mass Ornament* (2009), her most famous work to date, Bookchin collates hundreds of YouTube domestic videos of people dancing on camera to an added soundtrack, including Lullaby of Broadway from Busby Berkeley's *Gold Diggers of 1935* (1935) and the Wagner pieces featured in Leni Riefenstahl's *The Triumph of the Will* (*Triumph des Willens*, 1935).

Mass Ornament combines the frames not only consecutively through montage but also alongside each other in horizontal lines that recall YouTube's interface, an impression reinforced by the number of 'views' appearing underneath the frames. Frequently, the horizontal visual arrangement is orchestrated according to compositional analogy whereby similarly placed household objects in the frames – sofas, mirrors, personal computers, iron boards, Christmas trees – provide the connecting link. But in its emphasis on near-identical human gestures and dance moves across multiple videos, many of which are impersonations of Beyoncé's Put a Ring on It, *Mass Ornament* also reveals a process of human objectification (figure 3.12). That is, while each of the individuals may be answering YouTube's call to 'Broadcast Yourself' with the hope of expressing their unabashed uniqueness, *Mass Ornament* discloses these human gestures and moves to be in fact components of a larger thing: an ornament.

Mass Ornament lifts its title from Kracauer's eponymous essay, in which the choreography of the Tiller Girls, in its supremely coordinated symmetric patterns, is theorised as the formal expression of capitalism. Understood 'as parts of a mass, not as individuals', with movements that bear 'no meaning beyond themselves', the dancing girls constitute a 'mass ornament' by 'emptying all the substantial constructs of their contents'.[121] Kracauer's reflections on the mass ornament resonate with his essay on photography, written in the same year (1927): if photography underlined the 'mere surface coherence' of things, the Tiller Girls were for him 'surface-level expressions'.[122] But in the same way that Kracauer approached photography with ambivalence and curiosity, for him it was imperative not to dismiss the

121 Kracauer, *The Mass Ornament*, 76–77.

122 Ibid., 52, 75.

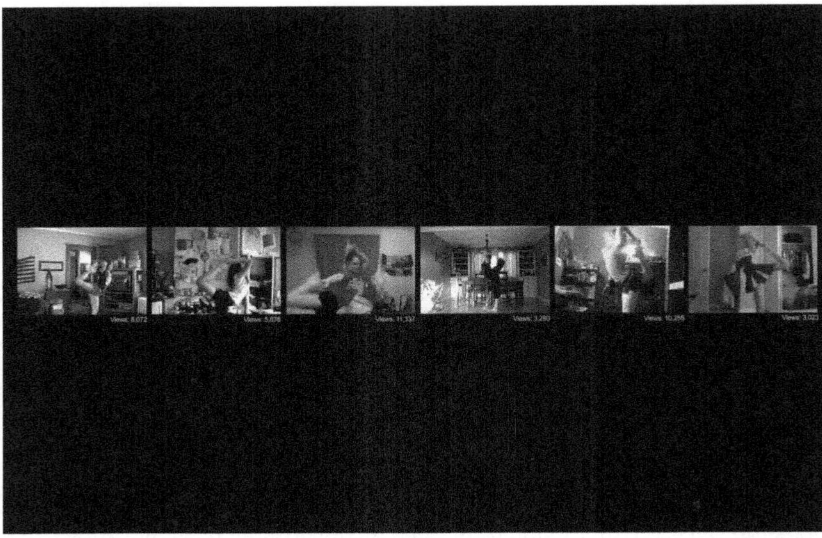

Figure 3.12 *Mass Ornament* (2009) reveals human gestures and moves to be components of a larger thing: an ornament. Courtesy of Natalie Bookchin.

ornament, since it was only by confronting it that one could gain 'unmediated access to the fundamental substance of the state of things'.[123] In other words, to understand one's time, one must face head on the products of that time.

Bookchin's work begs to be placed in relation to both Kracauer essays in its reflection on the mass ornament in connection with the blizzard of images stored on the digital database. By relating multiple and multiplying frames, *Mass Ornament* reveals the way individuals 'perform a partial function without grasping the totality'.[124] The assemblage of gestures in *Mass Ornament*, however, resists uniformity thanks to Bookchin's notable use of sound.[125] At points, amplified ambient sounds simultaneously emanate from different videos, while added 'found' noises such as thumbing, banging and shuffling are equally heard. By the same token, the corporeal movements do not congeal into an undifferentiated mass. The bodies are far from normative, their ethnicities are varied, and the delivery is often awkward: we are far

123 Ibid., 75.
124 Ibid., 78. In this respect, *Mass Ornament* resonates with Emma Sheffer's *Insta_Repeat* project (2018–ongoing), an Instagram account that posts grids containing nine pictures, all collected from Instagram, with the aim of highlighting their similarity in terms of location, visual content, framing and perspective.
125 For an illuminating account of Bookchin's sonic strategies, see Zoë Druick, 'Small Effects from Big Causes: The Dialogic Documentary Practice of Natalie Bookchin', *Camera Obscura* 31:2 (92) (2016): 1–25.

from the Tiller Girls' calculated geometry of lines and forms.[126] By retaining or highlighting individual details in each of the videos, *Mass Ornament* therefore eschews homogeneity by way of contingency, thereby casting a discerning glance at contemporary digital cultures. But if we are to follow its own aesthetic logic and place it in a bigger picture, *Mass Ornament*, like the contemporary projects examined earlier, is also a symptom of a quest to counter the ever-proliferating database through a repurposing of images into systematic patterns the aim of which is to assuage meaninglessness.

It is against this background that the global collaborative documentary *In Praise of Nothing* (2017), to which I now turn by way of concluding this chapter, stands out. Written and 'directed' by the Serbian filmmaker Boris Mitić, the film stitches together images filmed by 62 cinematographers across 70 countries over a period of eight years. Though, unlike the works explored above, *In Praise of Nothing* does not make use of personal and amateur images, it owes its existence to the Internet's participatory mechanisms. The film started as an online project through a call for contributions aimed at directors of photography around the world. At first contributors were asked to film 'Nothing as you see it, feel it or imagine it, without interference nor instructions'. Later they were required to upload the shots on to an online platform where they could see each other's images and brainstorm ideas. The final stage involved the commissioning of specific shots by Mitić as he put the film together.[127]

In Praise of Nothing may thus warrant comparison with *Life in a Day*. Not only did it harbour global aspirations from its inception, it was originally conceived as an assemblage of varied footage over which the director had little control at first. Yet the differences between both films are equally significant. The first relates to the fact that *In Praise of Nothing*'s images, filmed as they were by professional cinematographers, convey a more pronounced, controlled aesthetic dimension in terms of static framings and visual composition, with the film exuding a more deliberate quality that

126 A useful point of comparison is Pharrell William's *Happy Party* project (2015), made in partnership with the United Nations Foundation, where people were invited to create GIFs while dancing along to William's hit Happy to 'spread happiness, and demand climate action'. On the project's website, hundreds of GIFs are seen on a grid next to each other, all synchronically reacting to the beats of the song, which sacrifices multiplicity for the sake of oneness. See https://www.globalhappyparty.com/ (last accessed 7 August 2020). My thanks to Martin Roberts for drawing my attention to this project.
127 See call for contributions: http://dribblingpictures.com/nothing-1p.pdf (last accessed 01 August 2019).

may recall contemporary slow cinema's perceived penchant for moments where 'nothing happens', and I will return to this.[128]

Second, *In Praise of Nothing* eschews a life-in-a-day temporal structure and similarly avoids making recourse to the list as an organising device. While the film is framed around the absurdist idea of a weekend during which 'Nothing' visits Earth, and therefore unfolds over a specific time frame, there is no discernible chronology but only a loose tripartite structure involving the arrival, stay and departure of Nothing, given aural existence and male personification via Iggy Pop's voice. Doing away with celebratory humanism, *In Praise of Nothing* is instead an experiment in divesting images of meaning by asking the viewer to see 'the world according to Nothing, an all-encompassing vision', according to one of its promotional taglines.[129] The spectatorial contract the film proposes is therefore premised upon its insistence that we should, as Nothing tells us at the beginning, 'imagine that what you see is what I see. Or just enjoy the scenery and let your mind roam freely'.

Whether the film is successful in its attempt to confront a world of nothingness, however, remains open for debate. On the one hand, the narration remains largely within a nonsensical resister and is often unrelated to a seemingly unstructured visual track. On such occasions, *In Praise of Nothing* does offer a glimpse of what a parade of global images devoid of meaning and order might look like. At one point early in the film, for example, the following stationary shots appear: a hanging piece of clothing swirling with the wind; a dog under a bed next to a dirty sock; three topless men, drinking and on mobile phones, on a small boat at sea; the wrinkled hands of an elderly black man holding a cigar; a half-open window centralised in the frame. Whereas this sequence might recall earlier film symphonies' miscellaneous assortment of things, there is a difference related to the fact that, in such symphonies, things are, at the very least, made sense of within a temporal structure through which simultaneity and comparison can be verified in relation to a specific section of the day, which is not the case here. As *In Praise of Nothing* unfolds, establishing shots of landscapes are intercut with close-ups of objects, images of empty interior spaces alternate with external building façades, and glimpses of fortuitous occurrences

128 On slow cinema, see Tiago de Luca and Nuno Barradas Jorge (eds.) *Slow Cinema* (Edinburgh: Edinburgh University Press, 2016).

129 As found in the blurb for the website for the International Documentary Festival Amsterdam (IDFA) where the film can be purchased for streaming: https://www.idfa.nl/en/film/21dc6ed2-8ba7-4ffd-86cf-64c8289dd35e/in-praise-of-nothing (last accessed 2 August 2017).

abound, from a gyrating plastic bag through to barefoot women walking at night-time.

The striking thing about *In Praise of Nothing* when it comes to its allegiance to the contingency of the world would thus seem to reside in how it embraces its own ability to show anything and everything, from a dirty sock to wrinkly human hands, not as stuff in need of categorising, but rather, as nothing, that is to say, as the negation of meaning or even being in the world. As Brown notes, via Heidegger, '*thing* (*Ding*, *chose*, &c.) can designate merely something (*ein Etwas*) as opposed to nothing; it can refer to actions or conditions ("Let's get those things done now"; "things have been pretty shitty"); and it can name any quotidian object – a rock, a knife, or a watch'.[130] Like other works discussed in this chapter, both past and present, *In Praise of Nothing* hinges on film's ability to engage with this elastic notion of things. But the difference lies in how it sees itself as an aesthetic exercise to radicalise the medium's affinity for visual promiscuity by conceiving of that promiscuity not as in need of an overriding logic – whether, say, Vertov's productivist interdependence or Ruttmann's formalist analogy – but, on the contrary, as the gateway into a world without sense and significance.

In its promise of a world 'alienated from meaning', seen from the perspective of Nothing, *In Praise of Nothing* offers a reply to Kracauer's contention that photography may herald a world in 'its independence from human beings' and in which 'the original order is lost'.[131] Indeed, with Kracauer, we might say that *In Praise of Nothing* offers the possibility of a 'liberated consciousness' by 'destroying natural reality and scrambling the fragments'.[132] One could go so far as to say that the film attempts to offer a compendium of what Kracauer, in his 1960 *Theory of Film: The Redemption of Physical Reality*, listed as the medium's 'inherent affinities': 'the unstaged', 'the fortuitous', 'the indeterminate', 'endlessness' and 'the flow of life'.[133] If, for the Kracauer of the late 1920s the surface of things had to be confronted in order to be transcended, by the time he had published *Theory of Film* cinema was embraced on the basis of its capacity to cling 'to the surface of things' as an antidote to scientistic and utilitarian thinking.[134] In contrast with 'the abstractness of our approach to things in and about us', he maintained, film enables 'the experience of things in their concreteness'; it brings out

130 Brown, *Other Things*, 18.

131 Kracauer, *The Mass Ornament*, 61.

132 Ibid., 62.

133 Siegfried Kracauer, *Theory of Film: The Redemption of Physical Reality* (Princeton, NJ: Princeton University Press, 1997), 60–76.

134 Ibid., 285, 296.

'the qualities of things' and allows us 'to take away with us the objects and occurrences that comprise the flow of material life'.[135] *In Praise of Nothing*'s wandering gaze on a thingly world void of meaning resonates with these reflections.

On the other hand, however, on closer inspection the film does reveal structural, thematic and semantic directives underlying and regulating its textual makeup. One portion of the film, for example, is divided into geographical sections, as when Nothing sets out to describe with equal disdain the world's regions and its peoples, including the United States, Asia, Europe and the Arab world. Occasionally, a few images are joined up on the basis of graphic matches or around subject matter, examples including the shot of a hole in the wall followed by a mirror and a sequence comprising shots of empty streets at night. At the same time, the overpowering, unstoppable voiceover narration means that the images are never quite fully on their own, as phenomenal surface appearances in their own right as Kracauer would have it, meaning the viewer is often attempting to make connections between the visual and the aural tracks, even if to no avail.

More remarkably, as a film intent on capturing nothingness, it is telling that it does not allow its images to unfold for longer periods of time. Though, as noted earlier, *In Praise of Nothing* does toy with a contemplative aesthetics, it ultimately stops short of adopting a durational style, with takes running on average 7–10 seconds. This resistance to overstretched takes gains in significance when we consider that, in film studies, the idea of *Nothing Happens*, to cite the title of Ivone Margulies's seminal book (1996) on Chantal Akerman, is often related to lengthy shots whose duration exceeds representational and symbolic imperatives, as in the domestic chores on display in Akerman's *Jeanne Dielman, 23 quai du commerce, 1080 Bruxelles* (1975).[136] And it is a definition that has gained a new lease of life more recently thanks to the emergence of 'slow cinema', which, as Song Hwee Lim notes, often makes us question 'the very notion of "thing", and what counts as "nothing" within a film's narrative' by according durational importance to objects of undefined or indeterminate semiotic value.[137]

In the film's defence, *In Praise of Nothing*'s resistance to slowness, regrettable as it might be, seems intentionally designed to shun the accusations

135 Ibid., 296, 300.

136 Ivone Margulies, *Nothing Happens: Chantal Akerman's Hyperrealist Everyday* (Durham and London: Duke University Press, 1996).

137 Song Hwee Lim, 'Temporal Aesthetics of Drifting: Tsai Ming-Liang and a Cinema of Slowness' in *Slow Cinema*, 87–98; 90.

of affected aestheticism commonly levelled at slow cinema, with Nothing musing at one point: 'Make any shot too lengthy and get poetry?'. Moreover, the above should not stop us from appreciating the film both as an amusing parody of the global symphony and a sardonic existential commentary in a world dominated by images. Yet, as an attempt to reveal a world deprived of meaning, a no-thing, *In Praise of Nothing* is proof that cinema may never be ready to stop making sense of things.

4 The Face of the World

Abstract
This chapter examines the trope of faciality in relation to extinction discourse. It starts by considering the way photography was recruited to the typological project of classifying the faces of humanity according to nineteenth-century biological theories of race. I show how the head-on facial portraiture, ranging from carte de visite collections through to anthropometric photography, lent visual concreteness to the idea that supposedly inferior races were to disappear from the face of the Earth. I then examine the longevity of this trope in two recent documentaries: *Samsara* (2011) and *Human* (2015). Although both films try to distance themselves from evolutionary human hierarchies, their programmatic adherence to the codes of nineteenth-century still portraiture, I suggest, reveals the persistence of colonial modes of looking.

Keywords: face, race, anthropology, extinction discourse, *Samsara*, *Human*

Whether or not the subject is already dead, every photograph is this catastrophe.
Roland Barthes

We have explored so far how film and related visual media have been recruited to a world-seeing project aimed at disclosing the nonhuman world: be it the Earth itself as a singular planet, the life forms unavailable to perception or else removed from view due to geographical distances, or the things and objects crowding the visible. But this is only half of the story. For photography and cinema, as the remainder of this book will examine, have been equally put to the service of visualising, ordering and cataloguing a human collective on the globe.

In such a project one particular image has exerted a special allure: the human face. Whereas painted facial portraits had been the privilege of an elite, moneyed few, the appearance and subsequent popularisation of

de Luca, T. *Planetary Cinema. Film, Media and the Earth.* Amsterdam: Amsterdam University Press, 2022
DOI: 10.5117/9789463729628_CH04

photography widened this social practice. By the late 1850s, having one's own picture taken and sharing it with friends, relatives and acquaintances had become a widespread custom in many Western metropolises. The revolution that this signified cannot be overstated. As Hagi Kenaan observes, 'what serves as our most immediate *carte de visite*, our point of entry into the public visual sphere – the face – is something to which we ourselves do not have direct access. Our face escapes us'.[1] Carte de visite, French for 'calling card', was, in fact, the very name given to these nineteenth-century photographic portraits of oneself. Previously confined to its reflection in mirrors or reflective surfaces, or at most to a painting hanging on the wall, the face became a thing one could hold in one's hands and give to others, newly reproducible and endlessly circulating.

But is the face ever a 'thing', like those we excavated in the last chapter? Not for the philosopher Jean-Paul Sartre, who writes: 'We discover, among things, certain beings we call faces. But they do not exist like things'. Things, for Sartre, have no mobility and hence no temporality: 'Things are piled up in the present, shivering but never budging from their place; the face, on the contrary, 'projects itself ahead of itself in space and time': it is '*visible* transcendence'.[2] If we are to follow Sartre, this transcendental visibility may explain why faces immediately became privileged objects of attention for photography, which promised the fixing, via imaging, of this transcendence as subjects sat immobile for a long time in front of the camera.

But it does not fully explain why faces went global. For faces were held to exteriorise not only transcendence but also mental and intellectual traits which were themselves differentiated as they were mapped out across the planet and contrasted with the 'civilised', white European races. As Denise Ferreira da Silva argues in her influential *Toward a Global Idea of Race* (2007), nineteenth-century 'sciences of man', such as anthropology, instituted the racial as a global signifier through a 'grid of specification of human beings that identified each particular bodily form and the corresponding type of mind (mental functions) as coming from a particular global region or continent'.[3] In this context, 'the skull constituted the primary object of research. The reason was simple.

1 Hagi Kenaan, 'The Selfie and the Face' in Julia Eckel, Jens Ruchatz and Sabine Wirth (eds.) *Exploring the Selfie: Historical, Theoretical, and Analytical Approaches to Digital Self-Photography* (Cham: Palgrave, 2018), 113–130; 119, emphasis in original.

2 Jean-Paul Sartre, 'Faces' in *The Writings of Jean-Paul Sartre*, eds. Michel Contat and Michel Rybalka (Evanston, IL: Northwestern University Press, 1974), 67–71; 71, emphasis in original.

3 Denise Ferreira da Silva, *Toward a Global Idea of Race* (Minneapolis and London: University of Minnesota Press, 2007), 126.

Measurement of skulls would provide information about the volume and size of the brain'.[4] The presumed mental inferiority of the 'others of Europe', which found visible form in the shape of their heads, acted as global indexes of specific geographical places that, it was believed, directly conditioned such inferiority. 'What distinguishes this moment', Ferreira da Silva maintains, 'is a deployment of historicity that presupposes globality as a stage of human differentiation'.[5] In other words, globality was not a mere background for scientific race-thinking but the very ontological context that produced and enabled the classification and ranking of humans in the first place.

Mirroring this global spatialisation of humanity along hierarchic lines was its temporalisation according to evolutionist paradigms. For humans were not only assigned different intellectual attributes depending on their geographical provenance; they were also placed along a unidirectional temporal arrow that denied futurity to some humans according to their position on the evolutionary ladder. As a 'specific branch of the dual ideologies of imperialism and racism', this 'extinction discourse', as Patrick Brantlinger has shown, postulated 'the inevitable disappearance of some or all primitive races' in the face of 'civilisation' and 'progress'.[6]

Although not examined by Ferreira da Silva and Brantlinger, photography, and especially photographs of human faces (whether frontal or profile), were crucial disciplinary techniques deployed by the sciences of man to make visible planetary human differentiation and evolutionary extinction discourse. As I explore in this chapter, photography became not simply an appendage to anthropology but the very medium through which it validated its scientific authority as it measured and catalogued the heads of human beings while seeing as its mission the fixing for posterity of the 'always already vanishing others of Europe'.[7] My driving questions in this chapter will be: How did the technological reproducibility of faces provide the means whereby a global arena could be mapped out and rendered visible, and a global humanity hierarchised according to evolutionary parameters? How, in other words, did faciality intersect with globality, and how was that intersection itself imbricated with an overdetermined mortality whose internal functioning principle was racial differentiation?

4 Ibid., 121.

5 Ibid., 117.

6 Patrick Brantlinger, *Dark Vanishings: Discourse on the Extinction of Primitive Races, 1800–1930* (Ithaca, NY, and London: Cornell University Press, 2003), 1.

7 Ferreira da Silva, *Toward a Global Idea of Race*, 29.

To answer these questions, I first consider the mysterious allure of the human face at photography's dawn, its status as a special image for the lens of the camera, in relation to the daguerreotype and the carte de visite. I then proceed to investigate the deployment of facial and anthropometric photography in global compilations of 'race-types' that implicitly or explicitly proposed the disappearance of 'primitive races' from the face of the Earth according to social-evolutionist paradigms. My focus will be varied but concentrate especially on the British case, given its importance for the foundation of anthropology as a discipline. Case studies will include anthropological treatises and compendia from the 1850s onwards, but special attention will be given to 'popularisers' such as the Dammann brothers' 'atlas of types' *Races of Mankind: Ethnological Photographic Gallery of the Various Races of Man* (1876) and the illustrated serial *The Living Races of Mankind* (1901). This large temporal frame is necessary both because it enables us to see how photography became gradually valued in anthropological circles (in lieu of drawings of human faces) and because it allows us to evaluate how anthropology itself intermingled with popular visual culture, profiting from the carte de visite to then influence photographic serials and magazines at the turn of the century.

The final section of the chapter, in tune with this book's methodological approach, turns to contemporary documentaries that center on the human face as the common denominator of a differentiated humanity: Ron Fricke's *Samsara* (2011) and Yann Arthus-Bertrand's *Human* (2015). My rationale for choosing these documentaries is twofold. First, they programmatically adhere to the conventions of the portrait genre to the point that they subtract and/or downplay cinematic properties. Second, their galleries of humanity, not unlike their nineteenth-century counterparts, are equally concerned with human extinction, although in their case a disappearing humanity is no longer attributed to racial inferiority but to the prospect of global environmental catastrophe and population explosion. Nevertheless, as I hope to show, their allegiance to codes and features of the facial portrait reveals the persistence of a colonial gaze that is predicated on racial differentiation as a global signifier and qualifier.

The Inter-Face

The official genealogies of photography are well known. In 1837 France, Louis Daguerre recorded his first daguerreotype, a unique, 'positive' image formed on a metal plate coated with iodine vapours, which he showed publicly in

1839. In that same year, in England, William Henry Fox Talbot recorded a 'negative' image on paper, or a collotype, from which other images could be printed. The daguerreotype, however, remained the medium of choice for many in view of its superior qualities, and its preferred visual object was the human face. For Walter Benjamin, this was symptomatic. In his 'The Work of Art in the Age of Mechanical Reproduction' (1935), he noted:

> In photography, exhibition value begins to displace cult value all along the line. But cult value does not give way without resistance. It retires into an ultimate retrenchment: the human countenance. It is no accident that the portrait was the focal point of early photography. The cult of remembrance of loved ones, absent or dead, offers a last refuge for the cult value of the picture. For the last time the aura emanates from the early photographs in the fleeting expression of a human face. This is what constitutes their melancholy, incomparable beauty.[8]

Two things are worth noting here. For Benjamin, the uniqueness of the daguerreotype as a one-of-a-kind, glass-encased object finds itself replicated in the human face it represents, thereby preserving, for the last time, the aura of the image, its 'cult value'.[9] Unlike the reproducible images it would soon give rise to, photography is here still an artisanal and unrepeatable process. It requires subjects to sit immobile for a long time, and thus retains, as Noa Steimatsky suggests, 'a sense of intensified temporality' by capturing 'something of the subject's unfolding into the image', not unlike 'the traditional portrait's phenomenological constitution'.[10]

But equally important for Benjamin in this auratic process is the association of face with death, as it serves the purpose of remembering 'loved ones, absent or dead'. This ritualistic use of the face was not exactly new. In many ways, photography simply usurped the place that had once been occupied by the mummy portrait and the death mask: this, as is well known, was one of the main arguments put forward by André Bazin in his 'The Ontology of the

8 Walter Benjamin, *Illuminations*, ed. Hannah Arendt, trans. Harry Zohn (New York: Shocken Books, 2007), 225–226.

9 It must be noted that this is a revised position for Benjamin, who had written in 1931, in his essay on photography, that the daguerreotype could not entail an auratic process insofar 'as the camera records our likeness without returning our gaze'. He went on: 'To perceive the aura of an object we look at means to invest it with the ability to look at us in return'. Benjamin, *Illuminations*, 188.

10 Noa Steimatsky, *The Face on Film* (New York: Oxford University Press, 2017), 8.

Photographic Image', originally published in 1945.[11] Taking up Bazin, Laura Mulvey drives the point home by noting that the connection between death and photography was 'understood very quickly in the nineteenth century as people adopted photography into the rituals of mourning and memorials. The deathbed photograph came to replace the death mask'.[12] Already at its dawn, the photographic portrait thus latches on to the face both because of its promise of access into the subject's temporalised essence, a way of unlocking one's transcendence, and because of its immediate evocation of a time that is no more. To quote Roland Barthes: 'Whether or not the subject is already dead, every photograph is this catastrophe'.[13]

By the late 1850s, thanks to the glass negative wet-collodion process and new cameras, the technological reproducibility that, for Benjamin, washed away the aura of facial photography was already a reality in full swing. But even though sitting times had been significantly reduced and multiple photographs could be printed at once on to light-sensitive papers, the idea that a photographed human face could reveal the subject's uniqueness, his or her aura if you will, remained in place thanks to the appearance and popularity of the carte de visite, or calling card.

Patented by Frenchman André-Adolphe-Eugène Disdéri in 1854, the carte de visite was an inexpensive, cardboard-mounted photograph of oneself, typically measuring 6x10cm, that people exchanged with friends, acquaintances and relatives. As John Tagg has argued, insofar as portraiture had historically been an aristocratic practice and privilege, 'the rise of the photographic portrait' mirrors 'the rise of the middle and lower-middle classes towards greater social, economic and political importance'.[14] Facial portraits were, in this context, meant to assert individuality: a good portrait, in the tradition of the daguerreotype, was one that managed to reveal the subject's personality and essence through the face. But the advent of the carte de visite also meant, paradoxically, that individuality was submerged in uniformity, since studios often 'relied on a limited inventory of props, poses and backdrops' in order to secure the card's formal equivalence as an

11 André Bazin, *What is Cinema?*, volume 1, trans. Hugh Gray (Berkeley: University of California Press, 2005), 12.

12 Laura Mulvey, *Death 24x a Second: Stillness and the Moving Image* (London: Reaktion Books, 2006), 59.

13 Roland Barthes, *Camera Lucida: Reflections on Photography*, trans. Richard Howard (New York: Farrar, Straus and Giroux, 1981), 96.

14 John Tagg, *The Burden of Representation: Essays on Photographies and Histories* (Basingstoke, UK, and New York: Palgrave, 1988), 37.

industrially reproducible image.[15] And while, at first, technical limitations required the placement of subjects at a certain distance from the camera, often in rigid full-frontal postures because of still fairly long exposure times, soon the face became fuller and more centralised, sometimes highlighted with a vignetting effect.[16]

The carte de visite, in short, broadened the uses and reach of portraiture: not only was it now much cheaper to have one's picture taken, images of one's face and the face of others became circulating, interchangeable objects in a public sphere which, as recent scholarship has shown, can be deemed the precursor of contemporary social networks and the digital selfie.[17] As Kenaan tells us, it is unsurprising that the face lends itself so easily to mechanical reproduction and multiplication. As a 'modality of self-expression and self-presentation' in itself, the face is already an image, an 'inter-face' that is 'never just present but always already in the mode of self-presenting. It is an entity whose essence lies in its turning toward the eye, in facing a viewer'. Hence, 'in the age of "technological reproducibility", the image seems to be a natural vehicle through which the primary self-presentation of the face can expand its reverberation in the public sphere'.[18]

Though Kenaan's focus is on the selfie phenomenon, his remarks are applicable to the nineteenth-century carte de visite, the emergence of which institutes and formalises the face as a technologically reproducible mode of self-presentation. The impact of the carte de visite cannot be overestimated. For if the face is a special image because it is always already communicating and presenting even when seemingly it is not, the other, equally important, side of this process is that one's own face is always excluded from the field of visibility. We see other faces, which in turn see us, but we cannot see our own faces facing others. Sartre eloquently sums up: 'The trouble is that I don't see my face – or at least not at first. I carry it in front of me like a secret I am unaware of, and it is, on the contrary, the faces of others which teach me about my own'.[19] In this context, the carte de visite – and the myriad portrait and self-portrait genres that would follow in its wake – both

15 Deborah Poole, *Vision, Race, and Modernity: A Visual Economy of the Andean Image World* (Princeton, NJ: Princeton University Press, 1997), 111.

16 William C. Darrah, *Cartes de Visite in Nineteenth Century Photography* (Gettysburg, Pennsylvania: W. C. Darrah Publisher, 1981), 26.

17 See, for example, Annie Rudd, 'Victorians Living in Public: Cartes de Visite as 19th-Century Social Media', *Photography and Culture*, 9:3 (2016): 195–217.

18 Kenaan, 'The Selfie and the Face', 120.

19 Sartre, 'Faces', 68.

returns one's face to the field of visibility while formulating, via a doubling, the encounter with known others as a fundamentally inter-facial process.

The carte de visite, however, introduced faces beyond familiar faces. Seizing it as a tremendously lucrative business, studios published faces of celebrities, statesmen and members of royalty, while dispatching photographers to faraway corners of the world to record and gather the pictures of 'native' and 'primitive' races, often coded as such in terms of utensils, weapons, feathers, and body and face ornaments. A global network of images was then established, one in which cartes de visite, and later postcards and travel images, became items of transaction: they were purchased, collected and assembled in purposefully made albums that ordered a planetary humanity according to prevailing racist parameters. As Anne Maxwell observes:

> By the 1880s almost every middle-class household in Britain, the USA, Australia and New Zealand could boast a photographic collection that included images of indigenous peoples. But Victorian collectors did not simply amass things; taking their cue from scientists, they arranged postcards and photographs into series and sets that resembled the hierarchical taxonomies of race.[20]

In these collections, pictures of native peoples pertained to a different modality. As Deborah Poole explains in her *Vision, Race and Modernity* (1997), unlike 'the bourgeois cartes, whose imagery retained a residual "use value" as likenesses of friends and relatives, the colonial or "native" cartes de visite portrayed anonymous and historyless subjects'.[21] In other words, while people may have assembled pictures of natives because they may have been curious to see what 'others' looked like, these photographs were collected not because of the individual subjects they portrayed but because they embodied a racial and physiognomic *type* – a visual typology that was made possible by the very formal uniformity of the carte de visite as a reproducible image.

In large part, the necessity to order, differentiate and classify humanity can be attributed to the sensation caused by Johann Caspar Lavater's *Essays on Physiognomy* (1775–1778) in the first half of the nineteenth century, according to which the character of a person was an immutable God-given

20 Anne Maxwell, *Colonial Photography and Exhibitions: Representations of the 'Native' and the Making of European Identities* (London and New York: Leicester University Press, 1999), 11.
21 Poole, *Vision, Race, and Modernity*, 140.

attribution that could be read through his or her facial structures.[22] While Lavater's 'science' had illustrious precursors, such as the physiognomic treatises by Giovanni Battista della Porta and Aristotle, its popularity was as unprecedented as it was historically specific.[23] It constituted, as Stuart Ewen and Elizabeth Ewen have shown, a response to 'an expanding world, where interactions between European and non-European societies were intensifying', which in turn required 'a global map of physiognomic distinctions, a new hierarchy of inequality' in the face of an increasing proliferation of faces.[24] The popularity of the carte de visite was, in this context, intimately aligned with prevailing racial and physiognomic discourses. No wonder it was soon adopted as a fundamental tool by the emerging discipline of anthropology, which, energised by photography, took upon itself the task of producing a global taxonomy of human difference.

In a sense, as Roslyn Poignant points out, 'carte-de-visite portraits [...] appeared to be the ready-made constituents of a scientific narrative waiting to be pieced together. The extension of control over bodies implicit in anthropometric photography was but a short step to take'.[25] A measuring photographic system largely based on craniometry (and osteometry), anthropometry consolidated the Lavaterian idea that physical appearance revealed inner qualities of intellect and character along racial lines.[26] In the UK, two anthropometric systems were devised in 1869, both of which were used as disciplinary and measuring tools in the service of British colonial governance.[27] The first, created by John Lamprey, then secretary of England's Ethnological Society, displayed naked subjects in full-frontal and profile positions against a crisscrossed mesh made out of silk, the purpose of which was to delineate anatomical comparisons.[28] For the second, created by the anthropologist Thomas Huxley, naked persons were placed against

22 As Poole notes, by the time Disdéri patented his invention, Lavater's book 'had gone through hundreds of editions' in France. Ibid., 110.

23 For a succinct and useful history of physiognomy, see Therese Davis, *The Face on the Screen: Death, Recognition and Spectatorship* (Bristol and Portland: Intellect, 2004), esp. 25–40.

24 Elizabeth Ewen and Stuart Ewen, *Typecasting: On the Arts & Sciences of Human Inequality* (New York, Oakland, CA, and London: Seven Stories Press, 2008), 98.

25 Roslyn Poignant, 'Surveying the Field of View: The Making of the RAI Photographic Collection' in Elizabeth Edwards (ed.) *Anthropology and Photography 1860–1920* (New Haven and London: Yale University Press, in association with The Royal Anthropological Institute, London, 1994), 42–73; 57.

26 Ewen and Ewen, *Typecasting*, 98.

27 See Maxwell, *Colonial Photography*, 41–42.

28 It is also worth noting that albumen prints of Lamprey's system were the first photographs to ever appear in the *Journal of the Ethnological Society* in 1869.

a plain background and beside a measuring rod at a specific distance from the camera.

While shocking, these parametric standards were only the most extreme dehumanising version of what Elizabeth Edwards calls the 'scientific reference' of full-face and profile views, one that worked its way into personal portraiture as well as medical, military, forensic and penal uses of photography through the second half of the nineteenth century (and which of course persists in our time, not least in the guise of facial recognition technologies).[29] Here, the allure of the face is no longer related to the subject's elusive identity; it becomes instead a tool that permits the mapping out and surveillance of disposable bodies deprived of subjectivity and accordingly reduced to pure physicality. As one of the originary pillars of such a project, anthropology thus institutes faciality, as connected with craniometry, as the descriptor of racial difference. But, as Ferreira da Silva reminds us, faciality was in itself not enough for this project, which needed a global stage. Or, as she puts it: 'the racial constitutes an effective political-symbolic strategy precisely because the subjects it institutes are situated differently, namely, in globality'.[30] In this context, 'the scientists of man attempted to prove, to establish with certainty (as a "necessary" and "objective" fact), what eighteenth-century naturalists could only describe and the philosophers who framed man could only postulate', namely the inferiority of the racialised.[31]

One of the main instruments used for this empiricist purpose, as discussed in the following section, was photography, hailed precisely on the basis of its 'objectivity' in a number of British anthropological treatises and publications – in opposition to unreliable, because subjective, drawings.[32] But we will also see that anthropology did not merely want to draw a global map of human difference. As an epistemology that transformed individuals into specimens of race, anthropology's fascination with recording racial others across a differentiated global space was simultaneously informed by a social-evolutionist paradigm according to which the more mentally

29 Elizabeth Edwards, 'Evolving Images: Photography, Race, and Popular Darwinism' in D. Donald and J. Munro (eds.) *Endless Forms: Darwin, Natural Sciences and the Visual Arts* (New Haven, CT, and London: Yale University Press, 2009), 166–193; 189. On the aesthetics and uses of surveillance photography in the second half of the nineteenth century, see Tagg, *The Burden of Representation*, esp. 66–102. On facial recognition technologies, see Kelly A. Gates, *Our Biometric Future: Facial Recognition Technology and the Culture of Surveillance* (New York and London: New York University Press, 2011).

30 Ferreira da Silva, *Toward a Global Idea of Race*, 30.

31 Ibid., 120.

32 On the nineteenth-century discourse of mechanical objectivity, see Lorraine Daston and Peter Galison, *Objectivity* (New York: Zone Books, 2007), esp. 115–190.

inferior you were, the less time on Earth you had. Hence it became urgent to record these races before they vanished off the face of the planet, an urgency which conferred a new sense of mortality on the photographed human face.

Death's Head

In 1839 the British ethnologist and humanitarian James Cowles Prichard published his essay 'The Extinction of Native Races'. Denouncing 'the modern system of colonization' as 'the harbinger of extermination of the native tribes', he argued for the necessity 'to record the history of the perishing tribes of the human family'.[33] Prichard would take it upon himself to undertake such a task. In 1843, the same year he co-founded the Ethnological Society of London, he published *The Natural History of Man; Comprising Inquiries into the Modifying Influence of Physical and Moral Agencies on the Different Tribes of the Human Family.* This comprised dozens of sections on the physical and cultural differences between human races, and it reiterated the foundations of his ethnology as based on the principle of 'monogenesis', that is, the idea that all human races have a single and common origin, being thus 'of one species and one family'.[34]

Prichard's ethnological compendium would make history. In its wake came an explosion of similarly titled catalogues promising to compile the human races scattered across the globe. These included: Charles Pickering's *The Races of Man and their Geographic Distribution* (1850); Josiah Clark Nott and George Robins Gliddon's *Types of Mankind* (1854); Robert Knox's *The Races of Man: A Philosophical Enquiry into the Influence of Race over the Destinies of Nations* (1862); John G. Wood's *The Natural History of Man; Being an Account of the Manners and Customs of the Uncivilised Races of Man* (1868–1870); Robert Brown's multivolume *The Races of Mankind: Being a Popular Description of the Characteristics, Manners and Customs of the Principal Varieties of the Human Family* (1873–1876); C. and F.W. Dammann's *Races of Mankind: Ethnological Photographic Gallery of the Various Races of Man* (1876); and the serial magazine *The Living Races of Mankind: A Popular Illustrated Account of the Customs, Habits, Pursuits, Feasts and Ceremonies*

33 James Cowles Prichard, 'On the Extinction of Human Races', *Edinburgh New Philosophical Journal* 28 (1839): 166–170; 168, 169, 170.

34 James Cowles Prichard, *The Natural History of Man; Comprising Inquiries into the Modifying Influence on Physical and Moral Agencies on the Different Tribes of the Human Family* (London: Hippolyte Baillere, 1845), 546.

of the Races of Mankind Throughout the World (1901), also published as a two-volume book. Let us look at some of these projects in order to explore how anthropology became increasingly dependent upon the photographed human face so as to postulate a globally distributed and differently fated humanity.

The trope of the 'human family', as the titles above suggest, was pervasive in nineteenth-century anthropology. Yet the differences between these projects were equally significant, revealing the shifting and sometimes conflicting frameworks shaping the discipline in its complex rapport with popular visual culture. For example, unlike Prichard's culturalist *The Natural History of Man*, Nott and Gliddon's *Types of Mankind*, published in the US, advanced the hotly debated 'polygenetic' thesis according to which different human races originated from different, not the same, stock. Meanwhile, Wood and Brown were busy in their self-appointed mission as anthropological popularisers in the UK, even if, as Douglas A. Lorimer has shown, their *The Natural History of Man* and *The Races of Mankind* 'made limited use of physical anthropology and relied mainly on travel literature': the result was generalising and distorting treatises 'on the moral and psychological characteristics of various peoples'.[35] These two authors equally subscribed to the idea of a racial extinction whose outcome was as guilt-free as it was inevitable due to theology or 'progress'. Whereas, for Wood, 'the cause of extinction lies within the savage himself, and ought not to be attributed to the white man', for Brown, 'the decay of wild races', from Tasmanians to West Africans, was 'the effect of the collision of savage and civilised races [which] tends to the destruction of the former'.[36]

If the theory of polygenesis gave rise to a much more nefarious racism, insofar as it viewed human races as separate and fixed in their biological makeup, the many faces that the more enduring doctrine of monogenesis would assume was no less guilty of ranking humanity according to biological features that, it was believed, isomorphically evinced moral and intellectual traits. As Poignant explains:

35 Douglas A. Lorimer, *Science, Race Relations and Resistance: Britain, 1870–1914* (Manchester and New York: Manchester University Press, 2013), 125.

36 John G. Wood, *The Uncivilized Races of Men in All Countries of the World; Being a Comprehensive Account of Their Manners and Customs, and of Their Physical, Social, Mental, Moral and Religious Characteristics* (San Francisco: J.A. Brainerd & Company, 1883), 791; Robert Brown, *The Races of Mankind: Being a Popular Description of the Characteristics, Manners and Customs of the Principal Varieties of The Human Family*, volume III (London, Paris and New York: Cassell, Petter & Galpin, 1887), 219.

> By the 1860s an essentially historical Prichardian ethnology, which had attempted to establish a typology of the diverse races of mankind, was being gradually transformed by the Darwinian revolution in scientific thought into an anthropology that applied systematic methods of classification to produce developmental models of social evolution that were in essence hierarchical.[37]

This included the social-Darwinist paradigm according to which, as Brantlinger notes, 'nature's constant laws mandated the extinction of all unfit creatures and species to make room for new, supposedly fitter ones'.[38] It followed that it was urgent to record the faces of such races before their vanishing, a task which photography was to fulfil in a number of ways and contexts.

In fact, the human face haunts ethnological and anthropological accounts even before mechanical photographic printing. As Alison Griffiths points out: 'it was tales of what intrepid travelers, colonial officials, and missionaries had *seen* on their trips to the far corners of the earth in the form of drawings, sketches, engravings [...] that thrilled the general public and scientific societies alike'.[39] A case in point is Prichard's aforementioned *The Natural History of Man*. It included a number of steel and wood engravings of faces, many of them from the published accounts of the voyages of James Cook, Nicolas Baudin and Otto von Kotzebue, and which had, in turn, previously appeared in Johann Gottfried Schadow's 1835 publication *National-Physiognomieen*, thus confirming anthropology's foundational physiognomic proclivities.[40] Though not as heavily illustrated, Pickering's *The Races of Man* also included 13 pictures, including one based on a daguerreotype and eight original drawings by engraver Alfred Thomas Agate. Nott and Gliddon's *Types of Mankind* similarly relied on second-hand drawings and engravings. When there is an original sketch in the book, as in 'the likenesses of two Negroes', although the authors suggest it can 'be considered caricatured', this is dispelled by their following, racist qualification that 'one need not travel far, to procure, in daguerreotype, features fully as animal'.[41]

37 Poignant, 'Surveying the Field of View', 45.

38 Brantlinger, *Dark Vanishings*, 15. As Brantlinger notes, Darwin himself stressed the superior qualities of the civilised man in relation to savages doomed to extinction due to nature's selective process (164–186).

39 Alison Griffiths, *Wondrous Difference: Cinema, Anthropology, & Turn-of-the-Century Visual Culture* (New York: Columbia University Press, 2002), 88, emphasis in original.

40 Poignant, 'Surveying the Field of View', 57.

41 J.C. Nott and G.R. Gliddon, *Types of Mankind, or Ethnological Researchers* (Philadelphia: Lippincott, Grambo & Co; London: Trübner & Co, 1854), 260.

The recourse and references to the daguerreotype in these surveys evince the increasing investment in photography as the enabler of typological projects based on facial evidence. It is no accident that one of the very first books to include photographs had the human face at its center. Thanks to the heliotype, a multiple-printing mechanical process, Darwin's *The Expression of the Emotions in Man and Animals* (1872) featured 30 photographs, 19 by the photographer Oscar Rejlander and eight copied, with permission, from plates that had appeared in *Mécanisme de la physionomie humaine* (1862) by the French clinician Guillaume-Benjamin Duchenne de Boulogne.[42] Recognising the poor quality of some plates, Darwin justifies their inclusion by noting that 'they are faithful copies, and are much superior for my purpose to any drawing, however carefully executed'.[43]

By focusing on fleeting emotions rather than on 'the permanent form of the features', *The Expression of the Emotions in Man and Animals* attempted to distance itself from physiognomy.[44] Yet it pertained to the same discursive nexus around the legibility of the human face, which subtended a variety of practices and disciplines in the second half of the nineteenth century, including anthropology. For Darwin, facial pictures served the purpose of highlighting universality, or as he put it, that 'the same state of mind is expressed throughout the world with remarkable uniformity [in] all the races of mankind', even if pictures of 'racial types' are conspicuously absent in his book.[45] Conversely, in popular anthropological studies it was the opposite, that is to say physical difference, that was valued, as when Robert Brown, in the introduction to his 1882 *The Peoples of the World* (a republished version of *The Races of Mankind*) noted: 'Every individual is different in appearance. No two faces are exactly alike'.[46]

In any case, books such as *The Expression of the Emotions in Man and Animals* were still a rarity given the prohibitive costs of photographic printing. A more common practice, as noted earlier and stated in the preface to John G. Wood's *The Natural History of Man*, was 'for the countenances of

42 For an illuminating account of the use of photographs in Darwin's book, see Phillip Prodger, 'Photography and *The Expression of the Emotions*' in Charles Darwin, *The Expression of the Emotions in Man and Animals*, ed. Paul Ekman (New York: Oxford University Press, 1998), 399–410.

43 Charles Darwin, *The Expression of the Emotions in Man and Animals* (London: John Murray, 1892), viii.

44 Ibid., 1.

45 Ibid., 19.

46 Robert Brown, *The Peoples of the World; Being a Popular Description of the Characteristics, Condition, and Customs of the Human Family* (London, Paris and New York: Cassell, Petter, Galpin & Co., 1882), 1.

the people [to] have been drawn from photographs'.[47] But even when these drawings were based on photographs they still aroused suspicion. Thus, for the anthropologist and cultural evolutionist Edward B. Tylor, a central figure in the consolidation of anthropology as an academic discipline in the UK, the engravings in Wood's publication were still 'too picturesque and imaginative', and compared unfavourably with those featuring in Brown's *The Races of Mankind*, whose publication he accordingly welcomed in 1874.[48] Two years later, upon the publication of Carl Victor and Friedrich Wilhelm Dammann's photography book *Races of Mankind: Ethnological Photographic Gallery of the Various Races of Man* (1876; hereafter *Ethnological Photographic Gallery*), Tylor celebrated it as a watershed in racial studies. Its printed photographs, he maintained, evinced that 'most engravings of race-types to be found in books were worthless, either wanting the special characters of the race, or absurdly caricaturing them'. He concluded: 'Now-a-days, little ethnological value is attached to any but photographic portraits, and the skill of the collector lies in choosing the right individual as representatives of their nations'.[49] In other words, these portrayed individuals were valued not because of their individuality, as in the bourgeois carte de visite. Rather, they typified a particular human race that was itself globally demarcated: they were, in short, a racial type. To put it in even cruder terms, the human face was no longer cherished because of the inner transcendence that it exteriorised but it was, on the contrary, reduced to pure exteriority: the sizes and shapes of the head that denoted 'expressions of levels of "development" of mental functions'.[50]

Ethnological Photographic Gallery was a concise edition of C. Dammann's monumental *Anthropologisch-Ethnologisches Album Photographien* (1873–1874). The original album comprised 642 photographs, many of which were anthropometric photographs and cartes de visite collected, re-photographed and assembled by Dammann for the Berlin Society for Anthropology, Ethnology and Prehistory. Dammann's 'atlas of types' evinced not only the widespread use of the calling card in taxonomical projects but also the social-evolutionist bent of such projects as they segregated and hierarchised humanity on a global arena (figures 4.1 and 4.2).[51] As Teresa

47 John G. Wood, *The Natural History of Man; Being an Account of the Manners and Customs of the Uncivilised Races of Man* (London: George Routledge and Sons, 1868), v.

48 Edward B. Tylor, 'The Races of Mankind', *Nature* 9 (12 February 1874): 279–280; 279.

49 Edward B. Tylor, 'Dammann's Race Photographs', *Nature* 6 (6 January 1876): 184–185; 184.

50 Ferreira da Silva, *Toward a Global Idea of Race*, 124.

51 Another case in point is the attempt on the part of the British Association for the Advancement of Society (BAAS), in the late 1870s and early 1880s, to produce a survey of the origins of

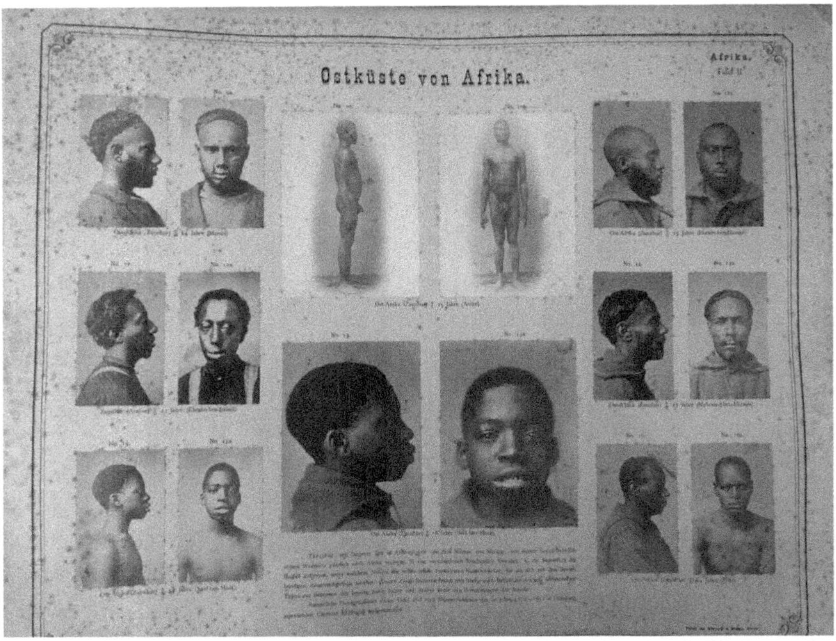

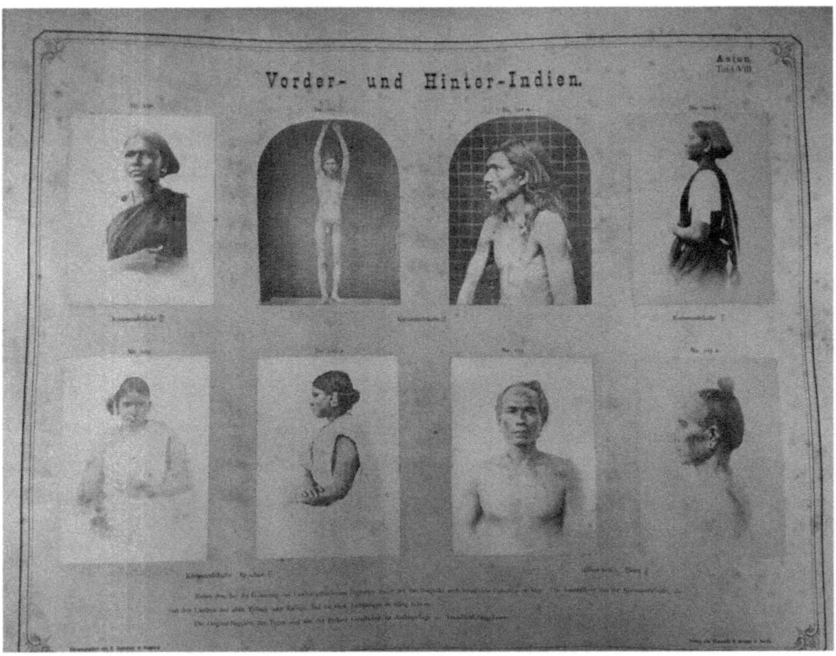

Figures 4.1 & 4.2 Pages from C. & F. Dammann's 'atlas of types' *Races of Mankind: Ethnological Photographic Gallery of the Various Races of Man* (1876) combine anthropometric photographs and cartes de visite, thus revealing their interchangeability in nineteenth-century galleries of humanity. Courtesy of the British Library.

Castro notes, it is striking that the book version starts with 'German or Teutonic Types', the only section featuring individualised, that is, named persons – and all of whom were men, including Prince Bismark and David Livingstone – and ends with the 'Australians Aborigines' in a 'clearly racist and sexist logic of progression'.[52] Racist clichés regarding 'savages', from the 'state of utter barbarism' of Amazon tribes through to the 'rudest barbarism' of Australians, equally abound throughout the book.[53]

As much as it concerned itself with human difference, however, the impact of *Ethnological Photographic Gallery* was not as unequivocal as perhaps envisaged, even within anthropological circles. As Jane Lydon observes: 'While mosaics of "types" had been compiled from the 1870s onwards [...] to define difference within social-evolutionist frameworks, even some of the most committed Darwinists acknowledged the complexity and ambiguity of applying concepts of biological race to humankind'.[54] For Lydon, evidence of this ambiguity is to be found in Tylor's aforementioned appraisal of *Ethnological Photographic Gallery*, which for him ultimately revealed likeness rather than difference: 'the real intricate blending of mankind from variety to variety'.[55] Yet the ranking of Australians in the album was certainly in line with the belief that they stood at the bottom of the evolutionary ladder and risked imminent disappearance, thus evincing the album's ties with extinction discourse.[56]

the British population through cartes de visite and cabinet cards collected across the country. See Poignant, 'Surveying the Field of View', 57–61.

52 Teresa Castro, *La pensée cartographique des images: Cinéma et culture visuelle* (Lyon: Aléas Editeur, 2011), 170.

53 C. & F.W Dammann, *Ethnological Photographic Gallery of the Various Races of Man* (London: Trübner & Co. Ludgate Hill, 1876), plates XXI and XXIII.

54 Jane Lydon, *Photography, Humanitarianism, Empire* (London and New York: Bloomsbury, 2016), 111.

55 Tylor, quoted in Lydon, *Photography*, 111.

56 Thanks in no small part to Darwin's *The Descent of Man* (1871), this belief generated a flurry of photographic series that attempted to salvage the race for posterity. See Edwards, 'Evolving Images', 179. These series included the well-documented set of Australian photographs of five surviving Oyster Cove Australians, taken by Charles A. Wolley in his own studio and circulated as 'The Last of the Tasmanians' (see Poignant, 'Surveying the Field of View', 45); the work of the German-Australian photographer John William Lindt who, from 1868 to 1876, took a number of pictures of aborigine peoples, in his studio in Grafton, New South Wales, carrying artisanal weapons and utensils as in a timeless *tableau vivant;* and the images taken by the French anthropologist Roland Bonaparte on the occasion of the French touring of the 'Australian Boomerang Throwers', described as the 'sole survivors of a party of nine Aborigines who were abducted by force in 1883 by the showman R.A. Cunningham' (see Maxwell, *Colonial Photography and Exhibitions*, 47).

Ethnological Photographic Gallery is historically significant as the first photographic compilation of a global humanity. It paved the way for colonial serials and magazines published at the turn of the century, when advancements in photographic printing enabled their mass production and circulation. In Britain, one such serial was *The Living Races of Mankind: A Popular Illustrated Account of the Customs, Habits, Pursuits, Feasts and Ceremonies of the Races of Mankind throughout the World*, published in 1900. Imperialist to its core, its existence is attributed in the introduction (signed by Reverend Henry Neville Hutchinson) to the fact that the 'half-clothed savage, just emerging from the brute condition' can turn 'into a customer for British trade'.[57] Equally important, its anthropological significance is exalted on the basis of its reliance on photography rather than drawings. Hutchinson goes on:

> Pictures, or wood-engravings, may sometimes be prettier, but they can never be so absolutely trustworthy as the products of the camera [...] Many standard works on Ethnology are disfigured by engravings which are far from accurate, and in some cases are nothing less than parodies of the people they profess to portray. Even when a woodcut is prepared directly 'from a photograph', it cannot always be trusted.[58]

Hutchinson's insistence that even photographic-based drawings are not comparable to actual photographs was probably a rebuke of Wood's and Brown's popularising efforts, examined earlier. He adds: the engraver 'has not studied anatomy, or the different types of human physiognomy, and to him there is very little difference between a Polynesian or a Papuan and an African negro'.[59] In short, only photographs could illuminate those very little differences that formed the epistemological backbone of anthropology's global cataloguing.

Whether everyone perceived differences, however, is another story. For the British author H.G. Wells, such photographs produced the opposite effect. Noting the 'extraordinary intensifications of racial definition', or what he disparagingly called the 'race mania' of his time, he enthused that at last

57 Henry Neville Hutchinson, 'Introduction' in H.N. Hutchinson, J.W. Gregory and R. Lydekker, *The Living Races of Mankind: A Popular Illustrated Account of the Customs, Habits, Pursuits, Feasts & Ceremonies of the Races of Mankind throughout the World*, volume I (London: Hutchinson & Co, 1901), i–iv; ii.
58 Ibid., iv.
59 Ibid.

> there exists to-day available one kind at least of unbiased anthropological evidence. There are photographs. Let the reader turn over the pages of some such copiously illustrated work as *The Living Races of Mankind*, and look into the eyes of one alien face after another. Are they not very like the people one knows?[60]

Wells's statement is striking. It provides evidence that the celebrated objectivity of photography could be used to bolster both human difference *and* sameness. But Wells's position is also symptomatic, since his universalism emerges as a response to the 'race-mania' discourse of difference, to use his own words. As Mary Ann Doane remarks, thanks to its multiplication in urban centers and via technological reproducibility, '[t]he face emerges as a particularly acute problem in modernity, raising the specter of an illegible strangeness and otherness. In such a context, the notion of the face as a universal language, leveling all differences, becomes especially attractive'.[61] The aforementioned statements and ideas by Darwin and Tylor also subscribe to this logic.

Yet, as we have seen, an even more pernicious and pervasive way of suppressing the threat of otherness was by ranking it low in the ladder of humanity – so low, in fact, as to be literally vanishing. *The Living Races of Mankind* wholly perpetuated this discourse. Its introduction is illustrated with the then highly disseminated pictures of Krao, a child from the forest of Laos, and the Mexican Juliana Pastrana, both hirsute and advertised as a 'missing link' with ape ancestors and whose placement in the serial, as Maxwell notes, served 'as evidence for a hierarchical concept of evolution in which Europeans were set apart from "primitive" racial types'.[62] Extinction is then underlined throughout as a natural and inevitable process. Opening the first chapter is the assertion that the Papuans 'have greatly declined in numbers since the white man brought them the vices and the diseases of civilisation'.[63] Later, the indigenous population of the Sandwich Islands is said to have been reduced from 300,000 to 40,000, while 'neither disease nor drunkenness offers a satisfactory explanation of the rapid dwindling'

60 H.G. Wells, *A Modern Utopia* (London: W. Collins Sons & Co, 1905), 228–229, 233. See also Lydon, *Photography*, 110–113.

61 Mary Ann Doane, 'Facing a Universal Language', *New German Critique* 41:2 (2014): 111–124; 122.

62 Maxwell, *Colonial Photography and Exhibitions*, 55. For an account of how these pictures were used to bolster evolutionism, see Edwards, 'Evolving Images', 177–180.

63 Hutchinson et al., *The Living Races of Mankind*, 1.

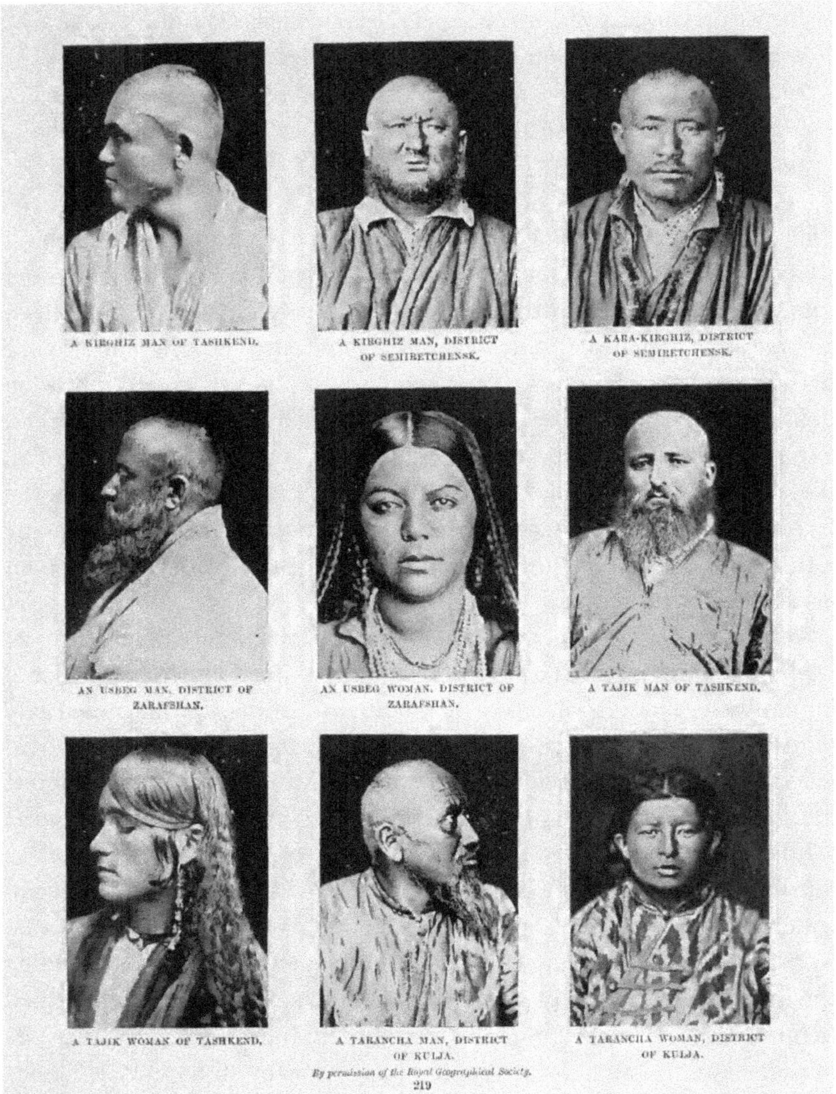

Figure 4.3 Frontal and profile facial pictures in *The Living Races of Mankind* (1901).

of the race.[64] In another section, the Australian 'dark-coloured natives' are relegated to 'the bottom rungs of the ladder of human progress', and their rapid disappearance attributed to the 'shameful' 'advent of the white man in Australia': 'Civilisation, alas! brought disease and vice in its train'.[65] Such

64 Ibid., 24.
65 Ibid., 51.

an admission of guilt, however, is quickly countered with the statement that '[i]nferior races must of course give way and make room for those that are more highly civilised'.[66]

We can perhaps conclude from these statements that the 'living races' of the publication's title was to be taken quite literally, that is, as the intimation that some races are alive and well and the premonition that others do not share such a fate: hence the importance of photography in memorialising them. With 329 illustrations in total, the first volume of *The Living Races of Mankind* comprised 148 individual portraits, 137 group photos and 43 pictures of pairs, with images variously collected 'from professional and amateur photographs at home and abroad', including 'the ethnographical collections of Paris, Leyden, Hamburg, Dresden and Leipzig besides Oxford and Cambridge'.[67] It features as a result an array of formats, sizes and styles, with pictures taken in situ, others in studios, and still others showing groups engaged in activities, dance and games. Yet, unsurprisingly, a significant portion of these images shows subjects in profile and frontal positions (figure 4.3) often isolated from their settings, whether through framing, a neutral background, or both.

Not all of these photographs were directly linked with extinction discourse, nor were all races portrayed as doomed. But it is fair to say that, by then, the frontal and profile facial portraits of nonwhites carried within themselves, that is, in their very formal structure and appearance, associations with death given the transactions and applications to which they had been subjected in the public sphere and anthropological domains. This was not the death denoted by pictures of deceased people adorning caskets or the more general sense of mortality attributed to photography per se, but a more specific, racialised death. Rather than standing by an individual who had passed away, such pictures instead suggested the disappearance of an entire human race. In place here, to cite Brantlinger, was 'the future-perfect mode of proleptic elegy [that] mourns the lost object before it is completely lost', that is, the 'work of cultural, national mourning occurs not because the aboriginals are already extinct but because they will sooner or later become extinct'.[68]

The longevity of facial portraiture cannot be overestimated when we consider the magazines that were to follow in the footsteps of *The Living Races of Mankind*, including, not least, *National Geographic*, founded in 1888 in the US by the National Geographic Society. As Catherine A. Lutz and Jane

66 Ibid.

67 Hutchinson, 'Introduction', iv.

68 Ibid., 4. That said, in some cases, such as the Tasmanians, total extinction did occur.

L. Collins point out, the society was set up, like its British counterparts, in the midst of an anthropological 'search for evidence of the evolutionary backwardness of subaltern peoples'.[69] The magazine supported such a project by 'position[ing] itself as a key actor in presenting "primitive" peoples for western perusal', with 'its encoding of hierarchy and power relations, and its projections of an inevitable outcome'.[70]

That outcome was disappearance. And while it was not until 1905 that photographs would become a common feature of the publication – and the magazine would nurture a range of photographic genres – the isolated frontal portrait of non-Western peoples would rank among its most favoured. As Lutz and Collins explain, *National Geographic* has reinforced ethnological codes by relying on a type of 'straight photography' that 'permitted the labors and point of view of the photographer and editor to recede into the background, thus encouraging the reader to see his or her contact with the photographed subject as unmediated, if necessarily indirect'.[71] Here, in line with the precepts of popular anthropology and physiognomy, head-on frontality enables evaluation 'of the other as a person or type', a 'close examination of the photographed subject, including scrutiny of the face and eyes, which are in common-sense parlance the seat of soul'.[72] Editors are as a result in constant 'search for the "compelling face" in selecting photos for the magazine'.[73]

On the face of it a simple recording of native peoples looking into the camera, this 'straight photography', as I hope to have shown, has nothing straight about it. Caught at the intersection of discourses of faciality, raciality and globality, photography's objectivity was swiftly put to the service of a planetary human taxonomy according to which the future did not belong to everyone. Faces of difference, these photographs of human others acted as the conduit through which an overdetermined logic of racial differentiation gained a global – and deadly – face.

'Don't Blink!'

When it emerged, cinema immediately joined photography in the project of making visible faces in the world. As Griffiths observes: 'The iconography

69 Catherine A. Lutz and Jane L. Collins, *Reading National Geographic* (Chicago and London: University of Chicago Press, 1993), 19.

70 Lutz and Collins, *Reading National Geographic*, 19.

71 Ibid., 29.

72 Ibid., 199.

73 Ibid.

of the "type" also found a home in early ethnographic film [...], functioning as a coded moment when film and photographic signification appear to collapse'.[74] In some travelogue films, as Jennifer Lynn Peterson has shown, 'the conventions of anthropometric photography, showing portraits of individuals frontally and in profile' are still evident, as in *Moroccan Shoemakers* (Pathé, 1915), 'where one witnesses a fascination with facial features that exceeds the film's attention to the local craft of shoemaking'.[75] A fascination with the face would similarly gain an illustrious place in early film theory, most notably in the writings of Béla Balázs and Jean Epstein, which, as Doane argues, tellingly revolved around 'the colonial encounter with otherness and its repercussions for identity'.[76] She writes: 'It is striking that a critical focus on the face and the close-up invariably seems to return to a history of origins structured by the primitive-modern divide and one that inevitably [...] invokes questions of race'.[77]

More recently, a focus on the racialised face has reemerged in early-cinema scholarship, which, according to Paula Amad, bespeaks an obsession with the 'return of the gaze' of film's colonial subjects. The aim, she notes, is to 'provide a textual compensation for the lack of photographic and cinematographic records made by peoples historically victimized by the camera', recovering facial looks and expressions as instances of resistance to the imperial gaze.[78] Film's time-based properties are fundamental for this recovery. As Gunning argues, early cinema 'allowed the drama of the look to develop a more dialogic relation to its filmed subjects, whose faces and gestures gained a further expressiveness and independence as they were filmed in time'.[79] Whereas, as discussed above, the facing gaze was the dominant convention of nineteenth-century racial portraits, in film the subject's look at the camera may not only emerge as an unintended element but also provide more sustained evidence of his or her feelings – discomfort, reluctance, animosity – at the time of registration.[80] As Steimatsky sum-

74 Griffiths, *Wondrous Difference*, 114.

75 Jennifer Lynn Peterson, *Education in the School of Dreams: Travelogues and Early Nonfiction Film* (Durham, NC, and London: Duke University Press, 2013), 166.

76 Doane, 'Facing a Universal Language', 123.

77 Ibid., 119.

78 Paula Amad, 'Visual Riposte: Looking Back at the Return of the Gaze as Postcolonial Theory's Gift to Film Studies', *Cinema Journal* 52:3 (2013): 49–74; 56.

79 Tom Gunning, 'Before Documentary: Early Non-fiction Films and the "View" Aesthetic' in Jonathan Kahana (ed.) *The Documentary Film Reader: History, Theory, Criticism* (New York: Oxford University Press, 2016), 52–61; 61.

80 That said, in a parallel gesture, scholarship on colonial photographs sometimes also turns to faces as a means of reading facial expressions as evidential (if often speculative) indications

marises: cinema, unlike photography, is often understood on the basis of 'its capacity to register, amplify, and perpetuate the face's expressive movements – the life of the face unfolding in time'.[81]

Stillness and mobility, photography and film: the reproduction of the gazing face is indelibly implicated in questions of race and issues of medium specificity. We now turn to two contemporary documentaries, *Samsara* (2011) and *Human* (2015), which provide ample materials for an exploration of such questions and issues. I argue that these films emerge as irresistible case studies to compare and contrast with the global compilations of humanity examined earlier – despite (or rather, because of) the fact that more than a century separates them. First, this is because they implicitly or explicitly organise a human collective against the background of human extinction, though extinction is here differently conceptualised. Second, both films deliberately attempt to sacrifice and subtract the defining feature of film as a medium – that is to say, movement – in order to adhere more firmly to the medium of photography, and more specifically the conventions of the studio-based, background-free facial portrait. My contention is that this intermedial borrowing is not fortuitous but that it evinces, on the contrary, the continued longevity of anthropometric and ethnographic codes of racialised portraiture when ideas of a global humanity and human extinction are at stake.

Brantlinger argues that the wiping out of the entire humanity was an implied, if often dormant, feature of nineteenth-century extinction discourse. Citing Malthus's ideas of population growth, the eugenics movement and publications such as Wells's *The War of the Worlds* (1898), he writes that anxiety 'about the future of all the races of mankind – indeed, of the entire human species – was always implicit in extinction discourse'.[82] Since the evolutionary paradigm postulated an ever-evolving human race, the idea that the white race 'would sooner or later vanish, experiencing the doom of the inferior races it had bested in the struggle for existence' was therefore always on the horizon.[83] Yet there is little doubt that ideas of human extinction are today more widespread thanks to a global environmental crisis

of the relationship between photographer and photographed at the moment of capture. See, for example, Maxwell, *Colonial Photography*, 45–46; Robert W. Rydell, 'Souvenirs of Imperialism' in Christraud M. Geary and Virginia-Lee Webb (eds.) *Delivering Views: Distant Cultures in Early Postcards* (Washington and London: Smithsonian University Press, 1998), 47–63; 60.

81 Steimatsky, *The Face on Film*, 6. See also Paul Coates, *Screening the Face* (Basingstoke: Palgrave Macmillan, 2012) for another study on cinema's facial visuality.

82 Brantlinger, Dark *Vanishings*, 15.

83 Ibid., 192.

that increasingly generates doubts about the future. In some predictive scenarios, the question is not so much if but for how long the conditions necessary for human life may still be maintained, which relativises humanity in vaster – the Earth's own – trajectories and frameworks.[84] As Déborah Danowski and Eduardo Viveiros de Castro summarise:

> The idea that our species is a newcomer on the planet, that history as we know it (agriculture, cities, writing) is even more recent, and that the energy-intensive, fossil-fuel based industrial way of life began only a second ago in terms of *Homo sapiens*' evolutionary clock all seem to point to the conclusion that humankind itself is a catastrophe: a sudden, devastating event in the planet's biological and geophysical history, one that will disappear much faster than the changes it will have occasioned in the Earth's thermodynamic regime and biological balance.[85]

But this idea is also a double-edged sword. On the one hand, it can be deployed to underline the pressing need for collective action in the face of an impending global catastrophe. As I will argue in chapter 6, projects such as *The Last Pictures* (2012) and *Homo Sapiens* (2016) visualise nonhuman futures in order to make us confront the world in the here and now. On the other hand, this same idea can reiterate the pernicious logic of inevitability that has often informed extinction discourses, even if from within a different conceptual space. To different extents, as my analysis will show, both *Samsara* and *Human* perpetuate the latter logic. It is therefore symptomatic that they make recourse to the face in their elegiac visions of a human world. Let us look first at *Samsara*.

Directed by Ron Fricke, *Samsara* (2011) can be deemed a variation of the 'world symphony' genre discussed in the last chapter and whose incarnations also include Godfrey Reggio's *Koyaanisqatsi* (1982) – on which Fricke worked as cinematographer – and Fricke's own *Baraka* (1992). *Samsara* features Fricke's signature style developed in those earlier films: no voiceover or interviews, dazzling aerial images, and time-lapse shots of natural and human-made settings. Shot on 70mm in 25 different countries over five years, *Samsara*'s global travelogue is structured around a cyclical notion

84 Christiana Figueres and Tom Rivett-Carnac, '"The Only Uncertainty Is How Long We'll Last": A Worst Case Scenario for the Climate in 2050', *The Guardian* (15 February 2020), https://www.theguardian.com/environment/2020/feb/15/worst-case-scenario-2050-climate-crisis-future-we-choose-christiana-figueres-tom-rivett-carnac (last accessed 1 January 2021).

85 Déborah Danowski and Eduardo Viveiros de Castro, *The Ends of the World*, trans. Rodrigo Nunes (Cambridge and Malden, MA: Polity Press, 2017), 15.

of life on Earth which is encapsulated in its title, a Sanskrit word meaning 'the turning wheel of life'. Here, as in *Koyaanisqatsi* and *Baraka* – the former meaning 'life out of balance' in the Hopi language and the latter a Sufi word for 'blessing' – the choice of a titular Hindu expression aims to position *Samsara* within the purview of Eastern spirituality and belief systems, as reinforced by its 'world music' soundtrack, composed by Lisa Gerrard, Marcello de Francisci and Michael Stearns. Nevertheless, as we shall see, *Samsara*'s gaze on the world, with its marked reliance on orientalist conventions and racialised tropes, is a flagrantly Western one.

The idea of cyclicality is mirrored by *Samsara*'s tripartite structure, with the film bookended by the image of a mandala, filmed at the Thikse monastery in India, being painted by monks at its opening and destroyed at its end. The first part, lasting roughly 30 minutes, combines images devoid of humanity, many filmed in ancient settings and natural environments, including: salt lakes in Bolivia, Naukluft National Park in Namibia, Antelope Canyon in Arizona and Mono Lake in California. The aim of these balletic images of sand deserts, dunes, lakes, glaciers and rock formations, often shot with a slow-moving camera in time-lapse cinematography, is plain and simple: the visualisation of a geophysical history where humans play no part.

But this quickly changes in the film's second part, where humanity rises as a planetary force to be reckoned with. Human faces and signs are first seen in a series of images of native villages in Ethiopia and Namibia, after which the film cuts to an aerial, night-time image of Los Angeles. What follows is a frenetic montage of sped-up images of crowded prisons, subways, factories and supermarkets. Filmed in bustling urban centers, such as Hong Kong, Tokyo, Shanghai and São Paulo, these images connote industrial modernity's intense and imbalanced relation with nature. In the film's last and briefest part (around 10 minutes), the prevailing themes are spirituality and religion, duly illustrated by images of Jewish synagogues, Catholic churches and Muslim temples. This includes an aerial shot of Mecca where the sea of people walking in circles down below transmogrifies into an undifferentiated mass through time-lapse. As the mandala is destroyed, the film then closes with human-free images of sand dunes.

The message is clear: humans, whether as agents or victims of a voracious and unrelenting modernity, are testing the limits of the Earth and its natural resources. The notion of impermanence, central to Buddhist and Hindu belief systems, is thus mapped on to the cycle of human existence, parenthetically situated within a much vaster planetary history. As Zoë Druick observes, here the viewer must either 'succumb to the inability to do anything but grit your teeth and experience the speed of life out of balance, or to realize that all this, too, will, in the cosmic scheme of things, pass'.[86] Indeed, there is not so much a sense of mourning for humanity's demise or a call to action in the face of socio-economic injustices and environmental ravages. Through the conceit of cyclicality, destruction is depicted as a perhaps undesirable but certainly inevitable process and, in doing so, *Samsara* recycles and updates nineteenth-century extinction narratives and tropes, including a programmatic adherence to the immobile human gaze.

Martin Roberts has argued that *Baraka* is illustrative of a 'coffee-table globalism', as its 'panoramic vision of global natural and cultural diversity [...] locate[s] it firmly within the humanistic documentary tradition of *National Geographic*'.[87] This also applies to *Samsara*, which claims its place within a photographic lineage even more overtly by framing motionless subjects without blinking when posing for the camera, a stilling of facial mobility that was duly publicised in promotional materials. On the official website dedicated to *Baraka* and *Samsara*, which includes an interactive world map where users can see the locations where the films were shot, the latter film is described thus: the 'filmmakers approach non verbal filmmaking with an understanding that it must live up to the standard of great still photography, revealing the essence of a subject, not just its physical presence'.[88] Elsewhere Fricke has noted: 'All the direction on the portraits was just, "Look into the camera and don't blink!" Because blinking portraits don't work too well. It was the stare we were after to connect everyone together. No matter who you are, where you are, you see that soul'.[89]

These statements cannot be overestimated. Not only do they seem lifted straight out of nineteenth-century physiognomic treatises and photographic guidebooks, they also evince how *Samsara*'s take on the ethnographic portrait emulates the lengthy exposure times of early photography, as a result of which subjects had to remain immobile looking intently into the camera. In total, *Samsara* features 65 subjects posing for the camera

86 Zoë Druick, 'On (Not) Falling from the Sky: Fly-Over Global Documentary as Capitalist Body Genre' in Steve Anderson and Christie Milliken (eds.) *Reclaiming Popular Documentary* (Bloomington: Indiana University Press, 2021). My thanks to Druick for sharing a pre-print version of her chapter.

87 Martin Roberts, '*Baraka*: World Cinema and the Global Culture Industry', *Cinema Journal* 37:3 (1998): 62–82; 67, 68.

88 'About *Samsara*', https://www.barakasamsara.com/samsara/about (last accessed December 17 2018).

89 The Creators Project Staff, 'Capturing the Stunning Cinematography in *Samsara*: Q&A with the Film's Director and Producer', *Vice* (6 September 2012), https://www.vice.com/en_au/article/9adewe/capturing-the-stunning-cinematography-in-isamsarai-qa-with-the-films-director-and-producer (last accessed 17 December 2018).

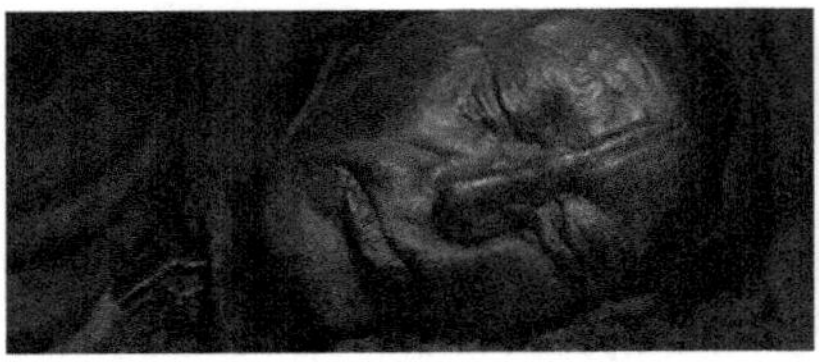

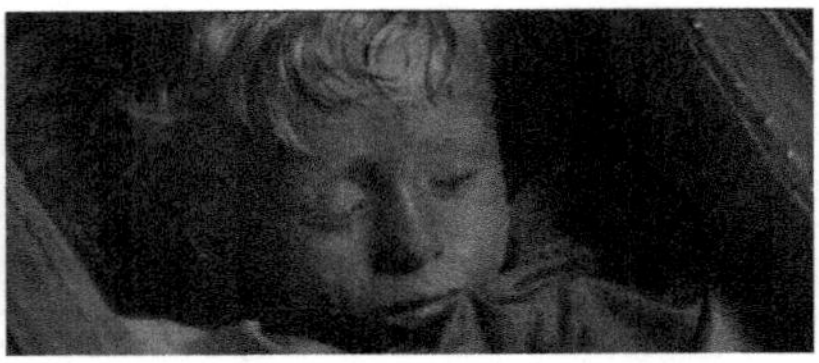

Figures 4.4 & 4.5 Images of mummified faces set up a connection between face and mortality in *Samsara* (2011). Courtesy of Magidson Films Inc.

during its running time.[90] These vary compositionally, with group shots, two shots and medium shots, yet facial close-ups dominate the film. A few subjects appear both in group shots and alone in short-distance shots, and the majority of these portraits, as per Fricke's instructions, display motion only in the form of a deliberate camera closing in on unblinking subjects. Whether or not these cinematic portraits reveal the soul of the face as intended by the director would be impossible to ascertain. But it is fair to argue that subtraction of movement here works in tandem with Barthes's famous contention that the unfolding presentness of filmic motion thwarts the connotations of pastness and mortality identified with photography.[91] Put simply, facial immobility in *Samsara* is deployed to evoke death.

This is confirmed already in the film's opening, which uncannily materialises a Bazinian genealogy by showing in succession the face of an embalmed corpse, a mummified face (figure 4.4) and an Egyptian death mask (the pharaoh King Tut, c.1341–c.1323 BC). As the first part of the film unfolds, this genealogy is reinforced via shots of the Mount Nemrut statue faces in Turkey (which are believed to surround a royal tomb dating from the first century BC), and facial close-ups of Rosalia Lombardo (figure 4.5), an Italian child who died in 1920 and whose mummified body can be visited in the Capuchin catacombs of Palermo. These embalmed, sculpted and mummified faces therefore confirm the anticipated spectatorship envisaged by the film when it comes to its (moving) images of (still) living humans parading thereafter: they are to be read as indexical signifiers of mortality.

At the same time, *Samsara* adheres to an evolutionary narrative of faciality that recalls nineteenth-century taxonomies of races while intimating the disfigurement of humanity, its increasing *unrecognisability*. This is what happens, for example, when the film transitions to the second part, which

90 Three of these shots include adults carrying babies, and I am not counting the babies. I am also not counting a montage sequence showing nine children being baptised in a catholic church in Brazil, given that they are not posing for the camera.
91 Barthes, *Camera Lucida*, 78–80.

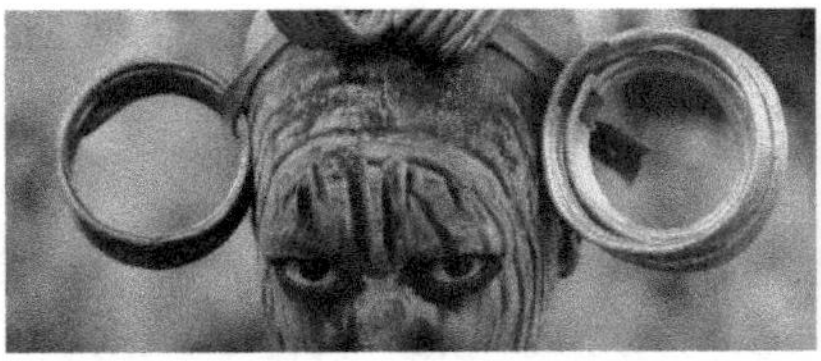
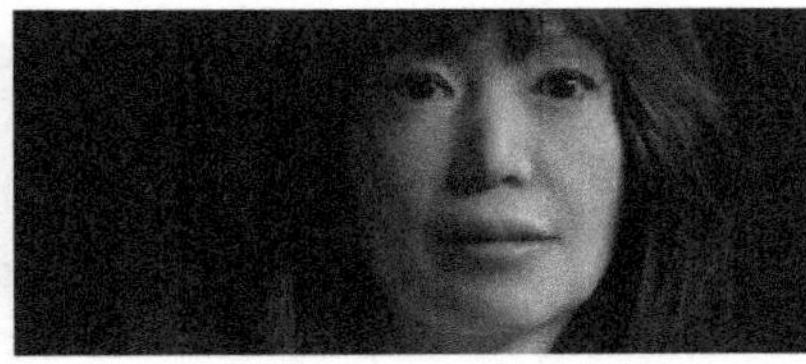

Figures 4.6 & 4.7 Pictures from a Mursi Village local and a Japanese android convey an evolutionary narrative of faciality that hinges on well-worn tropes of the primitive and the modern. Courtesy of Magidson Films Inc.

cuts from portraits of Mursi Village locals (Ethiopia) (figure 4.6) and Himba tribe (Namibia) members to a two-shot of Hiroshi Ishiguro, an eminent Japanese specialist in robotics, beside his lookalike android. A close-up of the android facing the camera appears next, followed by a close-up of a Japanese female android whose uncanny resemblance to a human face (figure 4.7) is then dispelled by her machinelike sounds. In a later sequence, a high-angle shot of a woman about to undergo facial plastic surgery gives way to a montage alternating faces of sex dolls and faces of strip girls. The sequence ends with a Japanese woman, in a geisha costume, staring into the camera while a tear drops down her face.

These juxtapositions are hardly unambiguous. Human faces, *Samsara* laments, are becoming gradually and dangerously indistinguishable from human-made ones. In another, preceding, sequence, this process of defacement is even more literal. It depicts, to a discordant soundtrack, the French artist and sculptor Olivier de Sagazan spasmodically applying layers of clay and paint onto his face, which turns as a result into amorphous and monstrous shapes. Therese Davis has argued that the 'image of a face becoming unrecognisable makes death visible', and this is why this filmed performance, which makes visible this process of unrecognisability, occupies such a central place in *Samsara*.[92] For Davis, the profusion of dead faces in contemporary media (especially television) has anesthetised us to death. Conversely, for her, the representation of a face becoming unrecognisable is a more 'viable site for the transmission of death' as it expresses 'the transient nature of human existence', making us confront 'finality, irreversibility, absolute otherness', in short, the passage of time.[93]

These remarks are applicable to de Sagazan's filmed performance. But they also raise questions relating not so much to the 'absolute otherness' of time but the 'absolute otherness' of the film's racialised faces. Indeed, there

92 Davis, *The Face on the Screen*, 4.
93 Ibid., 2.

is a noticeable lack of white countenances in *Samsara* (eight in total), and when these appear they are tellingly pierced, tattooed, and, in the case of US soldier Bobby Henline, disfigured by burning. In the film, then, the focus is on facial difference and a purported abnormality, one that tacitly assumes the white face – conspicuous by its absence – as the parameter against which 'others' can be measured. In this context, it is significant that it is a white face (de Sagazan's) that undergoes the process of transfiguration mentioned above. For this contradicts *Samsara*'s predilection for non-Western peoples (including subjects from Indonesia, China, Ethiopia, Namibia, Japan, the Philippines and Accra): from the wide-eyed, heavily made-up Balinese dancers performing the Legong dance at the beginning through to the braided, red-tinged Himba natives, *Samsara* consistently relies on figures of the exotic. This also applies to the modernity end of the spectrum, which in the film is related, predictably, to Japanese faces, robotic or otherwise. Another way of putting this would be to say that whereas a white face can express in the film the 'transient nature of human existence' through a performance that theatricalises such transience, the otherwise non-Western faces parading throughout are deemed unworthy of transformation and thus denied the passage of time, that is to say, existence itself. Immobilised in their unblinking stare, these faces are not exactly becoming unrecognisable so much as they already are the face of death.

Samsara's racialised gaze thus reveals a logic that reinforces whiteness as an invisible (because hegemonic) and qualitatively different ethnicity, not unlike previous catalogues of human difference. Given that this is a film whose extinction discourse is meant to encompass the whole of humanity, it is therefore striking that its collection of gazing subjects – themselves inserted within a history that has accorded them connotations of mortality according to evolutionist paradigms – hardly include white faces, who are not only assumed to be unworthy of the scrutinising gaze of the camera (presumably because they are behind it) but also, one wonders, to be immune to the global cataclysm that the film anticipates in its theme and structure.

The Face in the Crowd

In some respects, Yann Arthus-Bertrand's *Human* (2015) offers a corrective to *Samsara* by including faces from most parts of the world. What is more, not only do these have a share of screen time, but they are also given the opportunity to speak to the camera during personal testimonies. By eschewing a fetishising stillness and muteness of subjects, *Human* might thus be commended for reinstating facial movement and giving a voice to the subjects it interviews. On closer inspection, however, the film also resonates with colonial portraiture by subscribing to an anthropometric aesthetics of enumeration and uniformity, one that gains in significance in light of the film's investment in questions of overpopulation.

That *Human* reveals ties with portraiture is perhaps not surprising in view of Arthus-Bertrand's professional experience as a photo-journalist for *National Geographic*, *Geo* and *Life* magazines. That said, Arthus-Bertrand's visual brand is now predominantly associated with the aerial view technique, which he has obsessively employed in a number of photographic and cinematic projects over the last two decades. Examples include the bestseller *The Earth from the Air* (1999) and the documentaries *Home* (2009), *A Thirsty World* (2012) and *Terra* (2015). Often referred to, ironically, as France's answer to Al Gore, Arthus-Bertrand is a controversial high-profile figure: the customary target of criticism on the basis of his landscape aestheticising and costly helicopter shooting, he is the founder of the non-profit organisation GoodPlanet Foundation (behind many of his projects) and also Environment Goodwill Ambassador for the United Nations.

Though still peppered with his trademark views from above, *Human*'s most distinguished feature is its statistical reliance on the frontal face. Two years in the making, the film was shot in 60 countries and amassed over 2,000 interviews, largely with anonymous people (though exceptions include former Uruguay president José Mujica and actress Cameron Diaz). *Human* was part of a multimedia project comprising a number of audiovisual texts and artefacts, including: two theatrical films, making-of documentaries, educational projects, and promotional and photography books. The project's official release, on 12 September 2015, coincided with a public screening at the United Nations with Secretary General Ban Ki-moon in attendance, followed by a debate to mark the 70th anniversary of the United Nations. On that same day the film was made freely available as three volumes of 90 minutes each on YouTube and Google Play in a number of languages, with longer interviews also available as individual clips on an exclusive YouTube channel. At the time of writing, the film's first instalment has over 6 million views.

Human evolved from Arthus-Bertrand's web-based database project *7 Billion Others: Portraits of Humanity from Around the World* (formerly *6 Billion Others*). It is worth pausing for a moment to consider this project. During a six-year period starting in 2003, *7 Billion Others* compiled over 6,000 video testimonies from 84 countries, with participants answering 'universal' questions about fear, love, death and marriage. Featuring a boxy

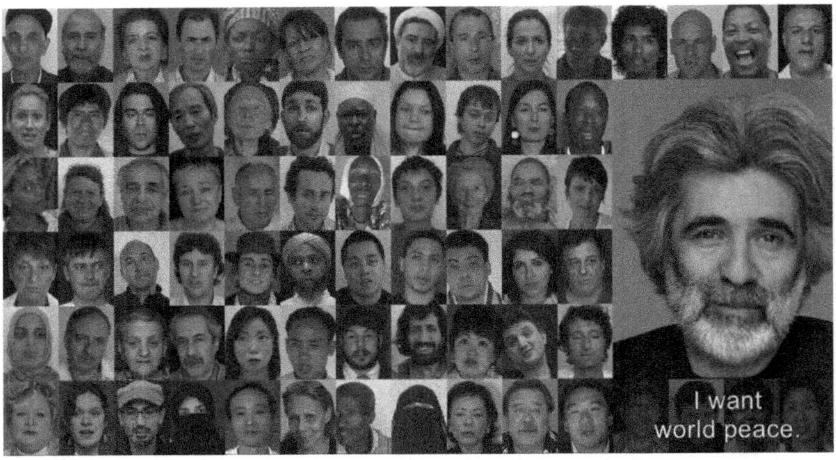

Figure 4.8 A multitude of faces in a mosaic video of *7 Billion Others*. Courtesy of GoodPlanet Foundation.

aspect ratio that encloses the face and eliminates the background, the anonymised participants speak directly to the camera while subtitles on the side translate or transcribe their testimonies. By framing faces according to a predetermined and unchanging model, *7 Billion Others* thus aims to render them equivalent and interchangeable. This is confirmed by the project's grid-like videos (figure 4.8), which combine a multitude of faces, mostly looking into the camera, while the face and voice of one participant is highlighted as he or she speaks.

As Druick notes, 'despite its reliance on the close-up, the ambitions of the *7 Billion Others* project are, as its title suggests, encyclopaedic and totalizing', with its database aesthetics evoking 'computational schemes to document and manage large populations'.[94] Indeed, it would not be a stretch to deem *7 Billion Others* the digital culmination of a project of population management that shares its genesis with that of anthropometric portraiture as a codifying and parametric mode of colonial knowledge. In this sense, the project's title, and the fact that it has changed so as to reflect an increase in the world's population, lays bare its concern with quantity and quantifiability. This is visually reiterated in the project's book version, the cover of which is made up of hundreds of tiny, barely legible facial portraits.

Previously relegated to the semiotic function of a *type*, head-on portraiture reappears here codified as a *number*, but one that still carries a sense of

94 Zoë Druick, '"A Wide-Angle View of Fragile Earth": Capitalist Aesthetics in The Work of Yann Arthus-Bertrand', *Open Cultural Studies* (2018): 396–405; 402, 401.

foreboding, now in relation to the ecological risks of population explosion. As Arthus-Bertrand explains: 'We're in the middle of destroying life on Earth. When I was born, we had a world population of 2 billion; now it's 7 billion [...] But there isn't enough for everyone. There isn't enough water, enough fish, enough wood'.[95] When placed alongside other Arthus-Bertrand projects, all of which have the Anthropocene as their narrative and epistemological substratum, *7 Billion Others* betrays in its title anxieties regarding unchecked population growth.[96]

Human rehashes the themes and features of *7 Billion Others*, with interviews structured around similar questions and the focus largely on the human face. Whether we consider the many versions and products it has generated, or the profusion of human faces it parades, *Human* thrives on accumulation and multiplicity. Like *7 Billion Others*, the film is quasi-anthropometric in its formal, compositional and structural rigour: against a black background, in frontal positions and looking into the camera, well-lit faces are positioned either on the left or right of the frame, with subtitles on the side transcribing their statements. But whereas the numerical drive of *7 Billion Others* is spatialised across mosaic compositions, in *Human* it is durationally stretched. At several moments, while one subject speaks to the camera and his or her testimony remains audible, the visual track is intercut with shots of other faces silently facing the camera, often unblinkingly and/or in slow motion. Although there is no voiceover narration, the film is organised around clearly defined themes, including happiness, love, marriage, sexuality, war, sickness, death and life. In the film's feature-length versions, subjects remain anonymised, but an option to include captions indicating their name and geographical provenance is available on the web versions and YouTube clips.

Like Fricke, Arthus-Bertrand justifies his portraiture-inspired formal choices as a physiognomic belief that 'faces' and 'looks' provide a 'powerful way of reaching the depths of the human soul. Each encounter brings you a step closer. Each story is unique'.[97] In this context, it is no accident that the film's second instalment includes an interview with Caleb, a US soldier who recounts his encounter with an Iraqi soldier thus:

95 Yann Arthus-Bertrand, in Rachel Segal Hamilton, 'Meet Filmmaker Yann Arthus-Bertrand Director of Woman' (n/d), https://www.canon.co.uk/pro/stories/woman-documentary-director-yann-arthus-bertrand/#:~:text=%22When%20I%20was%20born%2C%20we,%2C%20enough%20wood...%22 (last accessed 25 August 2020).

96 See Druick, '"A Wide-Angle View of Fragile Earth"', 397–399.

97 Yann Arthus-Bertrand, 'Introduction', *Human* Press Kit (n/d), available at https://www.human-themovie.org (last accessed December 21 2016), 4.

> There is a moment – and the reason his face is always going to be with me – there is a moment when he looked at me. Our eyes met and at that moment, it was like everything else disappeared. It was just two people looking each other in the eyes. For a moment connecting like two human beings in an event that is beyond any of their control. At that moment, he wasn't a terrorist, he wasn't an insurgent, he wasn't an Iraqi. He was a scared man and he was asking me for help. From that moment on, the war changed for me. It became a little more scary and it became a little more [...] I started to question decisions a little more because of that event. *It put a face.*

Caleb's testimony brings to mind the philosophy of Emmanuel Levinas, for whom the intersubjective encounter establishes an ethical relationship whereby the face of the other, in its ungraspable strangeness, commands acts of responsibility as a gateway into what Levinas terms 'the absolutely other', that is to say, infinity itself.[98] For Levinas, to look in the eyes of a face is to look at the face of all humans: the 'presence of the face, the infinity of the other, is a destituteness, a presence of the third party (that is, of the whole of humanity which looks at us), and a command that commands commanding'.[99] Although Levinas is no doubt concerned with the actual, indeterminate face-to-face encounter (as in Caleb's case), his remarks shed some light on *Human*, which not only believes in the powers of the destitute face of the 'other' in its appeal to morality, but also in the power of cinema to act as a vehicle for a mediated encounter with multiplying faces that together morph into the face of humanity.

Whether or not the film is successful at the task it sets for itself will depend on how much one subscribes to a Levinasian ethics of 'otherness'. To the credit of *Human*'s interviewers and editors, many of the stories told, in the manner they are told, are compelling: there is a sense these are genuine people telling their own stories with their faces, words and voices, an aspect even more pronounced in the longer individual videos available on the film's YouTube channel. On the other hand, if we are to side with Levinas's opponent Alain Badiou, as I do myself, *Human* is proof of the pernicious logic that supports an ethics of difference and 'its catechism of good will'.[100] This is the case especially when we turn to its privileging of emotional

98 Emmanuel Levinas, *Totality and Infinity: An Essay on Exteriority*, trans. Alphonso Lingis (The Hague, Boston and London: Martinus Nijhoff Publishers, 1979), 195.
99 Ibid., 213.
100 Alain Badiou, *Ethics: An Essay on the Understanding of Evil* (London and New York: Verso, 2012), 20.

testimonies and stories of suffering, which, to cite Badiou, reinforces a 'negative and victimary definition of man' to the detriment of a politics based on human affirmation, agency and action.[101] As Badiou sees it, the current orthodoxy of the 'rights for difference', which has its philosophical origins in Levinas, is at best commonsensical, at worst reactionary: 'in truth a tourist's fascination for the diversity of morals, customs and beliefs [...] Yes, the essential "objective" basis of ethics rests on a vulgar sociology, directly inherited from the astonishment of the colonial encounter with savages'.[102] Indeed, both *Human* and *7 Billion Others* not merely show this to be the case so much as they reinforce this ethic's reliance on tropes of faciality as a colonial inheritance.

One could go so far as to say that *7 Billion Others* and *Human* recall the 'systematic naming, categorization, and isolation of an otherness' of anthropometric photography as once defined by Allan Sekula.[103] The stress here is on quantity, informed by a statistical belief in comparison and physiognomic frames of reference and visuality – a belief that the face itself, alone, isolated, archived and multiplied, can reveal both the commonalities and divergences across humankind. Incidentally, Sekula noted, this numerical impetus was a condemnable feature of Edward Steichen's supremely popular exhibition *The Family of Man* (1955), described by Carl Sandburg in its book version as a 'multiplication table of living breathing human faces'.[104] For Sekula, *The Family of Man* was a site where 'arithmetic and humanism collide': 'The statistics that seek to legitimate the exhibition, to demonstrate its value, begins to carry a deeper sense: the truth being promoted here is one of enumeration'.[105] More recently, Janine Marchessault has treated the exhibition as a more complex project and asked us to reconsider it in terms of a contradiction 'between Steichen's desire to foreground commonality in the face of possible nuclear warfare that would end human life on the planet and the US government's new ideological war'.[106]

101 Ibid., 16.
102 Ibid., 26.
103 Allan Sekula, 'The Traffic in Photographs' in Allan Sekula, *Photography Against the Grain: Essays and Photo Works, 1973–1983* (Nova Scotia: The Press of the Nova Scotia, 1984), 77–101; 79.
104 Carl Sandburg, quoted in Sekula, 'The Traffic in Photographs', 90, 93. *The Family of Man* comprised 503 photographs taken in 68 countries by 273 photographers, attracting 9 million people in 69 countries.
105 Ibid., 93.
106 Janine Marchessault, *Ecstatic Worlds: Media, Utopia, Ecologies* (Cambridge, MA, and London: MIT Press, 2017), 24. Crucial to Marchessault's argument is what she sees as *The Family of Man*'s

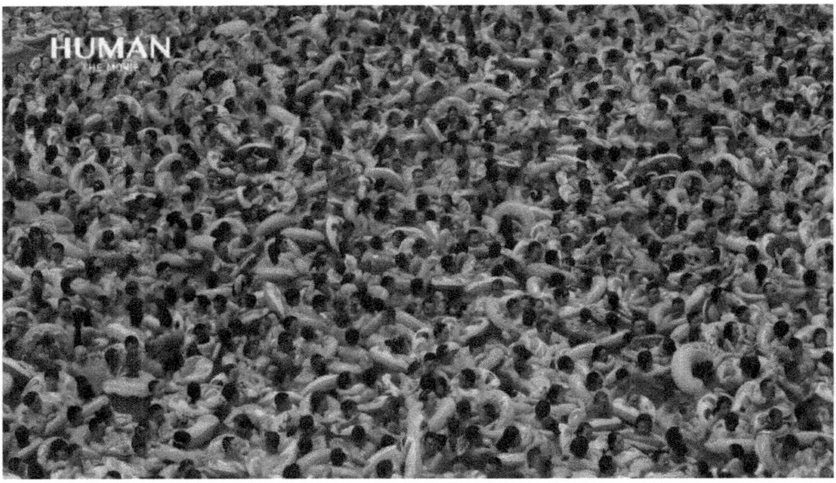

Figure 4.9 A sea of faces in *Human* (2015) makes visible anxieties over population explosion. Courtesy of GoodPlanet Foundation.

Human replicates both the arithmetic inflection of *The Family of Man* and its anxieties about the disappearance of the human race. Granted, the latter is not figured as a nuclear-induced threat, nor is the idea of a doomed human species explicitly thematised (as in *Samsara*). But it is an idea that lurks beneath the film, especially when we consider the inclusion of Arthus-Bertrand's trademark aerial shots of land devastation, endangered landscapes and heavy industry, all of which, in line with his other works, visualise a world running out of resources. More broadly, *Human*'s anxieties about the end of humanity are, as in *7 Billion Others*, visually formulated in relation to overpopulation. But whereas in that project population explosion is tamed via schematic grid-like visual models, in *Human* it is made visible in overcrowded shots where faces turn into tiny illegible dots as part of an undifferentiated mass. These include a wave pool in China (figure 4.9), a collective wedding in Japan, a football match in Hamburg and military parades in different countries – many of which are stylised and fetishised through slow motion and some of them recalling *Samsara*'s aforementioned time-lapse aerial shot of Mecca.

In tune with attendant physiognomic beliefs and anthropological discourses, nineteenth-century galleries of a global humanity constituted a response to the multiplication of faces increasingly crowding cities and accordingly reproduced and multiplied via technological means. The threat

much-overlooked centerpiece, 'a six-by-eight-foot glowing backlit color Milar transparency of a thermonuclear explosion', 38.

here, to cite Doane again, was 'the specter of an illegible strangeness and otherness', one that was in part countered by the discourse of 'the face as a universal language' and in part by the systematic differentiation of humanity according to evolutionary parameters.[107] These overcrowded shots in *Human* and *Samsara* demonstrate that such a treat has not so much gone away as it has been exacerbated in our time in the face of population explosion and ensuing catastrophic prospects. In this context, the ubiquity of the face evident in Facebook, the selfie and facial recognition technologies designed with the very aim of recognising the face in the crowd, all speak to this threat of illegibility and attempt, in different ways and with different socio-political and socio-cultural consequences, to reassert the face as a recognisable unit, whether as a mode of self-presentation or as a means of surveillance and control. *Human* and *Samsara* are part and parcel of this phenomenon.[108] Yet in light of their decidedly elegiac mode and their programmatic adherence to colonial modes of portraiture, their stilling and enclosure of the face is particularly striking. For it shows that whereas extinction discourse may have evolved in its causes and consequences, its face has remained remarkably intact.

107 Doane, 'Facing a Universal Language', 122.
108 Ibid., 123.

5 A Networked Humanity

Abstract
This chapter examines the discourse of film as universal language in relation to the narrative figure of the network. I contend that D. W. Griffith's *Intolerance* (2016) constitutes an illustrious precursor of the global network narrative genre. This analysis is then substantiated by my reading of Alejandro González Iñárritu's *Babel* (2006), a film that bears remarkable resemblance with *Intolerance* in its formulation of universality via a multinarrative textual design and focus on suffering as the lingua franca of humanity. The chapter closes with a consideration of Eduardo Williams's *The Human Surge* (2016), a film in which totalising ideas of humanity and globality are also mapped on to a networked narrative configuration. Here, however, miserabilism plays no part and the network is broadened into a wider web that includes the physical Earth itself.

Keywords: network narrative, universalism, humanity, *Intolerance, Babel, The Human Surge*

Ethics feeds too much on Evil and the Other not to take silent pleasure in seeing them close up.
Alain Badiou

In his book *Networking the World, 1794–2000* (2000), Armand Mattelart traces the genealogy of the network as a mode of international social organisation and integration since the turn of the eighteenth century. He notes: 'Networks, a leading symbol of progress, have also made their incursion into utopian thinking. The communication network is an eternal promise symbolizing a world that is better because it is united. From road and rail to information highways, this belief has been revived with each technological generation'.[1]

1 Armand Mattelart, *Networking the World, 1794–2000*, trans. Liz Carey-Libbrecht and James A. Cohen (Minneapolis and London: University of Minnesota Press, 2000), viii.

de Luca, T. *Planetary Cinema. Film, Media and the Earth*. Amsterdam: Amsterdam University Press, 2022
DOI: 10.5117/9789463729628_CH05

In short, the 'dream of reestablishing the pre-Babel "great human family" is present throughout the history of the imaginary of communication networks'.[2] Yet, as Mattelart hastens to add, 'networks have never ceased to be at the center of struggles for control of the world', motivated by interests as wide ranging and dubious as 'the universalism of a predestined civilization' and 'the pragmatism of the corporation'.[3] Whether in idealist or imperialist guises, networks are therefore inseparable from discourses of universalism which are themselves ridden with ideological contradictions.

The emergence of cinema was a milestone in the networking imaginary and accordingly duly lauded on the basis of its universalism. Mattelart continues: 'Cinema was to open the way for the mythology of universal communication in the "era of images", which became another symbol of the end of inequality between classes, groups, and nations'.[4] Film was hailed as a new language: one that was universal because it was visual; or, to cite one commentator, writing in 1914: 'Eye language is universal and direct – seeing is believing'.[5] Another writer exclaimed: 'The art of the film, alone, is unlimited and free. In sight we are all equal. The civilised man and the savage both have eyes and can see with them'.[6] And yet another critic enthused: 'Love of pictures is engrained in us, whatever our color, race, or age [...] Oratory has impressed thousands, the Press hundreds of thousands, the moving picture millions, and that in a way the others do not, reaching out amongst all races, stirring all grades from the old to the young'.[7]

As some of these statements already suggest, however, the universality of cinema, much like that of the galleries of humanity we explored in the last chapter, presupposed a segregation of humanity into the 'civilised' and the 'savage'. Film's universalism in this context was just another name for its purported 'civilising mission', as 'simple natives', one writer commented, could learn 'from the pictures how the white man lives in his home'.[8] Just as importantly, whereas cinema was seen as 'decidedly international in all its aspects [...] bounded neither by nationality nor prejudice', to cite another contemporaneous statement, there was no question that the nascent film companies, mostly

2 Ibid., 20.

3 Ibid., vii–viii.

4 Ibid., 19.

5 Joseph Best, 'Seeing the World by the Cinema', *Pathé Cinema Journal* 39:II (25 June 1914), 39.

6 'The Cinematograph Industry Round the World: Interview with Mr. A. C. Bromhead', *The Bioscope* 355 (31 July 1913), 335.

7 'The Lure of the Picture', *Moving Picture World*, reprinted in 'The World's Press', *The Bioscope* 211 (27 October 1910), 41.

8 Margaret I. Macdonald, 'The Trend of the Times', *The Bioscope* 256 (7 September 1911), 497.

based in Europe (especially France) and North America, hoped to benefit from this internationalism less as a means of establishing fraternal bonds between fellow humans than as a way to conquer the world *market*.[9]

In these early-cinema discourses and perceptions, universalism often revolved around the non-fiction travel film, or travelogue, which simultaneously showed and toured the world. The links between such films, or the places they depicted, were established either in the spectator's (and commentator's) mind, or at most flimsily forged by miscellaneous early cinema programmes organised under the theme of global travel, as we saw in chapter 3. Yet there was one film, released in 1916, which, directly inspired by the universal-language idea, explicitly attempted to create links, via editing, between different space-time worlds so as to concretise the idea of a global networked humanity. That film was D.W. Griffith's *Intolerance*, which, as this chapter will show, condensed both in its textual constitution and production history the contradictions and limitations of cinematic universalism outlined thus far.

As a canonical film in its own right, *Intolerance* has received voluminous critical attention, not least in relation to the discourse of film as a universal language. Here I hope to extend this important work by showing that the film's universalism is inseparable from a *networked* conception of the globe. A foundational figure when it comes to the application of editing as an expressive narrative device, Griffith's use of montage in *Intolerance* is revealing of a desire to make visible the material and immaterial links connecting humanity since times immemorial and across a global stage. Unlike the early photographs, films and projects examined in most chapters of this book, *Intolerance*'s claims of visible disclosure are therefore less dependent upon the recorded image's indexical properties. This is certainly a fiction film, mostly shot in purposefully built studio sets, and one that, indeed, freely reconstructed historical events. But, while existing in a different guise, an understanding of the film medium as a tool of visibility is still evident here, although crystallised around the notion of cinematic linkage, that is, the interplay between different images that can disclose or reiterate connections across time and space. In this context, as I hope to show, *Intolerance*'s figuration of a global network is not only established on the metaphysical level of its diegesis, as afforded by its (in)famous quadripartite, back-and-forth structure, but also quite literally in its modern narrative via recourse to images and tropes of railway, motorway and telecommunication networks.

9 'A Cosmopolitan Trade', *The Bioscope* 349 (19 June 1913), 1.

More specifically, I wish to argue, *Intolerance* constitutes an illustrious precursor of the contemporary 'network narrative', defined by David Bordwell 'as the dominant principle of offbeat storytelling' since the early 1990s, in tune with 'a deepening awareness of human connectivity on a global scale'.[10] Bordwell's terminological choice is indebted to 'a more formal theory of networks and "small worlds"' that emerged in the 1950s and 1960s, and which became popularised through 'the notion of "six degrees of separation"' and 'the butterfly effect'.[11] The idea that we live in a 'network society', made of 'networks in all the key dimensions of social organization and social practice', has also been advanced by Manuel Castells.[12] In the network film, this configuration often translates into a broadening of the film's geographical parameters, which can encompass: cities, be it the Paris of *Les Parisiens* (Claude Lelouch, 2004) or the London of *Wonderland* (Michael Winterbottom, 1999); countries, such as the China of *A Touch of Sin* (*Tian zhu ding*, Jia Zhangke, 2012); and indeed the world itself as characters navigate across countries and continents. Examples of the latter, which I will call here 'the global network narrative', include: *Traffic* (Steven Soderbergh, 2000), *Syriana* (Stephen Gaghan, 2005), *The Edge of Heaven* (*Auf der Anderen Seite*, Faith Akin, 2007), and most remarkably, the suggestively titled *Mammoth* (Lukas Moodysson, 2009), *360* (Fernando Meirelles, 2011) and *Babel* (Alejandro González Iñárritu, 2006).

While arguing for a reassessment of *Intolerance*'s universalism via the concept of the network, the contemporary focus of this chapter is, conversely, motivated by a quest to explore how the network film has rekindled the idea of film as a universal language in our time. Nowhere is this more visible than in *Babel*, which, as its title suggests and my analysis will demonstrate, channels the global network into the discourse of a universal language. Generally understood in terms of chance encounters and 'intertwining plotlines' that 'affect one another to some degree', to cite Bordwell again,

10 David Bordwell, *Poetics of Cinema* (New York: Routledge, 2008), 191–192; 198.

11 Ibid., 198. Although I will stick to 'network narrative', it must be noted that the genre has given rise to a number of critical terms, including: 'hyperlink', 'mosaic', 'multi-protagonist', 'complex' and 'multiplex'. On 'hyperlink' films, see Alissa Quart, 'Networked', *Film Comment* 41:4 (2006): 48–51. On the 'mosaic' film, see Patricia Pisters, 'The Mosaic Film: Nomadic Style and Politics in Transnational Media Culture' in Mieke Bal and Miguel Á. Hernández-Navarro (eds.) *Art and Visibility in Migratory Culture: Conflict, Resistance, and Agency* (New York: Rodopi, 2011), 175–190. On the 'multi-protagonist' film, see Maria del Mar Azcona, *The Multi-Protagonist Film* (Oxford: Wiley-Blackwell, 2010). On 'complex narratives', see Peter F. Parshall, *Altman and After: Multiple Narratives in Film* (Plymouth: Scarecrow, 2012). On 'multiplex cinema', see Hamid Nacify, 'Multiplicity and Multiplexing in Today's Cinema: Diasporic Cinema, Art Cinema, and Mainstream Cinema', *Journal of Media Practice* 1 (2010): 11–20.

12 Manuel Castells, *The Rise of the Network Society* (Malden, MA: Wiley-Blackwell, 2010), xviii.

the network narrative is often traced back to films such as *Grand Hotel* (Edmund Goulding, 1933) and directors such as Robert Altman.[13] I hope to demonstrate that even if *Intolerance* does not entirely fit into this description, to examine it alongside *Babel* makes sense because they are both the earnest products of a sentimental humanism that relies on comparative fatalism and human suffering, which in both films is moreover translated into graphic close-ups of bodies and faces in pain. By charting the continuities between these two films' take on the human with recourse to the philosophical theories of Richard Rorty, Alain Badiou and Hannah Arendt, my hope is to demonstrate the tenacity of the trope of the universal language as it has manifested itself at two crucial moments in film history, while evaluating its effectiveness and fault lines as both an aesthetics and a politics.

In the last section I turn to a more recent, small-budget film: Eduardo Williams's *The Human Surge* (*El auge del humano*, 2016), which, I contend, provides a meaningful counterpoint against which to compare the networked universalism of *Intolerance* and *Babel*. This is, first, because the film eschews human miserabilism in its networking of the planet, favouring instead sensory snapshots of everyday life in all its indeterminacies and hardships but also pleasures. Second, whereas the film critically thematises the network's capitalist subservience and triumphalist discourses on the Internet as a new form of universal communication, it equally shows the world, or the Earth, to be crisscrossed by other networks that escape and exceed the human, thereby injecting a much-needed ecological and material substratum into our global network society.

'One Common Flood of Humanity'

How could someone who had directed a blatantly racist film follow it up with another that claimed to speak on behalf of a universal humanity? The answer to this question is not as straightforward as it may seem but the fact is that *Intolerance*'s conception and execution is intimately bound up with *The Birth of a Nation* (1915). Controversial though it was, the latter was a hit upon its release,

13 David Bordwell, 'Lessons from Babel', *Observations on Film Art* (27 November 2006), http://www.davidbordwell.net/blog/2006/11/27/lessons-from-babel/ (last accessed 9 December 2016). On the lineages of the multinarrative film, see Bordwell, *Poetics of Cinema*, 194–197; Azcona, *The Multi-Protagonist Film*, 46–79; Parshall, *Altman and After*, 21–44. It must also be noted that Pisters, Azcona and Parshall do acknowledge *Intolerance* as a potential precursor of the multinarrative film, though they do so tentatively and in passing. See Pisters, 'The Mosaic Film', 175–176; Azcona, *The Multi-Protagonist Film*, 9; Parshall, *Altman and After*, 14.

making Griffith rethink the dimensions of his next film, which began to appear too small in comparison.[14] The director had already filmed *Intolerance*'s Modern Story, until then a single-narrative feature provisionally entitled *The Mother and the Law*, during the editing of *The Birth of a Nation*. Only later did Griffith come up with the idea of adding three more stories whose historical and religious weight were meant to evoke the 'European "historical-spectacle" films' he so admired, including *Quo Vadis* (Enrico Guazzoni, 1912) and *Cabiria* (Giovanni Pastrone, 1914).[15] The idea to organise the four stories around the thematic axis of 'intolerance' was also a rebuke to what Griffith considered the curtailment of artistic freedom by the Motion Picture Producers and Distributors of America (MPPDA). That is, *Intolerance* was never meant to rectify *The Birth of a Nation*'s racism but, on the contrary, arose out of Griffith's belief that he was being dishonestly persecuted by reformists.

In light of this, it is not difficult to agree with Slavoj Žižek, for whom Griffith's cry against intolerance 'is much closer to today's fundamentalists decrying the "politically correct" defence of universal rights of women as "intolerant" towards their specific way of life'.[16] Ryan Jay Friedman has similarly noted that *Intolerance*'s 'conceptual origins in Griffith's racism – not to mention his fatuous insistence on viewing himself as the victim of hatred – invalidates the film's putative humanitarian ideology, rendering it nonsensical from the start'.[17] On the other hand, Griffith did earnestly believe in his mission to fulfill the promise that cinema was the 'universal language that had been predicted in the Bible, which was to make all men brothers because they would understand each other', as once recounted by actress Lillian Gish.[18] *Intolerance* is therefore inseparable from the transnational discourse of film as a universal language against which its aims and claims must be accordingly assessed.

As Miriam Hansen has shown, *Intolerance* was one of the first attempts to 'put the universal language proposition into textual practice' through recourse to a multinarrative structure.[19] Hansen's authoritative analysis of

14 Melvyn Stokes, *D.W. Griffith's The Birth of a Nation: a History of the Most Controversial Motion Picture of All Time* (New York: Oxford University Press, 2007), 263.

15 Ibid., 78.

16 Slavoj Žižek, 'Against an Ideology of Human Rights' in Kate E. Tunstall (ed.) *Displacement, Asylum, Migration: The Oxford Amnesty Lectures 2004* (Oxford: Oxford University Press, 2006), 56–85; 63–64.

17 Ryan Jay Friedman, *The Movies as a World Force: American Silent Cinema and the Utopian Imagination* (New Brunswick, NJ: Rutgers University Press, 2019), 82.

18 Lillian Gish, with Ann Pinchot, *The Movies, Mr. Griffith, and Me* (Englewood Cliffs: Prentice Hall, 1973), 60.

19 Miriam Hansen, *Babel and Babylon: Spectatorship in American Silent Film* (Cambridge, MA, and London: Harvard University Press, 1991), 186.

the film, originally published in 1991, is both tacitly and overtly anachronistic. The former because it necessitates as its negative counterpart the theoretical notion of a classical cinema formulated many decades later against which the unusual narrative design of *Intolerance* can be positively assessed. The latter because, as she puts it, 'one cannot avoid reading *Intolerance* from a perspective of the present, along the vanishing lines of a tradition of alternative filmmaking that includes names like Dziga Vertov, Sergei Eisenstein, Jean-Luc Godard, Alexander Kluge and Chantal Akerman'.[20] In this section, my analysis will also be informed by a certain degree of anachronism, though my contention is that *Intolerance* appears as a fascinating precursor of the global network film examined in the next section in relation to *Babel*. This reading, I hope to show, finds substantiation not only when we look at *Intolerance*'s quadripartite structure, which revolves around fatalism and human suffering, but also when we consider its picturing of a networked society as the end of the circularity of history in its modern narrative. First, however, an exploration of some contemporaneous discourses of universalism is in order.

We saw earlier that the idea of film as a universal language shares its genesis with the new medium, ubiquitously featuring in the general press and trade magazines. From the early 1910s, universalism worked its way into early film theories developed in countries as disparate as Italy, Germany, France and the US.[21] Ricciotto Canudo's 1911 essay 'The Birth of a Sixth Art' offers a programmatic condensation of film's celebrated universality. It notes how cinema enacts a 'symbolic destruction of distances' and takes the 'spectacle of distant, enjoyable, moving and instructive things' to even 'the smallest human settlement'. It then cherishes film's appeal 'to all social classes, from the lowest and least educated to the most intellectual' in terms of a site-specific community in the cinema that evokes a worldwide community of spectators.[22] Similar ideas were reprised a few years later by the French film critic and director Louis Delluc, for whom cinema touched 'the unanimity of the masses' in a manner analogous to Greek theatre, which 'had to satisfy all classes of society'.[23] Delluc continued: 'The semicircle in which the cinema spectators are brought together encompasses the whole

20 Ibid., 132.

21 See Mattias Frey, 'Cultural Problems of Classical Film Theory: Béla Balázs, "Universal Language" and the Birth of National Cinema', *Screen* 51:4 (2010): 324–340.

22 Ricciotto Canudo, 'The Birth of a Sixth Art (1911)' in Richard Abel (ed.) *French Film Theory and Criticism, 1907–1929* (Princeton, NJ: Princeton University Press, 1988), 58–65; 60, 65.

23 Louis Delluc, 'From Orestes to Rio Jim (1921)', *French Film Theory*, 255–257; 257; Louis Delluc, 'The Crowd (1918)', *French Film Theory*, 159–164; 162.

world. The most separated and diverse human beings attend the same film at the same time throughout the hemispheres'.[24]

Other writers invoked linguistic models and metaphors even more explicitly to account for cinematic universality. For Vachel Lindsay, cinema was a 'moving picture Esperanto' whose mode of visual decoding aligned it with Egyptian hieroglyphics and whose emergence was as impactful as the dawn of print culture. He proclaimed: 'Edison is the new Gutenberg. He has invented the new printing'.[25] Nearly a decade later, the Hungarian Béla Balázs would similarly propose that cinema's 'impact on human culture will not be less momentous' than that of the 'printing press'.[26] But whereas for Lindsay moving pictures were still a form of picture-*writing* in their enmeshing of literal and symbolic meanings, for Balázs cinema held out 'the promise of redemption from the curse of Babel' and the conceptual instrumentalism of a word-inflected culture by making visible the 'true mother tongue of mankind', that is to say, the 'long-forgotten language of gestures and facial expressions'.[27]

These assertions of universality, however, were often fraught with contradictions. Thus, in a much-cited passage, Balázs reflects on the cinema as 'a machine that in its own way will create a living, concrete internationalism: *the unique, shared psyche of the white man*'.[28] In some of these theories, as Mattias Frey argues, the presumed universal legibility of the image also sat uneasily with the idea that 'film was an expression of national characteristics'.[29] In a way, these cinematic iterations of universalism reproduced what Rosi Braidotti defines as the Enlightenment's 'binary logic of identity and otherness as respectively the motor for and the cultural logic of universal Humanism', not infrequently as a way to bolster the national interests of nascent film industries on an emerging global market.[30] This was especially evident in the US upon the arrival of the feature-length

24 Delluc, 'From Orestes to Rio Jim', 257.

25 Vachel Lindsay, *The Art of the Moving Picture* (New York and Toronto: Random House, 2000 [1915]), 177, 224.

26 Béla Balázs, *Béla Balázs: Early Film Theory*; *Visible Man* and *The Spirit of Film*, ed. Erica Carter, trans. Rodney Livingstone (New York and Oxford: Berghahn Books, 2010), 9.

27 Balázs, *Béla Balázs*, 14; 10.

28 Ibid., 14, emphasis in original. This passage would be modified in the late 1940s, when Balázs revised and combined his two books into one single work, *Filmkultúra: A film müvészetfilozófiája* (translated into English in 1952 as *Film Culture: Theory of the Film*). Instead of 'the standard white man' and its 'shared psyche', Balázs now speaks of 'an international human type' in 'the development towards an international universal humanity'. See Carter, *Béla Balázs*, 15.

29 Frey, 'Cultural Problems of Classical Film Theory', 337.

30 Rosi Braidotti, *The Posthuman* (Cambridge: Polity Press, 2013), 15.

film, which, as Hansen observes, was frequently advertised in terms of 'a seemingly unmediated appeal to sentiment and sympathy' that 'reduced all social, cultural, and historical differences to an essential humanity'.[31] Although Hansen overstates that the 'celebration of film as a new universal language ultimately coincided in substance and ideology with the shift from primitive to classical modes of narration', given that universalisms existed since the advent of cinema, the idea certainly gained a new impetus with narrative cinema.[32] This was so especially during and after the war, when, as Friedman notes, 'the film-as-universal language rhetoric flourishes, in part, because it acquires a new political resonance, getting grafted onto discussions about the prospects for a "World State"'.[33]

Released in 1916, *Intolerance* was deeply implicated in these ideas, concepts and discourses. Although the film does not directly reference the Great War, it is peppered with historical battle scenes and ends with a suggestive extra-diegetic sequence where the image of a battleground is superimposed over that of angels in the heavens high above. This emphasis on human conflict was a reminder of the futility of warfare, used to bolster the film's pacifist message at a time when war was raging around the world and President Woodrow Wilson's policy of neutrality was beginning to lose public support in the US.[34] But *Intolerance*'s universalism, as we shall see, cannot be reduced to its anti-war rhetoric, channelling cultural expressions as disparate as Walt Whitman's poetry, universal exhibitions and millennialist discourse, all mashed up into a grandiose narrative of human suffering in close-up.

Intolerance's narrative was as unusual as it was strategically designed to up the film's game and cement Griffith's name on the world cinema map. Although the film performed relatively well upon its release and impressed some critics, it estranged many others, going down in film history as a 'magnificent failure'.[35] Crucial to *Intolerance*'s monumentality was the temporal enlargement of its textual edifice, which, encompassing over 2000 years, was mirrored in its long running time (roughly 3 hours).[36] As noted, the film's gestation started with the present-day narrative strand, or the Modern Story. Inspired by the news of a mine workers' strike in Colorado, this depicts the

31 Hansen, *Babel and Babylon*, 81, 80, emphasis in original.

32 Ibid., 79.

33 Friedman, *The Movies*, 13.

34 On *Intolerance*'s anti-war message, see Andrew Kelly, *Cinema and the Great War* (London and New York: Routledge, 2005), 13–23; see also Friedman, *The Movies*, 65–74.

35 Terry Reydan, cited in Hansen, *Babel and Babylon*, 130.

36 On the difficulty of ascertaining *Intolerance*'s original length, see Russell Merritt, 'D. W. Griffith's *Intolerance*: Reconstructing an Unattainable Text', *Film History* 4:4 (1990): 337–375.

plight of a working-class couple – the Dear One (Mae Marsh) and the Boy (Robert Harron) – as they face a series of fateful events: the couple's baby is taken away by reformists on the grounds that the Dear One is an unfit mother, while the Boy is wrongly accused of murdering a gangster. By contrast, the remaining strands, all added at a later stage, are reconstructions of historical events and religious myths: the fall of the Babylonian empire to the Persians in 539 BC; the Judeo-Christian story of the crucifixion of Jesus Christ; and the massacre of the Huguenots in France on St. Bartholomew's Night in 1572. What is more, *Intolerance* shuns away linearity by intermixing the four stories, turning 'from one of the four stories to another, as the common theme unfolds in each', as one of its first title cards announces.

The decision to situate the Modern Story within a transhistorical constellation was no doubt meant to solidify the film's universalist credentials through the depiction of a timeless humanity. In this respect, as William M. Drew has shown, a major influence was the poet Walt Whitman, from whom Griffith borrowed a 'total historical conception' by focusing 'on the continuity and unity of human experience through the centuries'.[37] Griffith even makes a direct reference to the 1855 edition of *Leaves of Grass* and the poem 'Out of the Cradle Endlessly Rocking', which appears in the film as the visual tableau The Woman Who Rocks the Cradle, famously played by Lillian Gish. Employed as a transitional device whenever the film jumps between historical periods and also referred to as the 'Uniter of Here and Hereafter' (another line in the poem), the tableau allegorises the 'cradle of humanity', as underlined in one publicity programme and confirmed by title cards: 'Today as yesterday, endlessly rocking, ever bringing the same human passions, the same joys and the same sorrows'.[38]

The publicity materials surrounding *Intolerance* similarly lauded its universal humanism. One programme commented on 'how the four stories alternate with one another until, in the climax of the last act, they seem to flow together in one common flood of humanity'.[39] Another noted that the

37 William M. Drew, *D.W. Griffith's Intolerance: Its Genesis and Its Vision* (Jefferson, NC: McFarland, 1986), 89.

38 Publicity programme for '*Intolerance*: Theatre Royal, Drury Lane', London (May 1917), held at the Bill Douglas Cinema Museum (BDCM), University of Exeter, Exeter. Item number: 18333. One of the anecdotes surrounding *Intolerance*'s cyclical conception of history is that it was down to Griffith's chancing upon a billboard stamped with the expression 'the same today as yesterday' on a trip to New York for the premiere of *The Birth of a Nation*. See Paul O'Dell (with the assistance of Anthony Slide), *Griffith and the Rise of Hollywood* (New York: Castle Books, 1970), 46. As O'Dell also notes, this expression appears as a sign in one shot at the beginning of the Modern Story, when millworkers are seen protesting.

39 '*Intolerance*: Theatre Royal, Drury Lane', BDCM.

switching back and forth between the narratives added to the suspense 'as though some great tornado gave wings to the very soul of all humanity, past and present'.[40] These ideas were then reiterated by Griffith himself. Thus, in the programme notes for the film's New York premiere, he explained that 'the purpose of the production is to trace a universal theme through various periods of the [human] race's history. Ancient, sacred, medieval, and modern times are considered'.[41]

On closer inspection, however, this all-encompassing human race reveals some fault lines and predilections. As Scott Simmon argues, the 'idea of a "race" that included Babylonians, Israelites, Frenchmen, and Americans requires a certain melting-pot ingenuity' and reinforced an 'Anglo-Israelite logic' according to which Anglo-Saxons descended from Jews 'whose ten tribes were themselves descended through the Persians and Babylonians'.[42] More broadly, *Intolerance* eschews a confrontation with race through its disregard for historical accuracy. An example is the Judean narrative, where, as Griffith's assistant Karl Brown recalls, there was no concern with 'the actual physical appearance of Christ during his life as a man on earth' but, rather, with what people knew about that era from recreations found in 'Bible pictures, Bible calendars, Biblical magic-lantern shows, Christmas cards'.[43]

Cracks also appear in *Intolerance*'s universality when one examines both the choice of the narrative strands and the uneven screen time they receive. As scholarship on the film has shown, the three historical narratives each emulated contemporaneous international film styles, with *Intolerance* textually reproducing the competition over which it attempted to assert its superiority.[44] Moreover, while the historical segments differ in tone and style, special importance is accorded to the Babylonian narrative. This was not coincidental. For it is this narrative strand that allows *Intolerance* to invoke the myth of Babel, according to which Babylonians had built a tower to reach God who, outraged by such an ambition, scattered the people all over the world and made them speak different languages. Thus,

40 Film publicity press book: *Intolerance*, n/d, BDCM. Item number: 18262.

41 Cited in Scott Simmon, *The Films of D.W. Griffith* (Cambridge, New York and Melbourne: Cambridge University Press, 1993), 143.

42 Ibid., 143.

43 Karl Brown, *Adventures with D.W. Griffith* (London: Faber & Faber, 1973), 135.

44 Whereas the French segment evoked the Films d'Art productions and the Judean Story the Passion Plays imported from France, the Babylonian narrative drew on Griffith's beloved Italian films. See Hansen, *Babel & Babylon*, 173; Ismail Xavier, 'On Film and Cathedrals: Monumental Art, National Allegories and Cultural Warfare' in Lúcia Nagib, Chris Perriam and Rajinder Dudrah (eds.) *Theorizing World Cinema* (London: IB Tauris, 2012), 21–44; 28.

a title card at the beginning of Act II indicates that the events portrayed are based on 'cylinders [that] describe the greatest treason of all history, by which a civilization of countless ages was destroyed, and a universal written language (the cuneiform) was made to become an unknown cypher on the face of the earth'. At the same time, the setting of Babylon – measuring 'over a mile in length and built to accommodate five thousand people' – materialised *Intolerance*'s quest to impress.[45] Partly a response to Pastrone's aforementioned *Cabiria*, from which Griffith borrowed its elephant motifs (as dramatised in the Taviani brothers' *Good Morning Babilonia*, 1987), the setting was equally indebted to the realm of world's fairs. As the story goes, Griffith would have come across, while location scouting, the 'newly risen spires' of the Panama-Pacific International Exposition (1915), whose craftsmen were then hired to construct the film set.[46]

Intolerance's connection with universal expositions is significant. Griffith is said to have declared that 'it would be a crime to let the [Panama-Pacific] exposition come and go without perpetuating it in photography. I don't mean ordinary photography, but something stupendous [...] I mean, in short, a film drama that will mark another forward leap as great as that of *The Birth of a Nation*'.[47] In turn, this invites a reading of the film's multistranded structure as the temporal translation of these fairs' hierarchical spatial organisation, which arranged stands, attractions and villages of human cultures according to colonial parameters and in such a way as to create the idea of a topographical journey through the world. The idea that *Intolerance* could take its viewers on a historical world tour was reiterated in its publicity. One programme noted how Griffith 'takes you from plain, home-like scenes of the present day to those in Judea in the time of the Nazarene. He carries you from Judea to the glories of mediaeval France. From France you go easily into the pomp and magnificence of ancient Babylon'.[48]

It is not a coincidence, in this context, that the film's promotional materials sometimes included global figures. To be sure, globes were not exclusive to *Intolerance*'s publicity. A case in point is a UK programme (figure 5.1) for Griffith's previous *The Birth of a Nation* (distributed in many European countries after the war), which features a global image, with the US prominently visible, and the caption 'Griffith's Mighty Spectacle', thereby

45 O'Dell, *Griffith and the Rise of Hollywood*, 63.

46 Richard Schickel, *D. W. Griffith: An American Life* (New York: Limelight Editions, 1996), 309.

47 Cited in Robert W. Rydell, *All the World's a Fair: Visions of Empire at American International Exhibitions, 1876–1916* (Chicago and London: University of Chicago Press, 1984), 231.

48 '*Intolerance*: Theatre Royal, Drury Lane', BDCM.

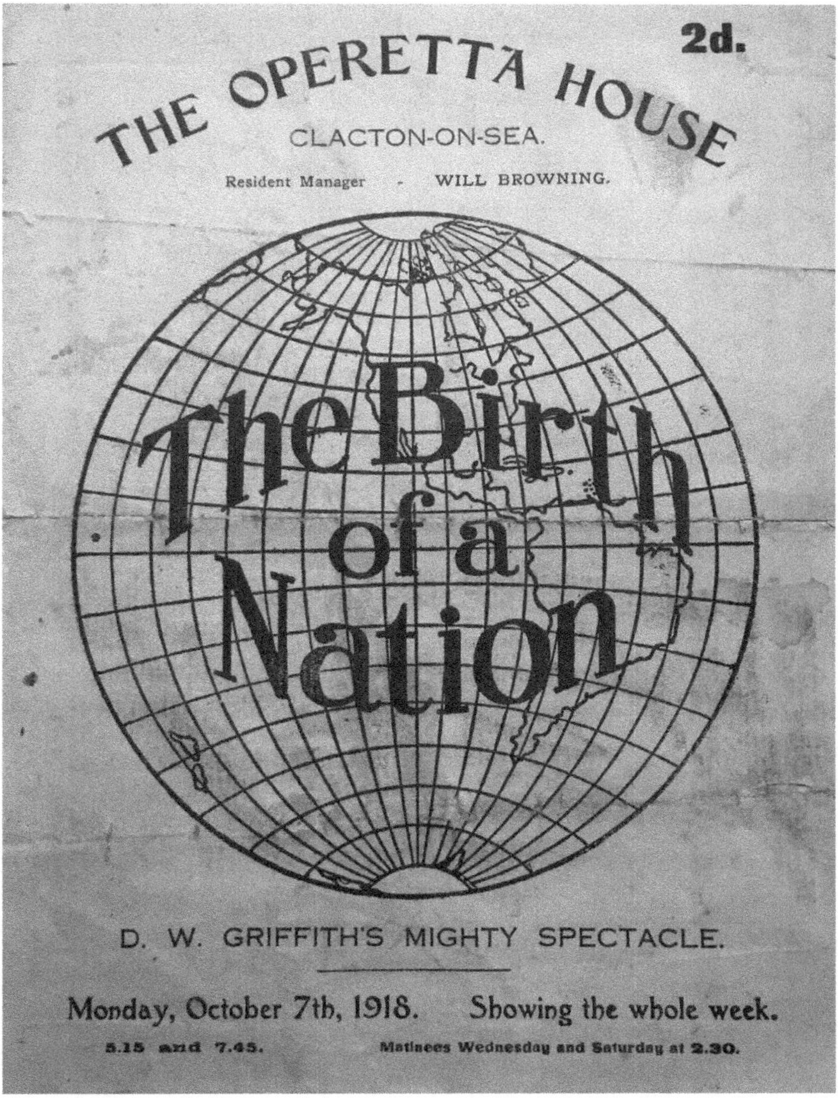

Figure 5.1 A programme for *The Birth of a Nation* (1915) reveals the global ambitions of its director. Courtesy of Bill Douglas Cinema Museum, University of Exeter.

confirming the global aspirations of this nation and this director.[49] But in the case of *Intolerance* globes mirrored its totalising diegesis and betrayed its unequal apportioning of the world. One poster (figure 5.2), for example, lauded 'Griffith's colossal spectacle', in which 'the four paralleled stories

49 Publicity programme for '*The Birth of a Nation*: Operetta House, Clacton-on-Sea' (October 1918), BDCM. Item number: 18710.

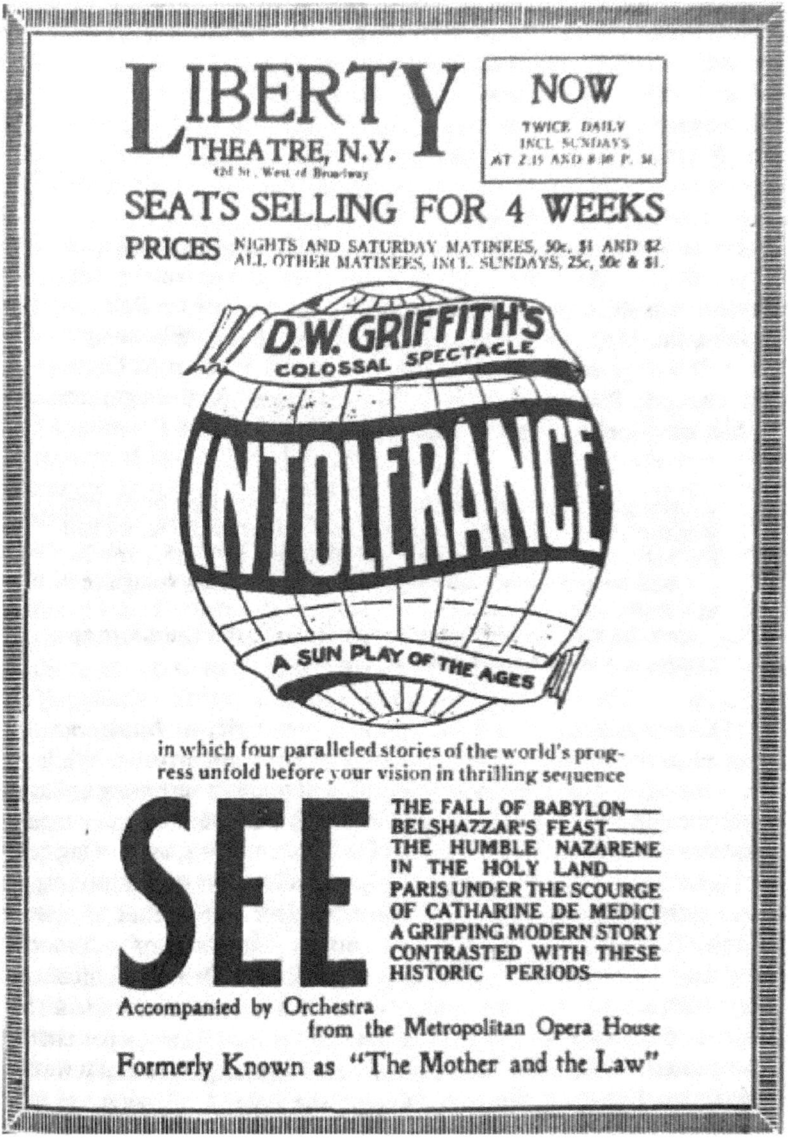

Figure 5.2 A poster for *Intolerance* (1916) reveals its world-encompassing aspirations and the unevenness of its world-making. Image reproduced in William M. Drew's *D.W. Griffith's Intolerance: Its Genesis and Its Vision.*

of the world's progress unfold in thrilling sequence', to then add that 'a gripping modern story is *contrasted*' with three different 'historic periods'.[50] These promotional lines spell out the film's uneven worlding, its gravitation

50 My emphasis. As quoted in Hansen, *Babel and Babylon*, 171.

towards the Modern Story. Not that this verbalisation was needed, though. In the film itself, the historical narratives, in spite of their differences, all come across as generalist and schematic when compared to the intricate design of the Modern Story, which instead displays Griffith's individualistic narrative inflection. Granted, some historical narratives are told through the lens of individuals: the unrequited love of the Mountain Girl (Constant Talmadge) for Prince Belshazzar (Alfred Paget) in the Babylonian narrative and the romance between Brown Eyes (Margery Wilson) and Prosper Latour (Eugene Pallette) in the French Story. Yet these individual stories are not fully developed, nor are these characters invested with the psychological nuances attributed to the Dear One or the Boy. The difference is even more glaring in relation to the Judeo-Christian narrative, which in the last instance remains a brief series of tableaus with no character or story development, confirming its placement by dint of its 'universal Christian appeal'.[51]

By emphasising 'the world's progress', these lines further reveal the film's conception of circular history to be a fallacy, traversed as it is by a teleological countercurrent, as Hansen has noted:

> The temporal succession of settings – pagan antiquity, Judeo-Christian period, Renaissance-Protestantism, and the Modern Age – corresponds to a geographical movement from the Orient via the Mediterranean and Western Europe to the United States – and thus to the millennial prophecy that was mobilized in the nineteenth century by the ideology of Manifest Destiny: 'Westward the course of empire takes its way'.[52]

In turn, the film's tension between circularity and teleology is mapped onto montage alternation and acceleration.[53] While *Intolerance*'s first two hours devote lengthy sequences to each individual story, often separated by intertitles and The Woman Who Rocks the Cradle, towards the end noticeably faster intercutting climaxes in the film's most memorable and commented-upon sequence: a montage tour-de-force of crowd and battle scenes – the fall of Babylon, the crucifixion of Christ and the Huguenot massacre in Paris – all of which are intercut with the rescue of the Boy in the Modern Story.

51 Schickel, *D. W. Griffith*, 312. This being said, the Christian narrative was not conceived to be as short as it turned out. Griffith was allegedly forced to cut it down after film executives deemed its depiction of Jews unfavourable. See O'Dell, *Griffith and the Rise of Hollywood*, 46.

52 Hansen, *Babel and Babylon*, 170.

53 For an illuminating metric analysis of *Intolerance*'s accelerating pace, see Yuri Tsivian, 'Beyond Comparing: The Internal Dynamics of *Intolerance*', *Cinemetrics*, http://cinemetrics.lv/tsivian.php (last accessed 4 September 2020).

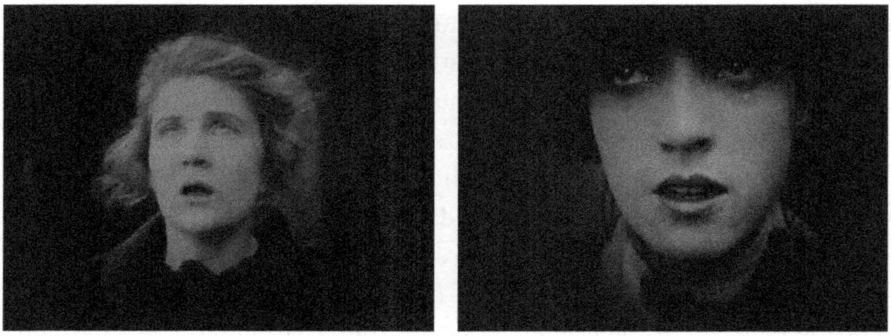

Figures 5.3 & 5.4 The universality of pain in close-up: the Dear One and the Friendless One in *Intolerance*.

As many commentators have noted, this montage sequence is revealing in that it culminates in the happy ending of the contemporary narrative, which announces the end of an eternally repeating history, itself materialised in the grief and gore closing the historical episodes. And indeed, despite the film's insistence that 'the same human passions, the same joys and the same sorrows' through the ages are depicted, there is little doubt that it is pain, in both its emotional and corporeal incarnations, that takes the upper hand in *Intolerance* as humanity's most universal attribute. As James Kendric summarises: 'Griffith gives us full view of two decapitations, bodies pierced by arrows, and an unedited shot of a character having a spear driven into his stomach, replete with gushing blood'.[54] Emotional pain is, for its part, conveyed through facial close-ups of the female characters. Although Griffith's self-ascribing claim as the inventor of the technique no longer stands up to scrutiny, *Intolerance* does offer expressive uses of the close-up to convey sorrowful mental states: the Dear One's despair as she thinks about the Boy's hanging (figure 5.3); the Friendless One's remorse at having killed the Musketeer of the Slums (figure 5.4); the Mountain's Girl's resignation when confronted with her own death; and Brown's Eyes' horror at the prospect of being killed.[55]

Intolerance's humanity is, then, a suffering humanity: a universal network of sorrows. On a narratorial level, this is articulated via the film's reliance on chance and fatalism both when considering its overall structure, which

54 James Kendrick, *Hollywood Bloodshed: Violence in 1980s American Cinema* (Carbondale: Southern Illinois University Press, 2009), 26.

55 On Griffith as the self-proclaimed inventor of the close-up and its connection with universalism, see Mary Ann Doane, 'Facing a Universal Language', *New German Critique* 41:2 (2014): 111–124; 111–115.

reiterates a ceaselessly self-repeating history, and in relation to the Modern Story, in which the Dear One and the Boy are repeatedly characterised as having no agency or control over their lives. Didactic intertitles drive the point home: the 'Loom of Fate weaves death for the Boy's Father'; 'Fate leads them all to the same district'; 'the Boy [is] wrongly convicted by some mischance of fate'. On the other hand, the Boy does get saved in the end, heralding the end of history's preordained cycles of human misery. What causes history to stop repeating itself after all?

One may argue it is simply chance as materialised in a plot contrivance: the unexplained appearance of the Kindly Officer who helps the Dear One to find the real culprit. Incidentally, this is in tune with Griffith's 'social-problem' films, such as *A Corner in Wheat* (1909), *The Usurer* (1910) and *What Shall we Do with Our Old?* (1911), in which, as Moya Luckett notes, 'resolutions emerge through chance, result from individual actions, or are simply unavailable'.[56] Also in line with Griffith's dramaturgy is the fact that the poor in *Intolerance* never become aware of the wider social structures dictating the course of their lives, nor is collective action presented as a possible means of bringing about change.[57] As Eisenstein once remarked:

> In social attitudes Griffith was always a liberal, never departing far from the slightly sentimental humanism of the good old gentleman [...] His tender-hearted film morals go no higher than a level of Christian accusation of human injustice and nowhere in his films is there a sounded protest against social injustice.[58]

This statement especially resonates with *Intolerance*, since the Boy is correlated with Jesus Christ at the film's end through crosscutting, which alternates his walk to the gallows with the crucifixion of Jesus in Golgotha. From this perspective, the prevention of the Boy's death appears not so much a product of chance but a miracle that announces the millennialist rebirth of Christian values.[59] This reading, in turn, finds substantiation in the film's closing sequence, where a battleground is superimposed over

56 Moya Luckett, 'Space, Gender, Oversight, and Social Change: Progressivism and the Films of D.W. Griffith, 1909–1916' in Charlie Keil (ed.) *A Companion to D.W. Griffith* (Oxford: Wiley Blackwell, 2018), 309–329; 311.

57 See Doyle Greene, *The American Worker on Film: A Critical History, 1909–1999* (Jefferson, NC, and London: McFarland & Company, 2010), 21, 29.

58 Sergei Eisenstein, *Film Form: Essays in Film Theory*, ed. and trans. Jay Leyda (San Diego, New York and London: Harcourt, 2011), 233.

59 Drew, *D.W. Griffith's Intolerance*, 20.

angels in the heavens, prisons turn into flowery fields and a luminous cross shines through, above in the sky.

Yet, as has been noted, the Modern Story also strongly implies that technology can change the course of things or, to use the film's own lexicon, that it can alter the twists of fate, especially when we turn to the film's climatic montage sequence. For Hansen, 'the teleology of history manifests itself in the progress of technology, the superior means of transportation and communication that make possible the last-minute rescue of the Modern story', since the Dear One stops the Boy's execution through recourse to cars, telephones and trains.[60] Following in Hansen's footsteps, Francesco Casetti has examined the sequence's 'breakneck speed' and Friedman the narrative and symbolic function of the automobile in the Modern Story as the bearer of such velocity.[61]

Inspired by these readings, I also want to close my analysis by drawing attention to the technological advancements figured in the Modern Story's ending. I want, however, to emphasise a different aspect, namely the fact that such advancements constitute, quite literally, a *network* linking geographically distant sites. This is highlighted through a number of shots showing railways and motorways (figures 5.5 and 5.6), as the Dear One chases by car the train carrying the Governor who can stop the execution, followed by images of officers calling the prison on a telephone, which delays the hanging. *Intolerance*'s focus on railroads perpetuates a fascination with the subject seen in preceding popular visual media in the US. As Brooke Belisle has shown, 'expanding networks of railroad lines offered a particularly good subject for nineteenth century photographic sets' not only because they helped visualise progress through a series of photographs implying a connecting line across vast expanses, but also because 'materially uniting the country became a pressing national challenge [...] weighted with ideas of Manifest Destiny'.[62]

Intolerance subscribes to these imperatives. But it also updates the network via film-specific devices designed to expand its reach beyond the nation and across time. While, as Belisle notes, photographic sets of railroads encompassed 'an extended scale of space and time' that defied human perception through omitted chunks and gaps, these sets were still

60 Hansen, *Babel and Babylon*, 170.

61 Francesco Casetti, *Eye of the Century: Film, Experience, Modernity*, trans. Erin Larkin with Jennifer Pranolo (New York: Columbia University Press, 2008), 117–123; Friedman, *The Movies*, 83–90.

62 Brooke Belisle, 'Picturing Networks: Railroads and Photographs', *Amodern* 2 (2013), https://amodern.net/article/picturing-nineteenth-century-networks/ (accessed 2 September 2020).

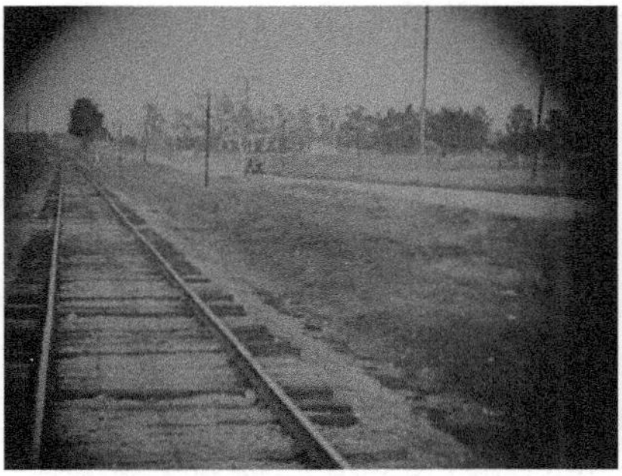
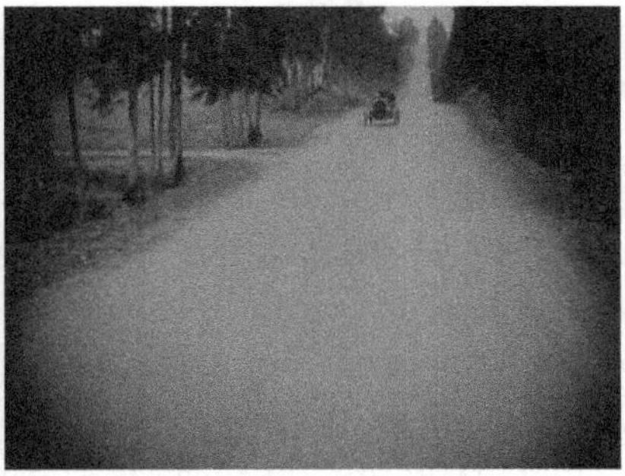

Figures 5.5 & 5.6 Railways and motorways in *Intolerance* constitute a physical network.

modelled on spatiotemporal unidirectionality.[63] Though this is also the case when it comes to the Modern Story, things become more complicated when we turn to the intermingling of geographical settings and time periods jumbled up in *Intolerance*'s final section via montage. Christian Metz, for example, speaks of a mixture that 'becomes a visual whirlpool' and gains 'the affective status of a *fusion*'.[64] Along similar lines, Friedman notes that 'the four stories ultimately *intersect*, resolving themselves into an image of unity that is decisive with respect to human history'.[65] For Friedman, this is corroborated by Griffith's own assertion, reproduced in a programme mentioned above, that 'in the climax of the last act, [the four stories] seem to flow together in one common flood of humanity'.[66]

Although the biblical imagery of this metaphor is far removed from urbanised networks, it produces a mental picture of divergent lines coalescing into one single point, not unlike a map or current digital modes of network visualising. At the same time, the four stories' fusion and intersection, emphasised via editing through graphic and compositional similarities (often between motorised and animal-based means of transport), gain in significance when examined next to Griffith's cherished crosscutting technique, which *Intolerance* radicalises in both form and content. Indeed, if parallel editing in Griffith's films was often employed to 'see more of the world' and 'bring together different worlds', as Luckett has suggested, then *Intolerance* takes this idea to new heights in its quest

63 Ibid.

64 Christian Metz, *Language and Cinema* (The Hague: De Gruyter, 1974), 107–109, emphasis in original.

65 Friedman, *The Movies*, 79, emphasis in original.

66 Ibid.

to bring together different worlds into a figuration, or networking, of the whole world.[67]

No doubt this commingling of 'different worlds' is miles away from the more recent concept of 'small worlds' according to which developments in one corner of the planet can have a demonstrably physical impact in another in the manner of a 'butterfly effect' traversing the vagaries of space and time.[68] From a metaphysical perspective, however, the small world of the Modern Story does have a retroactive impact on the worlds that temporally antedate it within the diegesis – not so much by changing the fate of those historical worlds in time-travel fashion, but by transfiguring the circularity of world history into the teleological linearity of progress.

That this linearity finds material configuration in highway, railroad and telecommunication networks is therefore significant insofar as it is this configuration that provides a physical grounding to the film's metaphysical networks. To use Griffith's own religious metaphor of floods and flows, the merging of the four stories into one single flow whose commanding steer is the contemporary narrative means that *Intolerance*'s transhistorical intersecting is effectively mapped onto the actual, intersecting lines of roads and cables leading to the saving of the Boy. And insofar as the Boy is directly associated with Christ, we may be forgiven for concluding that his saving signals, in the film, the dawn of a universal history which, as it turns out, is also that of a global network society.

A World of Strangers

If *Intolerance*'s spatial networks are restricted to one geographical locale and serve as the foundation onto which its temporally entangled, world-historical networks are grafted, in the contemporary global network narrative human flows are spatialised across the globe thanks to actual means of locomotion. Emerging on the world cinema map in the early 1990s, the network narrative has proved to be a genre uniquely suited to depict globalising processes related to the digital revolution, migratory displacements and tourism mobility. As Amanda Ciafone notes, via Jameson, the multinarrative can be understood as 'the filmic representation of the political unconscious of globalization' in its attempt to narrativise a world system whose vectors

67 Luckett, 'Space, Gender, Oversight', 318–319.

68 See Duncan J. Watts, *Small Worlds: The Dynamics of Networks between Order and Randomness* (Princeton, NJ, and Oxford: Princeton University Press, 2004).

and forces impinge upon the personal and the local.[69] Global multiplot films equally make visible what Arjun Appadurai calls our 'ethnoscape': 'the landscape of persons who constitute the shifting world in which we live: tourists, immigrants, refugees, exiles, guest workers, and other moving groups and individuals'.[70] Fernando Meirelles's *360* (2011), for example, follows the trajectory of more than half-a-dozen characters in Vienna, Paris, London and Colorado, with many scenes taking place in the 'non-places' of global modernity – airports, airplanes and hotel rooms – and recurrent shots of airplanes acting as a visual motif joining otherwise unrelated stories and characters.[71]

On an industrial level, the global network narrative also mirrors and is made possible by what Ciafone terms a 'networked industrial production (complex, temporary, and often transnational in financing, production, and distribution)' characteristic of our neoliberal age.[72] *360* is again a case in point. Directed by a Brazilian filmmaker, it brought together production companies from the UK, the US, Austria, Brazil and France, while featuring a vastly international cast from Czechoslovakia, Brazil, Russia, the UK and the US. Alejandro González Iñárritu's *Babel* was similarly financed by five production companies from Mexico, the US and France, and featured American, Australian, Mexican, Japanese and Moroccan actors and non-professionals. In these cases, as Paul Kerr observes, via Raymond Williams, the global narrative thus appears as 'a kind of "structural homology" to the film's mode of production and the social relations of its production'.[73]

In light of the above, we may conclude that the current network film is miles away from *Intolerance*'s time-bending, quasi-mystical globalism. And yet when we consider the latter as a foundational effort to transmute the universal-language idea into a multinarrative structure that crystallises a network of human suffering, then it appears as a paradigmatic film to bear in mind. For indeed, whereas literature on the multinarrative film has abundantly examined questions related to narrative mechanisms and storytelling devices, one of the most striking yet still underexplored features

69 Amanda Ciafone, 'The Magical Neoliberalism of Network Films', *International Journal of Communication* 8 (2014): 2680–2704; 2682, http://ijoc.org/index.php/ijoc/article/view/2452/1213 (last accessed July 29 2018).

70 Arjun Appadurai, *Modernity at Large: Cultural Dimensions of Globalization* (Minneapolis: University of Minnesota Press, 1996), 33, emphasis in original.

71 Marc Augé, *Non-Places: An Introduction to Supermodernity* (London: Verso, 2008).

72 Ciafone, 'The Magical Neoliberalism', 2682.

73 Paul Kerr, '*Babel*'s Network Narrative: Packaging a Globalized Art Cinema', *Transnational Cinemas*, 1:1 (2014): 37–51; 38.

of the genre is its quest to rekindle universalism through its situating of individual stories within a wider panorama.[74] Although I do not have the space to provide a survey of the genre, whose films can of course feature variations in theme, tone and style, I now turn to *Babel* as the most exemplary incarnation of this drive.

To glance over *Babel*'s mixed critical reception, littered with expressions such as 'incandescent self-importance', 'limitless ambition' and 'a grand wreck', is in fact to be immediately reminded of the canonisation of *Intolerance* as a 'magnificent failure'.[75] Iñárritu's third feature, *Babel* was the culmination of a meteoric international career. It followed on from his impressive Mexican debut *Amores Perros* (2000) and the international co-production *21 Grams* (2003). All scripted by Guillermo Arriaga, with whom the director publicly fell out after *Babel* over authorship issues, these three films ensured the two men would become household names, recognised for their multiplot narratives. But whereas *Amores Perros* was set in Mexico City and *21 Grams* in Memphis, *Babel* raised the stakes of the genre to giddy global heights.

As its title suggests, and not unlike *Intolerance*, *Babel* is the work of a skilful director who believes in the idea of film as a universal language and who wants to leave his mark on world cinema. The fact that Iñárritu's stature has only grown in importance since *Babel*'s release, with his last two films, *Birdman* (2014) and *The Revenant* (2015), earning him two consecutive Academy Awards for Best Director, suggests that he is heading in the right direction. These are different times and nearly a century separates *Intolerance* from *Babel*, with the latter film instead attuned to an 'ethics in a world of strangers', to cite the subtitle of Kwame Anthony Appiah's book.[76] Yet, as we shall see, it is striking that the religious myth of Babel makes an imposing resurgence, via a multinarrative architecture, in a film that similarly postulates universalism through bodies in pain and trapped by fate, thus confirming the obstinacy of cinematic tropes and figures of the universal across time.

Comprising four narrative strands, each set in a different country, *Babel* aims to paint a world beset by miscommunication, featuring six languages in

74 See footnote 11 in this chapter.

75 A.O. Scott, 'Emotion Needs No Translation', *New York Times* (27 October 2006), https://www.nytimes.com/2006/10/27/movies/emotion-needs-no-translation.html (last accessed 25 October 2018); Peter Bradshaw, '*Babel*', *The Guardian* (19 January 2007), https://www.theguardian.com/film/2007/jan/19/drama.thriller (last accessed 5 September 2020).

76 Kwame Anthony Appiah, *Cosmopolitanism: Ethics in a World of Strangers* (New York and London: W.W. Norton & Company, 2006).

total: Spanish, English, Japanese, Arabic, Berber and Japanese sign language. In Morocco, Abdullah (Mustapha Rachidi) gives a rifle to his goatherd sons Yussef (Boubker Ait El Caid) and Ahmed (Said Tarchani) to protect the goats from jackals. Unconvinced that the rifle can 'hit as far as 3km' as they are informed by its previous owner, the brothers shoot at a tourist bus atop a mountain. The bullet hits the neck of Susan (Cate Blanchett), an American tourist who is on holiday with her husband Richard (Brad Pitt), following the death of their newborn son. As a result of this incident, the Mexican nanny Amelia, who tends to the couple's children Debbie (Elle Fanning) and Mike (Nathan Gamble) in San Diego, finds herself unable to attend her own son's wedding in Tijuana, south of the US border. Failing to find care for the kids, she decides to take them to the wedding with her nephew Santiago (Gael García Bernal), leading to disastrous consequences. On their way back to the US, Santiago breaks through the border after a confrontation with a border patrol guard, leaving Amelia and the two kids in the scorching Californian desert. Meanwhile, in Tokyo, Chieko (Rinko Kikuchi) is a deaf-mute girl struggling with isolation and the suicide of her mother, unable to connect emotionally and sexually with other men. Eventually we find out that her father Yasujiro (Koji Yakusho) was the owner of the rifle sold to the Moroccan family, leading the Japanese police to start following her.

Babel's human networks, then, appear not only in the form of tourism mobility but also in relation to the themes of immigration, borders and border crossing. In particular, Amelia's back-and-forth trip to Mexico makes visible what Castells defines as a contemporary 'global process of urbanization' characterised by 'the formation of a new spatial architecture made up of global networks connecting major metropolitan regions', which he, tellingly, illustrates with the very geographical region depicted in *Babel*. He writes:

> Los Angeles is not the appropriate name for the actual spatial form of which it is only a component, because the relevant spatial unit comprises the entire Southern California Metropolis that extends from Santa Barbara to San Diego and Tijuana across the border [...] This is the undefined metropolitan region where 20 million people work, live, commute, and communicate [...], while retrenching in the polity of localities of a fragmented territory and identifying their diverse cultures in terms of ethnicity, age, and self-defined social networks.[77]

77 Castells, *The Rise of the Network Society*, xxxiii.

Babel entirely subscribes to this idea. Through its focus on immigration issues, it shows that the movements and flows enabled by this networked topography is not at all frictionless, but instead ridden with tensions and conflicts resulting from ethnic-cultural diversity and asymmetric relations of power.

On a global level, the connections between the four stories are only gradually revealed to the viewer, often via photographs as diegetic props. Thus, when waiting for the ambulance to arrive in the small village of Tazarine, Richard shows pictures of Debbie and Mike to the tour guide Anwar (Mohamed Akzham), whereas in Chieko's apartment the camera lingers for a moment on a picture on the wall showing Yasujiro holding the rifle. However, although the film alternates between the stories, which could lead one to infer they are simultaneous, this proves not to be the case. For example, *Babel* starts with Yussef and Ahmed shooting the rifle but then goes back in time to show Richard and Susan bickering in a restaurant. We are then made aware of this temporal disjuncture as the bullet incident is reprised, now from Susan's perspective on the bus. Similarly, at the film's beginning we hear Richard on the phone with his son the evening prior to the wedding only to realise, at the end, that the call has happened long after Susan's shooting. Another example is the Japanese narrative, which, though it is crisscrossed with the other stories through the film's running time, has taken place a few days after the Moroccan incident.

The gradual solving of this temporal puzzle reveals the two principles upholding *Babel*'s network narrative: interconnection and parallelism. If at the start of the film the links between the stories are unclear, by its end all dots are more or less connected. By intercutting the four stories rather than having them unfold autonomously, an overt narratorial hand is foregrounded and the film's emotional impact heightened. As Laura Podalsky has argued: '*Babel* reorders story events and juxtaposes sequences in particular ways as a means to reveal what it deems to be the true, affective connections between its characters'.[78] It must be noted, however, that *Babel* never blends all stories into one single whole via montage (as in *Intolerance*), and that the characters from different stories never brush against each other (a dominant trope of the multinarrative film). True, Gustavo Santaolalla's melancholy soundtrack does occasionally provide a bridge between transitions, some of which are based on graphic equivalences, as when poor kids running in Morocco turn into rich kids running in San Diego. Yet by and large transitions in the film

78 Laura Podalsky, *The Politics of Affect and Emotion in Contemporary Latin American Cinema: Argentina, Brazil, Cuba, and Mexico* (New York: Palgrave, 2011), 134.

are often jarring and abrupt, which is in line with the discrepancy between the look of each narrative achieved via different film stock (a common feature of Iñárritu's multiplot films, all photographed by Rodrigo Prieto).[79]

Babel's imagining of the global thus oscillates between two poles. If on the one hand there is a quest for narrative unity, the fact that the stories retain a certain autonomy means that on the other the film resists homogeneity and in so doing could be described, to cite Mads Anders Baggesgaard, as an 'exploration of the intimate relationship between location and globality'.[80] Indeed, location is emphasised through the film's style, which retains a documentary quality provided by handheld camerawork and fly-on-the-wall, street-life imagery, especially in the Japanese and Mexican narratives. But *Babel* is not merely interested in locality. The *locatedness* of one's place in the world is equally emphasised through its extensive use of point of view and hearing.

Thus, at points during Chieko's story, the film's aural track turns into barely audible vibrations so as to mimic her deaf-mute's perspective on the world. Likewise, point-of-view shots taken from inside vehicles convey what Deborah Shaw, in her analysis of the film, terms a 'tourist gaze' traversing foreign territories and confronting cultural difference. Examples include when Susan observes through the bus window burka-attired women walking in the Moroccan desert, or when Debbie and Mike stare with curiosity from the car at the bustling Mexican streets, and Mike tells Santiago that 'Mexico is a dangerous place'.[81] Yet the film also inverts the direction of looks and qualifies the safe distance of the tourist gaze as fragile. This is what happens, for example, when Michael and Debbie are stared at on their arrival at the party in Mexico, as well as when the American and English tourists become themselves an attraction for the people of Tazarine as the bus is forced to stop there. The seeing-through-glass motif reappears here, but now with the tourists depicted from inside the bus through the bullet-broken glass.

By foregrounding the phenomenology of one's being in the world via subjective hearing and looking, *Babel* calls to mind Hannah Arendt's concept of 'common world', which she defines thus:

79 Dolores Tierney, 'Alejandro González Iñárritu: Director without Borders', *New Cinemas: Journal of Contemporary Film* 7:2 (2009): 101–117; 105.

80 Mads Anders Baggesgaard, 'Picturing the World – Cinematic Globalization in the Deserts of *Babel*', *Journal of Aesthetics & Culture* 5:1 (2013): 1–10; 2.

81 Deborah Shaw, *The Three Amigos: The Transnational Filmmaking of Guillermo del Toro, Alejandro González Iñárritu, and Alfonso Cuáron* (Manchester: Manchester University Press, 2013), 141.

> For though the common world is the common meeting ground of all, those who are present have different locations in it, and the location of one can no more coincide with the location of another than the location of two objects. Being seen and being heard by others derive their significance from the fact that everybody sees and hears from a different position.[82]

Babel's replication of the same event from different perspectives – such as Susan's shooting and Richard's telephone call to Amelia – is another stance of this non-synoptic and multi-perspectival world-making. As Arendt notes: 'The end of the common world has come when it is seen only under one aspect and is permitted to present itself in only one perspective'.[83]

Clear-cut dichotomies – us and them, north and south, east and west – are further blurred in *Babel* by the fact that misunderstanding is also depicted between people from the same culture and/or speaking the same language, meaning the film's title is to a certain extent a red herring. No doubt linguistic miscommunication is a recurrent trope, as when Richard cannot be understood by a Moroccan driver when crying out for help; or when, on driving back to the US after the wedding, an intoxicated Santiago tells a confrontational US-Mexico border guard that Amelia is the kids' 'auntie', only for this to be disproved by Debbie. However, conflicts also arise between people speaking the same tongue. Episodes that substantiate this include Richard and Susan's bickering on the bus, the escalating row between Richard and the American and English tourists (who become impatient due to the ambulance's delay), Chieko's series of failed sexual and emotional encounters with other Japanese men, as well as her frustrating relationship with her father.

At least in principle, this effort at levelling humanity out beyond cultural and linguistic particularisms calls to mind Alain Badiou's indictment of a contemporary 'ethics of difference' (as also discussed in chapter 4). This Badiou defines as a 'commonsensical discourse defeated in advance', given that 'infinite alterity is quite simply *what there is*': 'There are many differences, say, between a Chinese peasant and a young Norwegian professional as between myself and anybody at all'.[84] It follows for Badiou that 'the real question – and it is an extraordinarily difficult one – is much more that of

82 Hannah Arendt, *The Human Condition* (Chicago and London: Chicago University Press, 1958), 57.

83 Ibid., 58.

84 Alain Badiou, *Ethics: An Essay on the Understanding of Evil*, trans. Peter Hallward (London and New York: Verso, 2012), 25–26, emphasis in original.

recognizing the Same'.[85] Universality in this account has little to do with a transhistorical human nature or an a priori Kantian morality, but what Badiou terms 'a truth', the possibility 'to think what might be in terms that break radically with what is', which is 'the obligatory starting point of all properly human action'.[86] He concludes: the 'coming-to-be of that which is not yet' that constitutes a truth and arises in particular, contingent events must be by necessity universalisable, that is, '*indifferent to differences*', since 'a truth is *the same for all*'.[87]

On the basis of its adherence to multiplicity, the multinarrative film would seem to appear as a privileged dramaturgic space around which notions of universality and alterity can be negotiated and contested, often to diverse ends. The Oscar-winning *Crash* (Paul Haggis, 2001) is a good example: featuring characters pertaining to ethnic minorities living in Los Angeles, such as Afro-American, Latino and Arabic communities, here is a film interested in bolstering a schematic discourse of difference through characters whose identities are defined by their ethnicity a priori, and who will de rigueur display racist attitudes towards other ethnic groups. Conversely, Badiou himself has examined Paul Thomas Anderson's fascinating *Magnolia* (1999), which interlocks the lives of nine characters during a rainy day in Los Angeles, as 'allegorical of humanity as a whole', with the film informed by 'a tension between the desire for multiplicity and a counter-tendency toward unity'.[88] For Badiou, *Magnolia*'s affirmative universalism is: '*humanity is love. When there's no love, then there's no humanity, there's only disconnection, in the strict sense of the term*'.[89]

To return to *Babel*, although the film grapples with these ideas, its thesis nonetheless proves to be removed from a Badiounian affirmation of human love and agency. The film instead locates its universality in the spectatorial recognition of pain and suffering. Iñárritu himself explains: 'I realized that what makes us happy as human beings could differ greatly but what makes us miserable and vulnerable beyond our culture, race, language or financial standing is *the same for all*'.[90] Such a statement, in turn, recalls the work of pragmatist philosopher Richard Rorty, for whom only witnessing the pain of others can activate a bond between fellow humans. For Rorty,

85 Ibid., 25, emphasis in original.

86 Ibid., 14.

87 Ibid., 27, emphasis in original.

88 Alain Badiou, *Cinema*, texts selected and introduced by Antoine de Baecque; trans. Susan Spitzer (Cambridge and Malden, MA: Polity Press, 2013), 178, 176.

89 Ibid., 183, emphasis in original.

90 Iñárritu cited in Shaw, *The Three Amigos*, 148, my emphasis.

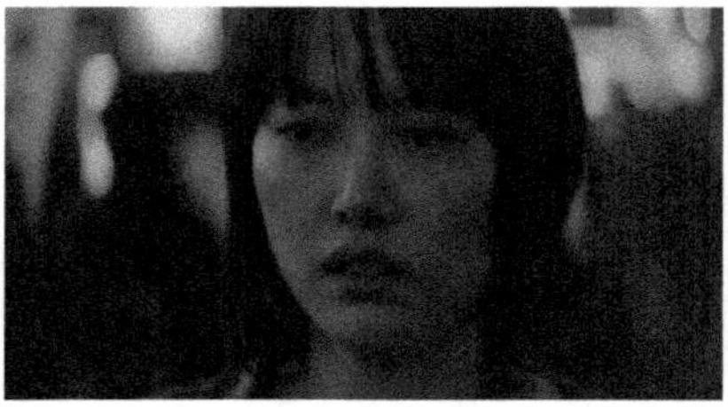
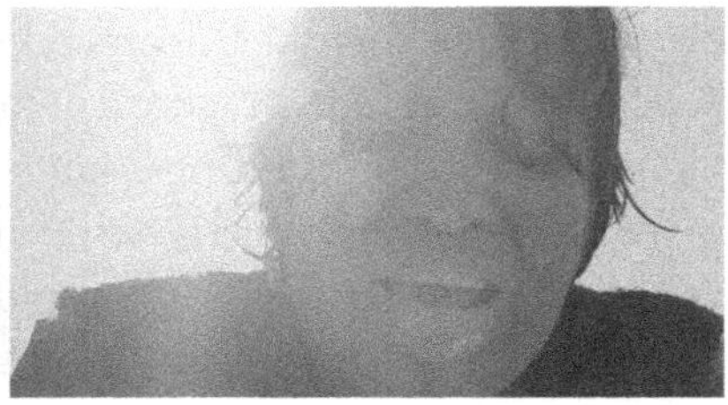

Figures 5.7 & 5.8 The universality of pain in close-up: grieving women in *Babel*. Courtesy of Paramount Vintage.

human solidarity should not be thought of 'as recognition of a core self, the human essence, in all human beings' but, rather, 'as the ability to see more and more traditional differences (of tribe, religion, race, customs, and the like) as unimportant when compared with similarities with respect to pain and humiliation'.[91]

Pain and humiliation are indeed the lingua franca of *Babel*, which revels in its own ability to convey these mental and physical states as viscerally as possible. Claustrophobic framings, out-of-focus shots, frenetic editing and erratic camerawork are all recruited to depict graphic events literally drenched in bodily fluids. And as in *Intolerance*, it is mostly women who predictably bear the brunt of suffering as universalising signifiers of martyrdom. We follow Susan's ordeal as she bleeds uninterruptedly, eventually pees herself and then has her gun-inflicted wound stitched by a veterinarian with no anaesthetic, all in extreme close-ups. We witness the comedown of Chieko's drug-fuelled night out as she walks, inconsolably, back home, her face all gloom and sweat (figure 5.7). And we are asked to participate in Amelia's plight in the scorching desert wearing a tight dress and high-heel sandals, her sunburnt face smeared with makeup (figure 5.8) and her feet bleeding in close-up. Tearful faces, twitching hands, fatigued legs: in *Babel*, the human body is truncated into synecdochal signifiers of pain and nothing escapes the attention of a camera eager to make the viewer feel intimately close with physical affliction as the conduit for universality.

In conformity with the network film, *Babel*'s rollercoaster of dehumanising situations is triggered by a chance event – an accidental bullet – that triggers other, increasingly harrowing events in an uncontrollable domino, or butterfly, effect. As Bordwell tells us, 'contingency replaces causality' as the motor of the network narrative, a staple of the genre that has been

91 Richard Rorty, *Contingency, Irony, and Solidarity* (New York: Cambridge University Press, 1989), 192.

positively appraised: either because, as Charles Ramírez Berg argues, contingency shows that 'agency is illusory and self-actualization risible'; or because, as noted by Todd McGowan in relation to *Babel*, as 'a film that bombards the spectator with the power of contingency', it proves 'that no underlying guarantee provides support for the sociosymbolic structure', hence the possibility of change.[92] On closer inspection, however, chance and contingency are not so much figured in the network narrative as the emergence of a Badiounian truth that can be seized as a catalyst for change, as they are channelled into determinism and fatalism. To use Jeffrey Sconce's blunt but useful expression: characters in the network film are often 'fucked by fate', and nowhere is this more evident than in *Babel*.[93]

It is true that in the end some characters in *Babel* are fucked by the wrath of fate more royally than others, whether because of their geographical provenance, social class or ethnicity. As in *Intolerance*, salvation may come for some, but certainly not for all. Although we can discern glimmers of connection and redemption in the stories of Richard and Susan and of Chieko and Yasujiro, this is hardly the case when it comes to Amelia or the Moroccan children. Indeed, I cannot follow McGowan's contention that the contingency of the shooting ultimately connects, in his words, 'Yussef and his family in Morocco, Richard and Susan visiting the country, Amelia and her family in Mexico, and Chieko and her father in Tokyo'.[94] While Amelia is informed that Susan and Richard have decided not to press charges against her, she is immediately deported back to Mexico, literally in rags, because she had been working illegally in the US. Tellingly, when she is informed of this decision, a close-up emphasises her fidgeting hands, which sits in contrast with the visual motif of holding hands (seen in the stories of Richard and Susan, and Chieko and her father) that McGowan treats as proof of the film's affirmative ethics. Yussef's future prospect could not be less promising either: the last time we see him he is sobbing in despair as his brother is killed by the police during a shootout.

92 Bordwell, *Poetics of Cinema*, 204; Charles Ramírez Berg, 'A Taxonomy of Alternative Plots in Recent Films: Classifying the "Tarantino Effect"', *Film Criticism* 31:1/2 (2006): 5–61; 40; Todd McGowan, 'The Contingency of Connection: The Path to Politicization in *Babel*', *Discourse* 30:3 (2008): 401–418; 403.

93 Jeffrey Sconce, 'Irony, Nihilism and the New American "Smart" Film', *Screen* 43:4 (2002): 349–369; 363.

94 McGowan, 'The Contingency of Connection', 413. McGowan's analysis substantiates this claim with examples only from Susan's and Chieko's stories (413, 415). See also Bert Olivier, 'Communication and "Real" Confusion: The Film *Babel*', *Communicatio: South African Journal of Communication Theory and Research*, 35:1 (2008): 1–16.

Granted, it could be argued that the film highlights the brutal fate that awaits these disenfranchised characters by subscribing to the rules of verisimilitude, that is, 'from a position of sympathy and identification with the residents of the Global South', as Dolores Tierney argues.[95] In this context, the televised departure of Susan from the Moroccan hospital watched by the Japanese detective just before the film ends, with the TV presenter highlighting the 'happy ending for the American couple', would confirm *Babel*'s knowing recognition of an unjust world system that benefits some to the detriment of many. The problem, in my view, is that the film risks perpetuating the very partition it aims to obliterate by characterising those residents of the global south solely as the recipients of pity. It thus reinforces what Badiou defines as a 'miserable moralism in the name of which we are obliged to accept the prevailing way of the world and its absolute injustice'.[96] Arendt similarly warns us about the dangers of a politics of pity, which, by requiring an unbridgeable distance between those who suffer and those who watch, 'can be enjoyed for its own sake, and this will almost automatically lead to a glorification of its cause, which is the suffering of others'.[97] This leads us back to the film's fatalism and the little agency it accords its characters, a feature of the network narrative that Berg defines as 'seriously subversive' in its deconstruction of the myth of American individualism.[98] And yet a world entirely devoid of human agency and purely governed by fate is just another name for nihilism, which, as Badiou notes, 'is a powerful contributor to subjective resignation and acceptance of the status quo'.[99]

Like *Intolerance*, *Babel* lauds cinema as a substitute for verbal language in connection with pain and suffering seen up close. In place here, to cite Rorty, is a belief that, within humanity, 'a nonlinguistic ability, the ability to feel pain, is what is important, and that differences in vocabulary are much less important'.[100] But *Intolerance* and *Babel* are also such revealing texts across the temporal gulf that separates them because they evince the limitations of a sentimental humanism. To construct a grief-stricken

95 Dolores Tierney, *New Transnationalisms in Contemporary Latin American Cinema* (Edinburgh: Edinburgh University Press, 2018), 91.

96 Badiou, *Ethics*, lv.

97 Hannah Arendt, *On Revolution* (New York: Penguin Books, 2006), 218. Karl Schoonover has explored Arendt's politics of pity in relation to the brutalised body of Italian neorealism. See *Brutal Vision: The Neorealist Body in Postwar Italian Cinema* (Minneapolis: University of Minnesota Press, 2012), xiv.

98 Berg, 'A Taxonomy of Alternative Plots', 40–41.

99 Badiou, *Ethics*, 32.

100 Rorty, *Contingency*, 88.

networked humanity, both films must rely on an unforgiving fate that can only produce marionettes for a history that has already been written or a reality that cannot be changed. Whereas *Intolerance*'s salvation comes through a miraculous conflation of theology and technology, *Babel* shows instead nihilistic resignation in the face of the ways of the world by succumbing to an aesthetic mythologising of the suffering body.

Networking the Earth

On first inspection, Eduardo Williams's debut feature *The Human Surge* (*El auge del humano*, 2016) could not be further removed from *Intolerance* and *Babel*. A hybrid between fiction and documentary directed by a young Argentinean, this small-budget film does not attempt to construct the ambitious Babelistic edifice on display in those films. Still, I want to argue in the remainder of this chapter that *The Human Surge* does reveal important ties with the network film. This is seen not only on a structural level, in the sense that the film interweaves three disparate stories across the globe, but also on a thematic one, with human connectivity an overarching topic. More importantly, *The Human Surge* is a useful text to examine against *Intolerance* and *Babel* because it offers a refreshing antidote to sentimental universalism. This is achieved both by its refusal to abide by the precepts of a suffering-based human community and the way it expands the network into more-than-human realms, which results in a grounded, earthly, indeed earthy perspective on the Earth.

Winner of the main prize in Locarno's Filmmakers of the Present competition in 2016, and a co-production between Argentina, Brazil, and Portugal, *The Human Surge* is divided into three parts, each of which episodically unfolds from start to finish. The film starts with the young Exe (Sergio Morosini) in Buenos Aires, Argentina, jumps to Maputo, Mozambique to follow Alf (Shine Marx), and ends on the Philippine island province of Bohol, with a few unnamed young characters. As in *Babel* and other multiplot films, these strands are formally differentiated by film stock: the first was filmed on 16mm, the second on a pocket camera whose images were then recorded from a monitor by a 16mm camera, and the third on a digital RED camera.[101] As far as its network narrative goes, however, one

101 Vadim Rizov, 'The Art of Walking: Eduardo Williams on *The Human Surge*', *Filmmaker Magazine* (18 January 2017) https://filmmakermagazine.com/101211-the-art-of-walking-2/#.W2Q_oBiZM_U (last accessed 3 August 2018).

of the film's original features resides in how it interrelates its characters, for the connections that bring them together, as discussed below, are not causal in terms of a 'butterfly effect'. Featuring non-professionals in all roles, and more than a dose of improvisation, the film moreover eschews depth of characterisation and story development: comprising long tracking shots trailing characters from behind, *The Human Surge* is more profitably understood as a collection of fragments of everyday life, including a focus on boredom and restlessness, but also on uncomplicated pleasures against the backdrop of natural settings.

One of the main features of the global network society, according to Castells, is a 'shift from traditional mass media to a system of horizontal communication networks organized around the Internet and wireless communication', which has ushered in 'a fundamental cultural transformation, as virtuality becomes an essential dimension of our reality'.[102] In *The Human Surge* this idea becomes one of the main axes around which the parallelisms and links between characters are established. Connecting all young men and women in the film, in the three different locations where the film is set, is their dependence on electronic gadgets resulting from an inability to cope with everyday reality and as a way to escape to the virtual world of text messages and online videos. Many scenes accordingly show characters holding their devices, in search of a wireless connection, walking and texting, or simply lying around glued to their phones. At one point in the Mozambican strand, Alf tells his friend Archie that 'it's the second time my telephone has been stolen', and later says he 'can't be without my phone'. Phrases such as 'I've just checked on the Internet' and 'I've seen it somewhere on the Internet' are similarly heard in chaotic group conversations.

That said, *The Human Surge* is no celebratory take on millennials, globalisation and technology, given its critical focus on issues of class. All characters in the film are uniformly disenfranchised and from the so-called global south, often living in modest if not shabby houses, and mostly unhappy in second-class jobs, when not jobless. While at the film's start we see the Argentinean Exe working in a warehouse supermarket, later the viewer learns that he has been fired 'yet again'. In another scene, in the Mozambican story, Alf confides, to his friends' amusement, that he is thinking of leaving his job. Cryptic, disconnected phrases like 'They need us to be in debt, really' and 'What are the millionaires going to do? We have to wake up early. I hope their fortunes never decline!', often dropped unexplainably, further underscore the social class to which these youngsters belong in the context

102 Castells, *The Rise of the Network Society*, xviii.

of an asymmetrical world system that feeds off inequalities. This point is pressed home in the film's closing sequence, filmed in Bohol. It shows a factory assembly line and workers putting together a tablet whose integrated circuit, with its tiny and interconnected resistors shown in close-ups, lends material form to the film's emphasis on wireless networks.

Unlike these sleek tablets, the devices used by characters in *The Human Surge* are old, broken or malfunctioning. Internet connection is similarly never a given for these young people. They are constantly in search of a free connection or, as in the case of the young Filipino woman in the last narrative strand, in urgent need of an Internet café for reasons that remain undisclosed. In short, the myth of a democratising and inclusive Internet as the new universal language of our time finds no place here. As Friedman notes:

> It remains commonplace to infer from the decentralized structure and easy accessibility that [social media networks] are threats to repressive regimes, tools of democratic self-expression that empower common people by enabling them to connect and interact. The speed with which certain words and images can spread across the network seems to realize the ideal of a universal plane of cultural experience.[103]

By showing such 'easy accessibility' to be not so easy after all, *The Human Surge* is sceptical of this new universalism. As Elena Gorfinkel rightly notes in her analysis of the film: 'For these youth, the virtues of the network form seem to be bound up in its promise, but in practical terms, it is in fact an architecture of failure, asynchrony, and unrequited desire'.[104]

If anything, *The Human Surge* is concerned with exposing the underbelly of the Internet via a focus on online pornography: an easy means of making cash for the destitute youth. This is depicted in the first episode, when Exe and a group of friends, not without nervous giggles, broadcast themselves live and semi-naked for a webcam, fiddling with their genitalia and performing fellatio on one another. While homoerotic, this scene also shows these sexual performances to result from the boredom that comes with being jobless and adrift. This is reiterated in the first transition between stories, which finds Exe on a computer displaying a video of young black boys playfully

103 Friedman, *The Movies*, 4.

104 Elena Gorfinkel, 'Introduction: Global Cinemas in a Time of Networks' in Elena Gorfinkel and Tami Williams (eds.) *Global Cinema Networks* (New Brunswick, NJ: Rutgers University Press, 2018), 1–20; 11.

staging anal and oral sex with a banana. The camera gradually closes in on the video, which then freezes due to a poor connection. When the image comes back to life, the viewer then realises that its visual source is no longer the computer monitor but the actual webcam filming in situ, which sets in motion the film's second strand in Mozambique. This transitional sequence thus functions as an interface that merges the digital and the real world into an image whose ontological status becomes momentarily indiscernible, thus resonating with a later shot showing Alf walking about town, which curiously displays a 'play' arrow symbol at the top-right of the frame.

However, if *The Human Surge* is populated by youngsters whose craving for virtual realities is unrelenting and palpable, the film's mise-en-scène and style act as a material antidote by focusing on embodied existence and the pulsing rhythms of the quotidian and the concrete. This is achieved via two interrelated formal strategies. The first is the durational quality of shots allied with a focus on mundane events. Whether we are following characters walking into densely tangled jungles, pottering about in a lagoon in the Philippines, wandering down the sinuous pathways of a hillside in Mozambique, or plodding through a flooded urban area in Argentina, *The Human Surge* foregrounds not only the instantaneous gratification of a connected world, but also the phenomenology of being-in-the-world, including its sensory and material pleasures. Miserabilism is here deterred through characters that, despite their precarious socio-economic status, are not defined as sorrowful victims. Against all odds, they enjoy themselves in the company of fellow humans and in natural settings, whether we think of Exe laughing away with his friends in a square in Buenos Aires or on the beach, or the extended sequence in the last narrative strand depicting a large group of people bathing in a swimming hole. Lasting eight minutes and filmed at water level, this merry sequence stays with these characters while they swim, chatter away, splash the water and joke around. While alluding to the asymmetry of the global network, *The Human Surge* nonetheless refuses to reduce its characters to bodily specimens of misery by underlining mundane pleasures in the natural world as both an affirmation of living in the physical world and as a counterpoint to the lure of the virtual world.

The second formal element worth noting is the ambulatory camerawork, which the film deploys when tracking characters from behind. Filmed without a Steadicam, these positively handheld shots reflect in the very constitution of the image the embodied contortions of the camera operator. This is especially the case when we consider the hilly, jagged and winding paths and alleys through which the camera traverses, with its erratic movements thus mimicking the phenomenological experience of traversing

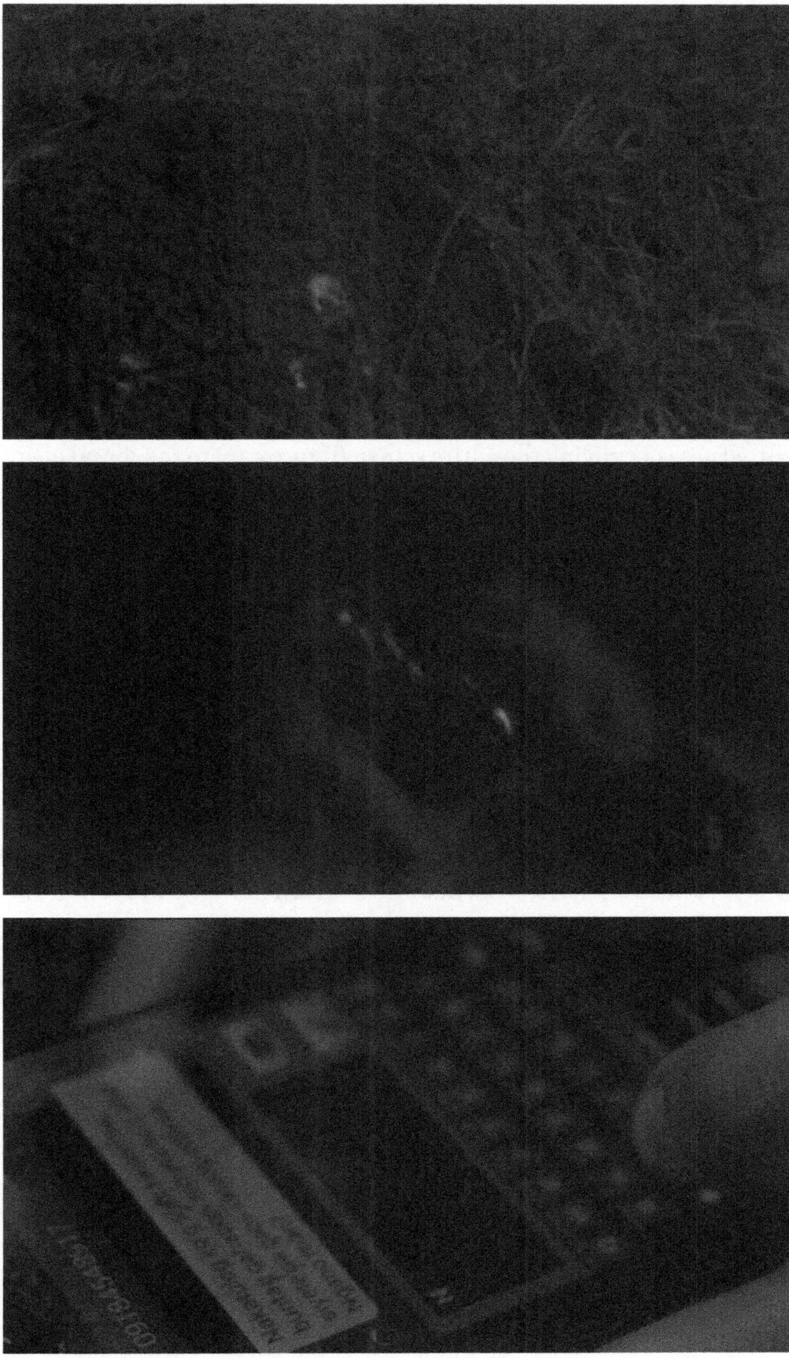

Figures 5.9, 5.10 & 5.11 In *The Human Surge* (2016), an ant colony provides the entry to an underground journey via the center of the Earth, which acts as a counterpoint to the idea of technology as the primary connective tissue of our network society. Courtesy of RT Features.

those settings. Although the resulting images are never coded as point-of-view shots, the camera seems to register an autonomous quality when decelerating, pausing or even changing its 'mind' as it stalks characters, not to mention the constant looks of passers-by, which underline further the corporeal I behind the camera's eye.

Yet if this view on the world seems to be coded as human, in the film's most memorable sequence this assumption no longer holds true. It shows Archie getting up at daybreak in an open field and urinating into an anthill, at which point the camera closes in on a hole on the nest (figure 5.9). This is followed by a wondrous, entirely unexpected interlude lasting over four minutes inside the nest. Instead of humans in full figure followed from behind – the film's main visual trope – the viewer is given the image of magnified ants occupying the entire screen (figure 5.10) to the sound of an atmospheric soundtrack. The ants then crawl out of the nest and on to a hand holding a smartphone, which brushes them off and begins typing in a message (figure 5.11), yet the geographical location is no longer Mozambique but the Philippines!

Rather than the digital interface connecting Argentina and Mozambique via wireless signals, the connection between Mozambique and the Philippines is instead forged through an underground journey across the center of the Earth, which acts as a counterpoint to the idea of technology as the primary tissue of our network society. That is, by halting a human-centric narrative to register an ant colony, the film relativises the world of characters not merely in relation to each other across the world (as is often the case in the global network narrative), but also in relation to the unseen world underneath our feet, and thus to a much wider, networked ecosystem that includes the natural environment and nonhuman animals.

Here, *The Human Surge* thus recalls not so much theories of network society but the 'actor-network-theory' (ANT), which, according to one of its most illustrious proponents, Bruno Latour, aims to trace the physical 'association between entities which are in no way recognizable as being social in the ordinary manner'.[105] Curiously, as Latour notes, the acronym ANT is 'perfectly fit for a blind, myopic, workaholic, trail-sniffing' mode of enquiry: defining his project as 'an ant writing for other ants', this expression takes on added meaning in *The Human Surge*, where an ant colony reconfigures the global network society into an earthly – and earthy – one. In turn, this leads us to take the film's title with a pinch of salt and similarly

105 Bruno Latour, *Reassembling the Social: An Introduction to Actor-Network-Theory* (Oxford: Oxford University Press, 2005), 65.

to take issue with its English translation, which does not fully grasp the ironic nuances of the original. The Spanish title reads *El auge del humano*, with *auge* literally meaning 'peak' or 'height'. A more faithful if nonetheless cumbersome translation would thus read as 'The Peak of Humanity'. While such a peak is thematised in the film in relation to connective technology, its focus on the pathways and trails of an infinitesimal animal simultaneously relativises and satirises the human network.

If *Intolerance* postulates the technical networks of urbanity as the ground on which to proclaim the metaphysical redemption of humanity, *The Human Surge* replies to that film, exactly a century later, by showing that ultimately technology is only one among many other networks of which we humans are also, and more humbly, a part. Worthy of note here is that the film is replete with holes, hollows and subterranean motifs that imply the existence of mysterious links beyond the nets and webs of technology. At one point, for example, Exe mysteriously walks into a cellar, while, in another scene, he and a group of friends enter into a hole inside a giant tree. In the Mozambican narrative, Alf unexplainably peeps into a cone-shaped object, while in the last strand, a man mentions he 'fell into a dark hole'. When placed in relation to the hole into which the camera plunges to reveal the ant nest linking different continents across the Earth, these episodes cumulatively gain in significance as they intimate worlds whose opening up into the 'real' world remain unseen but no less real.

In their condemnation of an ethics of pity, both Badiou and Arendt necessitate the nonhuman animal against which to laud human agency as a capacity for (re)invention as more than mere biological life. Whereas, for Badiou, we find the thinking human in 'an animal whose resistance, unlike that of a horse, lies not in his fragile body', for Arendt, as soon as people are 'forced to live outside the common world', as in concentration camps, 'they begin to belong to the human race in much the same way as animals belong to a specific animal species'.[106] And yet, as Kelly Oliver has recently argued, there is room for expanding Arendt's philosophy – and, I would add, Badiou's – into nonhuman territory in our uncertain environmental times. Recognising the distinction between Arendt's notion of the human world in contradistinction to the physical Earth, Oliver writes: 'Even if, with Arendt, we separate world and earth by associating world with the human world of meaning and earth with our given physical limitations or animal bodies,

106 Badiou, *Ethics*, 11; Hannah Arendt, *The Origins of Totalitarianism* (Cleveland, OH, and New York: Meridian Books, 1958), 302.

still both the human world of meaning and our existence on the physical earth are dependent upon not just human diversity but also biodiversity'.[107]

By reinstituting the networks of the physical Earth into the global network, *The Human Surge* calls to mind what Oliver terms 'earth ethics', which for her 'acts as a counterbalance to globalism insofar as it is grounded on the earth as a dynamic *network* of relationships through which each and all earthlings share the earth even if they do not share a world'.[108] She goes on:

> This means that individuals, species, and nations are fundamentally interconnected. It means that relationality is primary rather than secondary to who we are and what we can do. It means that if earth is an island, then we must reconceive islands as dynamic spaces constituted by their relationships to air, sea, and the elements that make them what they 'truly are'. In other words, it means that we must embrace the fact that we are limited creatures who are not just living on earth, but rather part of the biosphere that constitutes its very being.[109]

The Human Surge's contribution to this philosophical debate and the global network narrative genre resides both in its refusal to define the human as a suffering body and its resolution to open up the network – one that is as attuned to the phenomenology of being-in-the-world and an unjust world system, as it is to the Earth as the ground and planet we share with other human and nonhuman beings.

107 Kelly Oliver, *Earth and World: Philosophy after the Apollo Missions* (New York: Columbia University Press, 2015), 91.
108 Ibid., 40, my emphasis.
109 Ibid., 41.

6 A Disappearing Planet

Abstract
How lens-based media have confronted the task of recording disappearance, often via ruinous imagery, is the subject of this chapter. I first explore Albert Kahn's Les Archives de la Planète (Archives of the Planet, 1909–1931), a multimedia project whose raison-d'être was the capturing of disappearing realities owing to a sweeping global modernisation. Yet disappearance appears in another, equally significant guise in Kahn's Archive, which contains one of the most exhaustive collections of warfare destruction, as well as films of natural calamities, thus cementing a conception of planetary time as unpredictable rupture. These considerations provide the backdrop for my analysis of contemporary works interested in archiving the planet for a nonhuman future, including Geyrhalter's hybrid fiction-documentary *Homo Sapiens* (2016) and Trevor Paglen's *The Last Pictures* (2012) project.

Keywords: disappearance, Albert Kahn, ruins, Les Archives de la Planète, *Homo Sapiens*, *The Last Pictures*

To exist is to change.
Henri Bergson

We end with planetary imaginaries that have confronted the task of recording vanishing worlds for human and nonhuman futures. As noted in previous chapters, as much as the emergence of indexical media was tied to the promise of revealing novelty and the unseen, it was also inseparable from a quest to capture for posterity humans and nonhuman animals deemed on the verge of extinction. Yet in our time disappearance has gained new, decidedly global contours and dimensions. In the context of climate change, ocean acidification, coastal flooding, deforestation and other anthropogenic disturbances, processes that would normally unfold for humanly

de Luca, T. *Planetary Cinema. Film, Media and the Earth.* Amsterdam: Amsterdam University Press, 2022
DOI: 10.5117/9789463729628_CH06

imperceptible periods, if at all, are occurring at an unprecedented rate. Dystopian scenarios anticipated in the near future include the disappearance of entire towns and cities as a result of human-induced erosion and rising sea-rise levels, the Earth's sixth mass extinction, and even the eradication of the conditions that make human life possible on the planet.[1]

Of course, structural inequalities will not merely continue but be exacerbated during some of these scenarios, with poorer countries, regions and peoples likely the first to bear the brunt of disastrous environmental consequences, even if in the long run not even the rich may be spared. As the climate begins to act wildly and unpredictably, from extreme floods through to uncontrollable mega-fires, the idea that the 'end of the world' (meaning: the end of the world for humans) is nigh is no longer sci-fi territory but, as Bruno Latour notes, a global feeling shared by an unprecedented number of human beings inhabiting this Earth right now.[2] For Anna Lowenhaupt Tsing, it is imperative that we confront the 'terrors of indeterminacy' and our 'current precarity' as 'an earthwide condition': only then will we be able to come up with imaginative strategies of 'collaborative survival' in times of 'economic and ecological ruination'.[3]

A cursory glance over recent cultural and knowledge production confirms the prevalence of the trope of the ruin as a signifier of precarious and posthuman times. Bestsellers envisaging *The World without Us* (Alan Weisman, 2008) and *The Uninhabitable Earth* (David Wallace-Wells, 2019) now abound, while a number of visual projects have addressed the challenge of figuring a planet depleted of humanity.[4] In the last sections of this chapter I turn to

1 Stefano Valentino, 'World's Beaches Disappearing due to Climate Crisis', *The Guardian* (2 March 2020), https://www.theguardian.com/environment/2020/mar/02/world-sandy-beaches-disappearing-due-to-climate-crisis-study (last accessed 19 September 2020); Christiana Figueres and Tom Rivett-Carnac, '"The Only Uncertainty Is How Long We'll Last": A Worst Case Scenario for the Climate in 2050', *The Guardian* (15 February 2020), https://www.theguardian.com/environment/2020/feb/15/worst-case-scenario-2050-climate-crisis-future-we-choose-christiana-figueres-tom-rivett-carnac (last accessed 19 September 2020); Damian Carrington, 'Earth's Sixth Mass Extinction Event Under Way, Scientists Warn', The *Guardian* (10 July 2017), https://www.theguardian.com/environment/2017/jul/10/earths-sixth-mass-extinction-event-already-underway-scientists-warn (last accessed 19 September 2020).

2 Marc Bassets and 'Bruno Latour: "O sentimento de perder o mundo, agora, é coletivo"', *El País* (31 March 2019), https://brasil.elpais.com/brasil/2019/03/29/internacional/1553888812_652680.html (last accessed 29 April 2019).

3 Anna Lowenhaupt Tsing, *The Mushroom at the End of the World: On the Possibility of Life under Capitalist Ruins* (Princeton, NJ, and Oxford: Princeton University Press, 2015), 1, 4, 19.

4 See, for example, Alan Weisman, *The World without Us* (London: Virgin Books, 2008); Jan Zalasiewicz, with contributions from Kim Freedman, *The Earth after Us: What Legacy Will Humans Leave in the Rocks?* (New York: Oxford University Press, 2008); David Wallace-Wells, *The*

some of these – including the hybrid documentary *Homo Sapiens* (2016) and the photographic project *The Last Pictures* (2012) – in order to explore how they imaginatively face the prospect of archiving the world for nonhuman times ahead. I argue that their preservationist impulse to record the planet for the future is symptomatic of a marked sense of temporal disjuncture between an accelerated human time and the deep time of the Earth. But before we get there, I propose we first look at one of the most extraordinary projects in film history designed to fix the present for the future: Albert Kahn's Les Archives de la Planète, or Archives of the Planet (1909–1931). I suggest that Kahn's Archive, which set out to record humanity in the context of a disappearing and volatile world, can provide us with a more historically nuanced lens with which to assess the role of recording media as a visual repository of the planet against the background of temporal indeterminacy, precarity and instability.

As its title indicates, the Archive's ambition was planetary: to record the whole world, its places and its peoples, with the novel technologies available at the time. During its existence over two decades the Archives de la Planète amassed 183,000 metres of film, 72,000 colour photographs, or autochromes, and more than 6,000 stereographs, all taken in over 50 countries across four continents. For Kahn, an internationalist and liberal pacifist, the archiving of the world by way of evidentiary media could provide the inhabitants of this planet with the visual proof of human commonality. Yet there was another, equally important goal underwriting the project, the mission of which, in Kahn's own words, was to 'fix, once and for all, the aspects, practices and modes of human activity whose *disappearance is but a question of time*'.[5] As Kahn saw it, his Archive was to respond to a sweeping globalisation that threatened local cultures and rural lifestyles, thus safeguarding 'the development of life' at the dawn of the twentieth century 'to the benefit of future Humanity'.[6] As described by its appointed director, the geographer Jean Brunhes, the Archive, as a 'portrait of real life

Uninhabitable Earth: Life after Warming (New York: The Duggan Books, 2019); Clive Hamilton, *Defiant Earth: The Fate of Humans in the Anthropocene* (Sydney: Allen & Unwin, 2017).

5 Kahn, cited in letter from Emanuel de Margerie to Jean Brunhes, reprinted and included in Mariel Jean-Brunhes Delamare and Jeanne Beausoleil, 'Deux témoins de leur temps: Albert Kahn et Jean Brunhes' in *Jean Brunhes, autour du monde: regards d'un géographe/ regards de la géographie* (Boulogne-Billancourt: Musée Albert Kahn, 1992), 91–106; 92, my emphasis. Unless otherwise stated, all translations in this chapter are mine, with precious help from Alastair Phillips and Lúcia Nagib. My thanks to both.

6 Kahn, cited in Paula Amad, *Counter-Archive: Film, the Everyday, and Albert Kahn's Archives de la Planète* (New York: Columbia University Press, 2010), 149.

in our time', would 'preserve planetary occurrences that are going to die out' just prior to their vanishing, that is to say, 'while there was still time', as Kahn noted himself.[7]

As I explore below, this resolve to record the world before it changed irrevocably was itself indicative of a historically specific formulation of time as threat. As Philip Rosen has shown, whereas 'pre-Enlightenment historical temporality was not exclusively directional, but reversible, recursive', one of the prevailing conceptions of time in modernity is that it is 'an irreversibly linear, directional stream, any instant of which is unstoppable and unrepeatable. Irreversibility is constant loss, always threatening decay, death, disappearance'.[8] It is against this background, Rosen explains, that 'a preservative impulse or obsession in the modern West' emerged in the nineteenth century across a number of intellectual and cultural domains, including history, architecture, geology and visual arts.[9] These gave pride of place to the indexical trace – a crumbling building, a fossil, a historical artefact, a photograph – as the material conduit for a spectator to enter into phenomenological rapport with pastness, often discursively 'converted from death and lifelessness to life: nonbeing is called back into being'.[10] Though Kahn's Archive, which was never public during its time, compiled images for a future spectator, its preservationist credentials must be examined within this prevailing idea of time as a 'threateningly dynamic force'.[11] The obsessive recourse to figures of life and death made by Kahn and Brunhes are therefore not coincidental but symptomatic: for them, the image was to preserve living phenomena in the context of an irreversibly dissipating reality that could, nonetheless, be relived in the future.

On the other hand, as I also hope to show, the Archives de la Planète contains one of the most exhaustive collections of images of destruction and devastation, that is to say, death. As recent French scholarship on the Archive has emphasised, the outbreak of World War I interrupted its inaugural internationalism, with photographers now dispatched to document a destroyed France in line with Kahn's unconcealed patriotism

7 Jean Brunhes, 'Ethnographie et Géographie humaine', *L'Ethnographie: Bulletin de la société d'ethnographie de Paris* (1913): 29–40; 38; Delamare and Beausoleil, 'Deux témoins de leur temps', 91.

8 Philip Rosen, *Change Mummified: Cinema, Historicity, Theory* (Minneapolis and London: University of Minnesota Press, 2001), 105, 84–85.

9 Ibid., 6, 46.

10 Ibid., 130.

11 Ibid., xi.

and the pictures put to the service of propaganda.[12] In contrast with the films and photographs teeming with human life 'just before' their demise, the war images instead figure disappearance as a perceptually verifiable phenomenon rather than as an invisible temporal force: ruinous buildings and bombed-out towns depleted of human figures. These being ruins formed not through the continuous passage of time but instead the products of mechanised warfare, however, they likewise testify to the Archive's adherence to a modern conception of time as entropic rupture, one that, as my analysis will consider, was also evident in films and photographs of natural calamities produced and acquired by the Archive. In turn, these images of transformation and destruction gain in significance when considered in relation to Brunhes's proto-ecological 'human geography', which, by focusing on the reciprocity between human action and the environment, prefigures Anthropocene discourses and concepts in significant ways.

An archive that remained relatively outside the spotlight through much of the twentieth century, the Archives de la Planète saw its cultural and intellectual status change since the late 1990s both on its native soil and abroad. In France, this was partly due to a number of anthologies published by or in conjunction with the Musée Albert-Kahn (MAK), which administers and disseminates Kahn's projects since 1990.[13] This includes the digital storage of materials in a database available for researchers – the FAKIR, or Fonds Albert-Kahn Informatisé pour la Recherche – and the museum's opening up to the public. At the time of writing, most autochromes can be viewed and downloaded by anyone, anywhere in the world on the museum's website, while the Kahn films are soon expected to join the photographs. Parallel to this, and no doubt as a response to some of these developments, the Archives de la Planète has received growing scholarly attention, with a number of studies illuminating the project's tensions and contradictions between still and moving images, cosmopolitanism and nationalism, imperialism and universalism.[14]

12 See Valérie Perlès and Anne Singaud (eds.) *Réalités (In)Visibles: Autour d'Albert Kahn, les archives de la Grande Guerre* (Paris: Musée départemental Albert–Kahn, Bernard Chaveau Édition, 2019); Isabelle Marinone (ed.) *Un monde et son double: Regards sur l'entreprise visuelle des Archives de la Planète (1919–1931)* (Perpignan: Presses Universitaires de Perpignan and Institut Jean Vigo, 2019).

13 This includes *Jean Brunhes*; and more recently, *Albert Kahn, singulier et pluriel* (Boulogne-Billancourt: Musée départemental Albert-Kahn, 2015); Valérie Perlès with Manon Demurger (eds.) *Les Archives de la Planète* (Paris and Boulogne-Billancourt: Musée départemental Albert-Kahn and Lienart, 2019).

14 Teresa Castro, *La pensée cartographique des images: Cinéma et culture visuelle*, (Lyon: Aléas Éditeur, 2011); Jay Winter, *Dreams of Peace and Freedom: Utopian Moments in the 20th Century* (New

In particular, Paula Amad's indispensable book-length study, *Counter-Archive: Film, the Everyday and the Archives de la Planète* (2010), has argued for the centrality of the concept of the everyday to account for Kahn's Archive. Looking predominantly at the film collection, Amad argues that the Archive's impetus to 'record and store the raw data of routine experience, transient details, uneventful moments, ordinary gestures, and casual occurrences' furnished it with the potential to destabilise from within the notion of the archive understood as a positivist apparatus of historicist knowledge.[15] As she notes: 'the uniqueness of [Kahn's] films resides in their capturing of the spatial and temporal webs of daily life – eating, working, playing, walking, reading' – ordinary moments that unsettled traditional epistemologies of history understood as the concatenation of significant happenings.[16] Yet Amad knows she is treading on precarious ground, since as she herself admits, Kahn's Archive is in fact teeming with milestone historical events, meaning the everyday life on show in the films is not 'separate from or neatly opposed to political or economic history'.[17] Just as importantly, as she also notes, the Archive is 'based on a profound sense of the afterlife of the planet, or more precisely, the apocalyptic feeling about the world that existed as it entered and exited the First World War'.[18]

With its connotations of mundanity, uneventfulness and repetition, the everyday is therefore, in my view, insufficient in itself to account for the tumultuous background against which quotidian life often unfolded in Kahn's films and autochromes. Accordingly, I want to suggest that if the everyday is to remain a useful category for examining the Archive, then we might need to recruit the help of its conceptual opposite and frame it as the oxymoron 'extraordinary everydayness', or 'ordinary life in extraordinary times' – to borrow the subtitle of Sheila Fitzpatrick's book on Soviet life in the 1930s.[19] My hope in doing so is to home in on the idea

Haven, CT, and London: Yale University Press, 2006), 11–27; Katherine Groo, *Bad Film Histories: Ethnography and the Early Archive* (Minneapolis and London: University of Minnesota Press, 2019), 43–104; Trond Lundermo, '"The Archives of the Planet" and Montage: The Movement of the Crowd and "the Rhythm of Life"' in Eivind Røssaak (ed.) *Between Stillness and Motion: Film, Photography, Algorithms* (Amsterdam: Amsterdam University Press, 2011), 207–224; Trond Erik Bjorli and Kjetil Ansgar Jacobsen (eds.) *Cosmopolitics of the Camera: Albert Kahn's Archives of the Planet* (London: Intellect, 2020).

15 Amad, *Counter-Archive*, 5.

16 Ibid., 9.

17 Ibid.

18 Ibid., 23.

19 Sheila Fitzpatrick, *Everyday Stalinism: Ordinary Life in Extraordinary Times: Soviet Russia in the 1930s* (New York: Oxford University Press, 2000), 1, 2.

of disappearance underwriting the Archive and to demonstrate that its obsession with recording perishable phenomena was variously visualised and negotiated alongside categories of disturbance, turbulence, upheaval, disruption, catastrophe and turmoil: sociopolitical and environmental, natural and human induced. Given the Archive's monumentally heterogeneous holdings, the following is a brief, modest and, in some places, even arbitrary analysis. Yet I hope to show that the sense of planetarity formulated by the Archive via human geography cannot be dissociated from ideas of precarity, instability and temporal variability, which asks us to reconsider the current doxa that these categories are exclusive to our time as a global condition.

A Human Planet

Albert Kahn was born Abraham Kahn in 1860 in Alsace, northeastern France. Following the Franco-Prussian War in 1871 and the annexation of Alsace-Lorraine to Germany, Kahn's Jewish family opted for French citizenship and Kahn moved to Paris at the age of 16. There he began his career as a bank clerk and eventually became a trader on the foreign stock market. Over 20 years Kahn amassed a sizeable fortune by speculating on emerging markets, especially gold and diamond mining in South Africa. Astonishingly, by 1897 Kahn was the owner of his own bank, investing in Japan through international loans while working as a financier and economic adviser in the Japanese Imperial Court.

Kahn harboured ambitions other than finance, however. As a result of his growing fortune, he designed and supported a number of projects with utopian and cosmopolitan aspirations, including: a garden at the back of his mansion on the outskirts of Paris in Boulogne-Billancourt, today open to the public, which combines plants, soils and ornaments from different parts of the world; the Autour du Monde (Around the World) scholarships, founded in 1898, which awarded male (and later female) students grants to travel abroad and whose cosmopolitan foundations formed the basis of the Société Autour du Monde (Around the World Society, 1906–1949); and the Comité National d'Études Sociales et Politiques (National Committee for Social and Political Studies, CNESP, 1916–1931), the aim of which was to foster international political and economic collaborations.

The Archives de la Planète must therefore be seen as one, though certainly crucial, branch of an interconnected web of projects that promoted global thinking and instigated transnational debate, exchange and cooperation.[20] No doubt Kahn's cosmopolitanism was in some measure informed by the imperial imaginary prevailing in the France of his time, and while the Archive itself was never a colonial enterprise, it benefitted from the structures and networks of colonialism; indeed, the largest portion of the countries recorded by Kahn's operators were French colonies or connections.[21] As Jay Winter summarises: though 'Kahn never succumbed to simple racial stereotypes', in the end he 'was a French republican imperialist, and could not escape the contradictions in the ideas he held'.[22] As we shall see, such contradictions came into particularly sharp relief as soon as the war broke out.

It is worth mentioning that Kahn was also a close friend of the French philosopher Henri Bergson, whom he met while taking philosophy classes just after moving to Paris. The confluences between Bergson's ideas and Kahn's Archive are well documented.[23] Here I want to simply draw attention to Bergson's well-known concept of durational flow – or *durée* – as yet one more manifestation of what Rosen defines as a specifically 'modern conception of historical temporality' noted earlier.[24] Indeed, one would be forgiven for assuming that Rosen is quoting in full Bergson's theory of *Creative Evolution* (as per his 1907 book) when he mentions that, for modern historicity, 'life becomes a process of unending transition to a different future, for the present is the perpetual change of old into new'.[25] In response to modernity's wholesale rationalisation of time, Bergson proposed the idea of a qualitative, indivisible duration as the motor of life (*élan vital*): 'a

20 For an overview of these different projects and how they interconnected and resonated with each other, see Anne Sigaud, 'The Archives of the Planet: Between Cosmopolitanism and Patriotism' in *Cosmopolitics of the Camera*, 178–205. My thanks to Sigaud for sharing her chapter with me before publication.

21 See Sigaud, 'The Archives of the Planet'; Paula Amad, 'Experimental Cosmopolitanism: The Limits of Autour du Monde-ism in the Kahn Archive' in *Cosmopolitics of the Camera*, 133–154.

22 Jay Winter, 'Pacifist Photography: Seeing the Face of Humanity' in *Cosmopolitics of the Camera*, 116–132; 117.

23 See Sophie Coeuré and Frédéric Worms (eds.) *Henri Bergson et Albert Kahn, Correspondances* (Strasbourg and Bolougne: Desmaret/Musée départemental Albert-Kahn, 2003); Kjetil Ansgar Jacobsen, 'Henri Bergson and Albert Kahn: The Cosmopolitan Method', *Cosmopolitics of the Camera*, 206–235; Kjetil Ansgar Jacobsen, 'Bergson's Aesthetics? Autochroming a World of Memories' in *Cosmopolitics of the Camera*, 261–291; Amad, *Counter-Archive*, 96–132; Winter, *Dreams of Peace*, 18–19.

24 Rosen, *Change Mummified*, 106.

25 Ibid. It is curious that Rosen mentions Bergson in passing only once in his study, not only because Bergson constitutes a major influence on Bazin's formulation of time but also because Bergson's *durée* was itself a symptom of the nineteenth-century conception of temporality which is the focus of Rosen's book.

history that is gradually unfolding'.[26] He noted that temporal 'succession is an undeniable fact, even in the material world' and as such, organic and inorganic entities cannot sidestep 'the irreversibility of time'.[27]

Though perhaps with an added dose of anxiety, Bergson's universal temporalisation of reality informed Brunhes's human geography and the Kahn Archive in their battle against time to capture vanishing phenomena. Yet unlike the philosopher, who famously decried recording media due to their transfiguration of the ever-dynamic *élan vital* – 'the continual *change* of form' – into immobile snapshots, Kahn and Brunhes had no qualms about regarding the same media as the perfect means by which to preserve life as a soon-to-be past that could be relived indefinitely in the future.[28]

The origins of the Archives de la Planète hark back to a round-the-world trip on which Kahn embarked with his chauffeur Albert Dutertre in 1908–1909, which covered the United States (notably New York and San Francisco), Japan, China, Singapore, Sri Lanka and Italy. Dutertre documented the trip via film, photography and especially stereography, the last a popular recording medium among travellers at the time due to its small plate sizes, which required little light input and short exposure times.[29] Upon his return, Kahn hired his first professional camera operators: Stéphane Passet and Auguste Léon, the last a former postcard photographer who subsequently accompanied Kahn on a two-month trip to South America in 1909. Stereography was dropped as the medium of choice in favour of the autochrome, an expensive colour photography that used plates coated with dyed starch grains. Patented and commercialised by the Lumière brothers in 1907, the autochrome required an atypically lengthy exposure time, which, as seen in the Kahn picture collection, made it suitable only for immobile objects and required individuals to pose for the camera for long periods so as to prevent blurs.

The official beginnings of the Archives de la Planète, however, coincide with the appointment of Jean Brunhes as its director in 1912. In exchange for accepting the role, Brunhes, a disciple of Paul Vidal de la Blache's 'human geography', was given a Chair at the Collège de France, personally funded by Kahn himself. Film was then recruited as the twin recording technology of the project to compensate for the autochrome's inability to capture and

26 Henri Bergson, *Creative Evolution*, trans. Arthur Mitchell (New York: Random House, 1944), 9.

27 Ibid., 12, 21.

28 Ibid., 328, emphasis in original.

29 See Michel Poivert, 'Les yeux d'Albert Kahn: les photographies d'Albert Dutertre durant le voyage autour du monde (1908–1909)' in *Les Archives de la Planète*, 358–373.

inhibition of motion. Kahn's initial idea for the role of director was the physical geographer Emmanuel de Margerie, who, when approached, suggested Brunhes given his use of photography as a geographical tool for empirical and field observation, as seen in publications such as *Human Geography* (*La Géographie humaine*, 1911) and *Photographic Atlas of Terrestrial Relief Forms* (*L'Atlas photographique des formes du relief terrestre*, 1908–1914). As de Margerie confided to Brunhes, he had been sure to show Kahn the latter book and to tell him that, in the field of photography, Brunhes was a 'highly experienced practitioner and, moreover, a friend of Mr. Gaumont' (founder of the film company).[30] De Margerie added: 'it is above all the living and human element that interests [Kahn]', thus emphasising the 'ethnographic [as opposed to geographic] dimension' of the project, even if, he concluded, Brunhes should 'banish all fears' and accept the invitation.[31]

In some important ways, however, the 'living and human element' was already a substratum of Brunhes's own human geography, although, admittedly, that discipline was more concerned with human *traces* on the Earth's crust than with humans themselves. Already in the first edition of *Human Geography*, Brunhes invites the reader to 'go up some hundreds of yards above the earth in a balloon or an aeroplane, somewhat after the fashion envisaged by the Swiss [*sic*] geologist Suess at the beginning of his great book *The Face of the Earth*'.[32] Brunhes goes on: 'From this imaginary point what shall we see? Or, better still, what are the human facts that a photographic plate would register just as well as the retina of the eye?'.[33] As Jeanne Haffner has shown, in the first decades of the twentieth century, aerial photography was embraced by an array of disciplines such as social sciences, architecture, history, ethnography and human geography thanks to 'its promise to offer new insight into the study of humankind and its relationship to the environment'.[34] In this context, as Marie-Claire Robic notes, Brunhes's approach must be viewed within an 'ecological' rather than 'chorological' French geographical tradition, one that 'inscribed geography in the problematic of the interaction between the biosphere and the physical environment', and attempted to measure '"the traces of man" on the face

30 Delamare and Beausoleil, 'Deux témoins de leur temps', 92.

31 Ibid., 92.

32 Eduard Suess was Austrian. Jean Brunhes, *Human Geography*, abridged edition, trans. Ernest F. Row (London: George G. Harrap, 1952), 36.

33 Ibid.

34 Jeanne Haffner, *The View from Above: The Science of Social Space* (Cambridge, MA, and London: MIT Press, 2013), 20–21.

of the earth': 'the destruction of nature, the attacks on the integrity of the planet'.[35]

Brunhes's geographical programme anticipated two underlying principles of the Kahn Archive. The first is the Bergsonian temporal postulate that both human and physical forces are mutually reinforcing 'in a state of perpetual change'.[36] Brunhes writes: 'there is constant progress and recession [...] human phenomena, like all terrestrial phenomena, never remain the same'.[37] In fact, as Robic notes, the idea to record the force of time on material reality goes back to Brunhes's early work on the morphogenesis of rivers and water flows, as well as his study of 'perishable forms' in human habitation as 'the mark of the ephemeral'.[38] These included 'swiftly abandoned earthen ruins' and 'the gradual disappearance of wooden structures', both of which he recorded when travelling through Egypt in 1897–1899.[39]

Second, human geography's axiom of global interconnection prefigured the Archive's planetary aspirations. Brunhes notes: 'It is not enough to study separately various set of phenomena, for they are not really separate: they are closely connected with one another'.[40] Here, it is 'the great savant, observer, and innovator Alexander von Humboldt', discussed in chapter 1, that looms large as an influence.[41] Brunhes speaks of the planet in the Humboldtian sense of 'a system of harmonious order', 'the close interdependence of the different parts of the earth', and 'the concept of the "terrestrial whole", or the unity of the entire globe'.[42] As Brunhes saw it, human geography's intervention in this field was to place a greater emphasis on human actions as 'part of a wider whole' in terms of a dynamic interaction between human and geographical elements: 'we are entitled and obliged to add to those material forces which we have seen continually in action this new force, not entirely of a material nature, though it shows itself in material effects – the force of human activity'.[43]

35 Marie-Claire Robic, 'La géographie dans le mouvement scientifique' in *Jean Brunhes*, 52–65; 56, 64.

36 Brunhes, *Human Geography*, 20.

37 Ibid.

38 Marie-Claire Robic, 'Administrer la preuve par l'image: géographie physique et géographie humaine' in *Jean Brunhes*, 221–233; 229; Marie-Claire Robic, 'Un laboratoire: l'Egypte' in *Jean Brunhes*, 164–180; 175.

39 Robic, 'Administrer la preuve par l'image', 229.

40 Brunhes, *Human Geography*, 23.

41 Ibid., 26.

42 Ibid., 21, 26.

43 Ibid., 28.

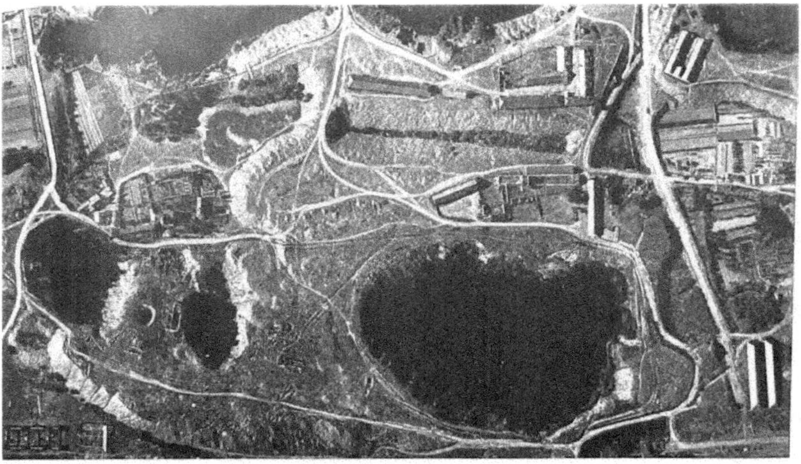

FIG. 96. SLATE-QUARRIES AT TRÉLAZÉ, NEAR ANGERS

Man's exploitation of minerals is marked by great holes in the earth—the mines they dig, the oil-wells they bore, and the quarries they work. Here are seen the gaping holes made by the extraction of slate. Man takes wealth from the earth without replacing it: that is what we have called 'destructive economy.' This photograph illustrates the point particularly clearly.

[*See pp.* 38, 147.]
Photo Marcel Chrétien

Figure 6.1 Aerial view of a 'gaping hole' in *Human Geography* bears an uncanny resemblance to contemporary environmental imagery, as seen in *Earth* (2019) discussed in chapter 1 (figure 1.6). Image reproduced from item held at the British Library.

It is hard not to think of current Anthropocene discourses when reading these lines, even if one must be wary of anachronisms and remember that the geological periodicity of the Anthropocene as a scientific designation is obviously missing in Brunhes's proto-ecological explorations of the relationship between humankind and the planet. Moreover, as Trond Erik Bjorli and Kjetil Ansgar Jakobsen note, Brunhes distinguished between 'nature as such [including glaciers and the atmosphere] and nature as product of human action', a distinction which, according to the Anthropocene, is no longer tenable given that human action is now widely considered a natural force in a geological sense.[44] Nevertheless, in its levelling out of geophysical, morphological and human phenomena as reciprocally influential and co-constitutive, human geography prefigures Anthropocene tropes, concepts and figures in important ways.

Brunhes classified human activities on the Earth according to three criteria: 'A) those connected with the unproductive occupation of the soil', such as the construction of houses and highways; 'B) things connected with the conquest of the plant and animal worlds', such as the cultivation of fields

44 Trond Erik Bjorli and Kjetil Ansgar Jakobsen, 'Introduction' in *Cosmopolitics of the Camera*, 1–23; 16.

and the domestication of animals; and finally, 'C) the destructive economy', as seen in 'mineral exploitation', where 'men have removed the stone for their own use, without making any restitution', and the 'destruction of plants and animals', including 'trees cut down, and forests burned'.[45] In order to illustrate these types of activities, the third edition of *Human Geography*, published in 1925, makes recourse to a number of panoramic and aerial photographs, some of which uncannily foreshadow contemporary environmental imagery. A case in point is the aerial picture of a slate quarry with 'gaping holes' (figure 6.1), which bears a striking resemblance to the satellite shots opening individual episodes in Geyrhalter's *Earth* (explored in chapter 1).[46] Whether through its concern with the dissipation of matter as a result of time's entropic force, or with the degradation of the natural environment as the product of human-induced change, the world envisaged by human geography was therefore always already disappearing: hence the need for empirical observation and the employment of recording technologies to meet that need.

Yet the Archives de la Planète had a reciprocal impact on Brunhes. Foremost among these, as anticipated by de Margerie, was the ethnographic dimension that Kahn's Archive brought to bear on human geography.[47] Perhaps in an attempt to justify what might have been regarded as disciplinary promiscuity, in 1913 Brunhes published an article in the journal *L'Ethnographie*, entitled 'Ethnographie et Géographie humaine', in which he argues that rather than being opposed to each other, both disciplines in fact 'complemented [...] and must help each other'.[48] This conflation allows Brunhes to extend the ecological understanding of 'reciprocal actions between the natural environment and humanity' to the idea of a globalisation that, in tune with Kahn's views, leads to cultural homogenisation and eventually disappearance.[49] Brunhes writes:

> At the moment when, as a result of the intensity and multiplicity of communications, all parts of the universe are violently brought into relation

45 Brunhes, *Human Geography*, 37–38.

46 Ibid., plate XXXIV, fig. 101.

47 That said, both projects also retained differences in their visual approaches. As Didier Mendibil notes, when compared with the autochromes, the 'framing of the photographs of *Human Geography* is not as close and the "landscape views" are twice as frequent'. See Didier Mendibil, 'À la recherche de l'effet Brunhes' in *Les Archives de la Planète*, 130–139; 133. See also Franziska Scheuer, 'Autochromes in Service of Human Geography: Jean Brunhes and the Aesthetics of The Archives of the Planet' in *Cosmopolitics of the Camera*, 238–260.

48 Brunhes, 'Ethnographie et Géographie humaine', 32.

49 Ibid.

> with one another, humanity is tending towards a certain uniformity of customs, needs and daily habits. The economic and geographic fact of circulation (the *Verkehrskultur*, as the Germans say) plays such a predominant role that, in the global competition, the former small economic units which lived their own life, with complete independence, producing roughly all they needed, are facing an increasing threat to their existence on a daily basis and will soon disappear.[50]

In this context, Brunhes goes on, the aim of the Archives de la Planète was to establish a 'record of humanity captured in the midst of life at the dawn of the twentieth century'.[51]

On the one hand, Brunhes's repeated invocation of figures of vitalism to describe the Archive is connected with the medium of film, which, thanks to its properties of motion, permits 'an exact and perfect register of that which belongs to the domain of life'. Thus, he goes on, the Archive's aim is to 'to capture and preserve planetary occurrences that are going to die out', thus offering a 'portrait of real life in our time [...] for those who will come after us'.[52] Whereas ethnographic museums and herbariums displayed work utensils and 'dead plants' cut off from their environments, film enabled these elements to be 'reassembled and brought to life, as they are in reality'.[53] On the other hand, Brunhes devotes equal attention to the Archive's other medium, the autochrome. He justifies its application on the basis of its 'rigorously exact' rendering of colour (as seen in traditional garments) and the idea that it can capture 'the truth' of human behaviour in their contextual environment, in contrast with the isolated human figures and types in 'wax casts' or 'posed photographs' in ethnographic museums.[54]

It is worth pausing for a moment over Brunhes's reflections. First, his acknowledgement that the Archive is for 'those who will come after us' reveals that it comes into full fruition only when considered from its point of reception, that is, an envisaged spectator in the future. Second, the static autochrome was for him not to be set against the motion picture, for both technologies put *life* into the reproduction of reality: the one because of movement and the other because of colour. Finally, Brunhes's pejorative allusions to anthropometric photography, wax casts and dead plants evince

50 Ibid., 37.

51 Ibid., 38

52 Ibid.

53 Ibid.

54 Ibid., 39. For an analysis of the role of traditional vestments in the Kahn Archive, see Valérie Perlès, 'Les types en costume: de l'ethnologie à la politique' in *Les Archives de la Planète*, 182–197.

his concern with situating the Archives de la Planète in what Stephen Bann has defined as 'a continuum of forms of historical representation' in nineteenth-century media culture. For Bann, these forms, which included painting, taxidermy and dioramas, promised 'effects of resurrection' that betrayed a formulation of time as threat.[55] Whereas tropes of resurrection and afterlife received a new boost with the advent of cinema and were recurrent and even commonplace in early film culture and theory, what is striking in Brunhes's remarks is the self-consciousness with which he locates the Archives de la Planète within a wider cultural history in order to bolster, by contrast, the increased realism of its own films and colour photographs in their vivifying effect.[56]

Yet how was the 'domain of life' visualised, after all, in the Kahn films and photographs? At the project's dawn, a notable concern on Kahn's part was to record rural communities due to the view these were the first to vanish in the age of large-scale industrial capitalism.[57] Even before Brunhes's appointment, Auguste Léon had been dispatched to villages in rural Sweden and Norway in 1910, with some of the autochromes depicting the country dwellers in local garments posing individually or in groups outside their traditional houses and workplaces. One photograph, for example, shows five women sitting and standing outside the hotel where they worked (online catalogue entry: A286).[58] Another portrays five children on the stairs of their house, with imposing pillars behind (A458). Bearing in mind the diversity of images in the Kahn Archive, the visual set-up of locals posing frontally outside their homes or workplace, in traditional costume and staring fixedly into the camera is a dominant one across the pictures of peasant life. Given the long exposure required, the autochrome process demanded immobile poses not unlike the earlier daguerreotype, in whose resulting portraits Benjamin detected an 'inhuman, one might even say deadly' quality thanks 'to the (prolonged) looking into the camera'.[59] Something similar can be discerned in many of Kahn's autochromes, and their 'deadly quality' gains further significance in the knowledge of what motivated their taking.

55 Stephen Bann, *The Clothing of Clio: A Study of the Representation of History in Nineteenth-Century Britain and France* (New York: Cambridge University Press, 1984), 136.

56 For an account of these tropes, see Amad, *Counter-Archive*, 139–141.

57 Michel Lesourd, 'L'appropriation du monde' in *Jean Brunhes*, 15–51; 36.

58 Number in parentheses after films and autochromes throughout this chapter refer to their catalogue entry on the online collection: http://collections.albert-kahn.hauts-de-seine.fr/.

59 Walter Benjamin, *Illuminations*, trans. Harry Zohn and edited by Hannah Arendt (New York: Shocken Books, 2007), 188.

In their intimation of a world that had remained unscathed by the forces of modernity, Kahn's images of pastoral life suspended in time recall what Rosen defines as the 'just before' temporality of so-called living history and museum villages in their reconstruction of a rural past that is about to be swept away. Looking at Colonial Williamsburg in Virginia and Greenfield Village in Michigan, conceived by John D. Rockefeller Jr. and Henry Ford respectively in the 1920s, Rosen notes the discrepancy between the 'historical "just before"' capitalism figured in these museums and the fact they were supported by 'major industrial and finance capitalists'. For Rosen, this begs analogies with 'the "just before" in Freud's much-cited account of fetishism' as a threatening knowledge that must be continuously averted through recourse to an 'obsessive overvaluing' of an object, in this case the past itself.[60] The prevalence of pre-industrial temporalities in Kahn's Archive, itself brought to fruition by a finance trader, participates in this problematic and activates similar defensive mechanisms.

However, the Kahn Archive cannot be reduced to its penchant for rurality. Equally prevalent in its holdings is modern, street life captured on film in urban settings, especially Kahn's beloved Paris. Trond Lundermo has examined the prevalence of crowds in the Archive films, which often register the hustle and bustle of cities, or the 'rhythm of life', via static long takes, panoramas and tracking shots.[61] Indeed, unlike the contemporaneous, highly trimmed 'actualities', the majority of Kahn's films are unpolished rushes that relied on on-camera montage and unbroken shots. In turn, this preference for 'uninterrupted location shooting' has been queried by Amad in terms of a correspondence with 'Bergson's *durée réelle* (time as human awareness experiences it)', and consequently André Bazin's 'own Bergsonian [...] "cinema of *la durée*"'.[62] For Amad, however, 'the idea that Kahn's films somehow materialize the continuity of *la durée*' must be approached with caution, given that 'they focused just as much on newly emerging rhythms, ruptures, and experiences specific to modernity as on the inevitable disappearance of more traditional modes and practices of human activity'.[63]

And yet, from a different angle this dual focus does not so much attest to a dichotomy as evince that, within the Archive's conception of time, both phenomena – the new experience of modernity and the disappearance of

60 Rosen, *Change Mummified*, 71, 73.

61 Lundermo, '"The Archives of the Planet" and Montage', 211–216. See also Teresa Castro, 'Vues panoramiques et travellings: Les Paysages en mouvement des Archives de la Planète' in *Les Archives de la Planète*, 390–399.

62 Amad, *Counter-Archive*, 104.

63 Ibid.

traditional cultures – are two sides of the same temporal coin: a 'history that is gradually unfolding', to cite Bergson himself.[64] In this sense, Bazin's famous definition of film as 'change mummified', according to which 'for the first time, the image of things is likewise the image of their duration', is a useful concept to bear in mind.[65] As Rosen has shown, for Bazin, 'reality is definitionally temporalized, in the sense that it always involves change', a foundation of his theory that must be treated itself historically, that is to say, in the context of 'a preservative impulse or obsession in the modern West' that started to gain traction in the nineteenth century.[66] In many ways, the Archive was both part and parcel of this impulse to capture *change* as it happened in flux, and both Kahn and Brunhes agreed on the 'urgency of the work to be accomplished' in order to safeguard 'the memory of a reality in the process of disappearing'.[67]

This urgency resulted not only from technological transformations but also from a sense of instability related to continuous war and a rapidly changing world order. As Winter argues, Kahn's liberal brand of internationalism, according to which peace was the precondition for free trade, cannot be dissociated from a 'backdrop of worldwide violence', with the 'decade from 1895 to 1905 [...] marked by armed conflict in every continent'.[68] At the same time, Kahn's awareness that 'the machinery of progress [could] all too easily become engines of destruction and disaster' meant that it was imperative to record geographic areas on the brink of cataclysmic sociopolitical change, as proved by the Archive's first missions under Brunhes's direction.[69]

In 1912 Passet was dispatched to China and Morocco, the former because of the rise of the nationalist movement of Sun Yat-sen, the latter due to the fact it had become a French protectorate.[70] In that same year Brunhes himself travelled to Bosnia-Herzegovina in his first mission with Léon and in May 1913 to Macedonia, a few months prior to the outbreak of the second Balkan war. The aim of these missions was to document the ethnic diversity of the region before it was too late, as nationalist factions had begun to scramble for sovereignty. The resulting autochromes, which were used in

64 Bergson, *Creative Evolution*, 9.

65 André Bazin, *What is Cinema?*, volume 1, trans. and edited by Hugh Gray (Berkeley, Los Angeles and London: University of California Press, 2005), 15.

66 Rosen, *Change Mummified*, 6, 46.

67 As noted by Marie Bonhomme (with Mariel Jean-Brunhes Delamarre), 'Le champ du monde' in *Jean Brunhes*, 181–194; 181.

68 Winter, *Dreams of Peace*, 12.

69 Ibid., 18.

70 Bonhomme, 'Le champ du monde', 182.

Brunhes's first illustrated lectures at the Collège de France and attested to the co-existence of peasant Muslims, Jews and Christians, soon became visual records of a reality that vanished months after the moment of its registration.

As soon as the war ended, it was Passet's turn to travel to the region, in September 1913, to document the ravaged countryside and the displacement of thousands of Muslims. One especially poignant picture, taken in today's Bulgaria, shows a group of women, some carrying babies, posing for the camera amid the rubble, with only the skeleton of a house in the background, two thin walls barely standing (A3904). Other films and autochromes document a nefarious refugee camp near Salonika: wooden carts, covered with rags, turned into makeshift houses (A3847); a man drinking a cup of coffee seated on the ground (A3863); a woman cooking on the floor, with blurred children looking into the camera in the background (A3869). Documents of everyday life in unprecedented and deadly times, these images of mundane occurrences against the background of forced displacement and conflict foreshadow the Archive's painstaking coverage of WWI on French soil in the years to come.

An Inhuman Planet

The Great War constituted a watershed during the Archive's existence. As Valérie Perlès tells us: 'Between 1914 and 1918, 50% of the production of images was devoted to the war and the subject matter was recurrent until 1930, to the point of representing today 20% of the entire collection'.[71] This turn to the national was not inconsequential or neutral, as recent French scholarship (Perlès's work included) has noted in relation to the many roles the Kahn Archive played in generating public support for the Allies. As Anne Sigaud has shown, from December 1914 to August 1916, thanks to Kahn's contacts in the army, nearly 1,500 war-related photographs were produced in missions in northeastern and central France, in two of which Brunhes himself participated and used the pictures in lectures at the Collège de France in order to garner public support against the Germans.[72] In 1917, Kahn struck a deal with the Section Photographique de l'Armée (SPA), recruiting the army photographers Fernand Cuville and Paul Castelnau,

71 Valérie Perlès, 'Des images pour quoi faire?' in *Réalités (In)Visibles*, 18–26; 25.

72 Anne Sigaud, 'Entre documentation et propagande: vocation et usage politiques des Archives de la Planète' in *Les Archives de la Planète*, 280–291; 282.

who started taking two plates for each autochrome, one for the Archive and the other for the army.[73]

That a pacifist such as Kahn would collaborate with the army and become involved in war propaganda raises some thorny questions, some of which can be illuminated by a pamphlet he wrote between March 1917 and July 1918. Published in December 1918 and entitled *Des droits et des devoirs des gouvernements* (On the Rights and Duties of Governments), it argued for a supranational federation to safeguard peace for future generations, in many ways prefiguring the creation of the League of Nations (1920), of which Kahn became an ardent enthusiast. As Frédéric Worms has shown, Kahn's manifesto has the imprint of Bergson all over it, both in its rhetorical recourse to vitalist figures and in 'the opposition between France and Germany', which is mapped on to 'an opposition between Spirit and Matter, Life and Death'.[74]

Whereas Kahn's distress over the war is palpable – 'the storm that shaking the Earth touches every living being' and 'destroyed all dreams of humanity' – humanity itself is measured against the Germans, in turn defined as 'the negation of all morals, of all justice'.[75] Kahn asks: 'The question we must answer is: will it be the German sword or humanity that will triumph?', then adding: 'For the first time, a group of individuals set their sights on the entire human race; for the first time, all of humanity, at the same time, is threatened with the same fate, a fate that we dare not reveal'.[76] Kahn's allusion to total death formulates a 'negative universalism' that is, paradoxically, segregational from the outset in its exclusion of Germans.[77] In the following decade, the Archive's cosmopolitan credentials would suffer other blows, with Brunhes increasingly aligned with imperial ideology and interests and Kahn himself embracing some aspects of French colonial expansion by the late 1920s.[78]

73 Anne Sigaud, 'Des Archives de la Planète aux archives de la guerre: "Réalités invisibles, notre raison doit les discerner"' in *Réalités (In)Visibles*, 30–54; 34; Perlès, 'Des images pour quoi faire?', 23.

74 Frédéric Worms, '"Agir en homme de pensée, penser en homme d'action?": L'action e la pensée de Kahn et Bergson à l'épreuve du siécle' in *Henri Bergson et Albert Kahn*, 111–126; 120.

75 Albert Kahn, *Des droits et des devoirs des gouvernements* (Paris: Imprimerie de Vaugirard, 1918), 11.

76 Ibid., 3, 24.

77 I borrow the term 'negative universalism' from Dipesh Chakrabarty's 'The Climate of History: Four Theses', *Critical Inquiry* 35: 2 (2009): 197–222; 207.

78 See Marie-Claire Robic, 'Regard géographique et intelligence du monde: Jean Brunhes, la géographie humaine et les Archives de la Planète' in *Les Archives de la Planète*, 84–96; 93–94; Sigaud, 'Entre documentation et propagande', 284, 286, 288. See also Amad, *Counter-Archive*,

In her study of the Archive's involvement in war propaganda, Sigaud points out that the 'photographic motifs are perfectly consistent thematically and stylistically with the numerous illustrations in the contemporaneous pamphlets denouncing the German "barbarism"', with France placed 'in a position of legitimate self-defence against a Germany presented as solely responsible for the war, deliberately destructive'.[79] Taken with the aim of boosting morale, the war autochromes were put to the service of patriotism, which in turn meant that certain subjects, such as the tens of thousands of corpses produced by warfare, are often nowhere to be found. By the same token, soldiers frequently appear in dignified and pensive positions; in groups denoting camaraderie in the trenches and the camps; or else engaged in quotidian errands: seated at a table having lunch (A005935); sleeping inside a tent (A011030); writing in a notebook (A010999) – all photographed during Passet's mission in 1915–16.[80]

The majority of the pictures in the war collection, however, are of dilacerated towns and ruinous buildings. These too can be linked to propaganda, given that destroyed national monuments and especially cathedrals played on national and religious sentiment as symbols of French cultural heritage.[81] Yet, as I explore in the remainder of this section, the prevalence of ruins in the Kahn Archive also begs to be examined in relation to its foundational concept of disappearance, which was now visualised in concrete reality and as the absence of humanity rather than as an invisible temporal force traversing the 'domain of life'. Moreover, these images of destruction are significant when explored next to the films of natural catastrophes acquired and produced by the Kahn Archive, which, I suggest, further confirms its attachment to a conception of time as unpredictable rupture.

To be sure, the Archive's ethos of extraordinary everydayness still prevailed during the war. Many autochromes show residents posing outside their house remnants, and a busy Paris was similarly the sustained object of filmic attention – no doubt, as Teresa Castro has astutely shown, as a way of proclaiming in patriotic mode that life defiantly went on in the capital.[82]

265–268 for an account of Kahn's Archive and its varied if not necessarily direct connections with colonialism.

79 Sigaud, 'Des Archives de la Planète aux archives de la guerre', 38.

80 For an account of the Archive's autochromes from WWI, see also Emmanuelle Danchin, 'The Archives of the Planet and the First World War' in *Cosmopolitics of the Camera*, 98–113.

81 See Claire Maingon, 'Dans l'oeil de la Grande Guerre: Esthétique et culture visuelle' in *Réalités (In)Visibles*, 157–167; 164.

82 Teresa Castro, 'Le Quotidian est politique: Le cas des films 1914–1918' in *Réalités (In)Visibles*, 188–192.

At the same time, human-free landscapes and cityscapes are anything but unusual in the Kahn Archive, many of which can be linked to the precepts and influence of human geography.[83] Yet the specific configuration of ruins and lack of humanity prevalent in the war pictures is especially noteworthy, thus requiring some unpacking in relation to questions of temporality and the rhetorical figures of life and death so dear to both Kahn and Brunhes.

As Rosen notes, ruinous buildings were favourite objects of contemplation in nineteenth-century culture because they were deemed to crystallise in their very structure the 'sheer continuousness of the passage of time', hence the common recourse to romantic, organic metaphors to characterise those buildings as indeed 'living' in time.[84] This idea is perhaps best exemplified in Georg Simmel's famous 1911 essay 'The Ruin', which he defines as 'the site of life from which life has departed'. Simmel adds: 'In the case of the ruin, the fact that life with its wealth and its changes once dwelled here constitutes an immediately perceived present'.[85] This statement, in turn, resonates with Bazin's aforementioned concept of 'change mummified', as noted by Johannes von Moltke in relation to what he theorises as 'cinema's and the ruin's common function to visualize time and history in modernity'.[86]

During the war, however, the Kahn Archive mostly produced still rather than moving images of ruins. Moreover, when it comes to these autochromes of bombed-out towns there is a further difference related to the fact that they are not the product of the slow passage of time privileged by nineteenth-century romanticism or even Brunhes's own geographical interest in 'perishable forms' discussed earlier. Rather, they are evidence of an event that had produced a radical disjuncture in the course of time. For Simmel, ruins derive their aesthetic pleasure and epistemological appeal from the fact 'that here the work of man appears to us entirely as a product of nature', much like the 'forces which give a mountain its shape through weathering, erosion, faulting'.[87] By contrast, the ruination obsessively documented in the Kahn autochromes cannot be dissociated from human engineering and the industrial barrage of artillery, thus making it less akin to the workings of time on geographical forms and more in tandem with the

83 See Mendibil, 'À la recherche de l'effet Brunhes', 130–139.

84 Rosen, *Change Mummified*, 50.

85 Georg Simmel, 'Two Essays', *The Hudson Review*, 11:3 (1958): 371–385; 384–385.

86 Johannes von Moltke, 'Ruin Cinema' in Julia Hell and Andreas Schönle (eds.) *Ruins of Modernity* (Durham, NC, and London: Duke University Press, 2010), 395–417; 396.

87 Simmel, 'Two Essays', 381.

type of 'destructive economy' outlined by Brunhes.[88] And indeed, Kahn's war collection holds such crucial historical importance today because it comprises the first wide-ranging documentation of a mode of destruction the scale of which had never been catalogued or experienced before. Here, Simmel's definition of the ruin thus necessitates a qualifier in the form of an adverb: it is the site of life from which life has *abruptly* departed.

Yet this does not mean that these images are removed from the aesthetic fascination historically identified with ruin gazing.[89] Indeed, as Julia Hell and Andreas Schönle note, via Benjamin, as much as ruins 'invite reflections about history: about the nature of the event, the meaning of the past for the present', the 'aestheticization of ruins is unavoidable'.[90] Beautifully framed and symmetrically arranged, these are remarkable images of evacuated towns and empty streets where houses, buildings, churches and shops have become material vestiges of their former selves: roofless skeletons, windowless façades, sheer rubble (A4841 and A5408). Human presence is conspicuous by its absence, as in the many wooden carts unattended in the middle of streets and presumably left behind, a recurrent visual motif denoting rushed departures.[91] Of course, the photographers may well have arranged objects in the same way that the absence of human figures was sometimes an aesthetic choice. An example is a couple of near-identical autochromes of a wrecked house in the commune of Crévic, taken by Georges Chevalier in 1915, with the only difference between the two pictures being that one features its likely dwellers sitting on a pile of rubble in the foreground (A005361) and the other no humans (A073109). But human evacuation was also a reality, for thousands of civilians had indeed died or fled the German offensive, an example being the town of Reims, where only 14,000 inhabitants remained out of 120,000 and whose near-obliteration was documented by Cuville and Castelnau in 1917.

Whatever political, aesthetic or propaganda purposes these human-free autochromes served, they therefore also need to be understood in relation to Kahn's concern that 'for the first time, all of humanity, at the same time, is threatened with the same fate'. Initially conceived as a 'record of humanity captured in the midst of life at the dawn of the twentieth century', during

88 As also noted by François Amy de la Bretèque, 'Les Archives de la Planète "Régions Dévastées", un regard humaniste et sa place dans l'histoire du cinéma' in *Un monde et son double,* 235–248; 245.

89 See, in this respect, Brian Dillon, *Ruin Lust: Artists' Fascination with Ruins, from Turner to the Present Day* (London: Tate Publishing, 2014), 5–12

90 Julia Hell and Andreas Schönle, 'Introduction' in *Ruins of Modernity,* 1–14; 1.

91 See, for example, A004797; A005309; A005366 in the online catalogue.

the war the Archives de la Planète becomes a visual laboratory of what a world without humanity might look like.[92] As Amad has noted, the war violently reminded Kahn of the 'unpredictability of the present', lending 'an even more apocalyptic urgency' to his archival project.[93] Though Kahn unequivocally attributed responsibility to the Germans in his pamphlet, his idea of a supranational 'world union' was also meant to counter what he called the forces of 'arbitrariness and chance', which could potentially erupt and endanger the prospect of planetary peace.[94]

Kahn's anxiety over these forces is in tune with modern notions of temporality according to which time is always already change, instability, disorder: whether a continuously gradual or a brutally unexpected one. As Mary Ann Doane observes, 'modernity is [...] strongly associated with epistemologies that valorize the contingent, the ephemeral, chance', which become complex sites of pleasure and anxiety, since '[c]hange becomes synonymous with "newness," which, in its turn, is equated with difference and rupture'.[95] The centrality of WWI ruins in the Kahn Archive cannot be dissociated from this conception of historical rupture and attendant notions of time as an unceasingly differentiating and potentially disordering force, especially when we consider that the ruins were occasioned not by natural processes but modern industrial warfare.

This fascination with ruins persisted in the Kahn Archive until its end, with Lucien Le Saint and Camille Sauvageot intermittently recruited to film and photograph the extent of war destruction on French soil and the reconstruction until the early 1930s. In 1922, Sauvageot was dispatched to Turkey with Frédéric Gadmer to register the harrowing aftermath of the Greco-Turkish war and the Great Fire of Smyrna, which extinguished much of the port city and forcibly expelled thousands of Greeks and Armenians. As with the WWI collection, the resulting films and autochromes depict a quasi-lunar landscape of demolition and dispossession. In the 5-minute film *Casaba et Smyrne* (Casaba and Smyrne, AI90577), for example, the utterly destroyed city of Smyrna is recorded in circular pans, with the odd human figure or dog glimpsed amid the rubble. Similarly, in the hundreds of autochromes taken by Gadmer, human presence is often nowhere to be seen and the impression is that of a world after the apocalypse, as in the striking

92 Brunhes, 'Ethnographie et Géographie humaine', 38.

93 Amad, *Counter-Archive*, 149, 200.

94 Kahn, *Des droits*, 9.

95 Mary Ann Doane, *The Emergence of Cinematic Time* (Cambridge, MA, and London: Harvard University Press), 20.

picture of a derailed train standing diagonally over a destroyed railroad bridge, following on from the Greek Army's retreat from Turkey (A037214).

These images of human-induced destruction acquire further significance when placed in relation to those of natural phenomena and calamities in the Kahn Archive. In some respects, the presence of the latter can be attributed to Brunhes's human geography, one of the pillars of which, as we saw earlier, was the reciprocal relationship between human activity and nature. Although ethnographic matters did get the upper hand over more properly geographical ones in the Archive, soon after his appointment in 1912 Brunhes trained a camera operator to record geographical phenomena, while dispatching his student Cesare Calciati to an excursion across the US organised by the American Geographical Society, with two resulting films depicting geysers in action in the Yellowstone National Park and the Niagara Falls.[96]

Yet even before Brunhes the Archives had already showed a propensity for natural disturbances. One needs only consider the 1906 San Francisco earthquake ruins extensively photographed by Albert Dutertre during his trip with Kahn. Just as remarkably, the very first film catalogued in Kahn's Archive in 1910, titled *Inondations des quais devant la Cercle Autour du Monde* (Flooding of Docks Outside the Autour du Monde Society, AI48261), records in an unbroken shot the Autour du Monde members leaving Kahn's mansion on small boats, because the Seine had overflowed. In turn, this film makes a companion to the longer, nine-minute *Inondations* (Floods, AI 48953), filmed by Léon or Dutertre in that same year. *Inondations* documents the floods in the wider Boulogne region, complete with shots of crammed chariots and carts wheeling through the water, small boats coming and going, and agglomerations in critically flooded areas. Floods in Paris and surrounding regions also appear in many films produced for the Archive, as do glaciers. One film, shot in 1924 by Sauvageot and titled *Crue de la Seine* (The Seine Flood, AI48955), begins with the remarkable shot of lampposts almost entirely submerged in water.

Natural disasters are equally the topic of a select number of actualities acquired by Kahn, many from Gaumont, namely: *Tremblement de terre* (Earthquake, AI93248), filmed in Santa Barbara, US; *Tremblement de terre sur l'île de Shikoku* (Earthquake in Shikoku, AI12613), in Japan; *La Ville détruite par le cyclone* (The Cyclone-Ravaged City, AI93244), in Miami, US; *Eruption du Mauna-Loa* (Mauna-Loa Eruption, AI93255), in Hawaii; and

96 Valérie Perlès, 'La création d'un corpus filmique inédit aux origines des Archives de la Planète' in *Les Archives de la Planète*, 102–115; 108.

Eruption de l'Etna (Mount Etna Eruption, AI45301), in Italy. Unlike the Kahn films, these are highly edited and short, and yet they are similarly intent on showing wreckage in the wake of natural catastrophes – and in the case of the *Eruption du Mauna-Loa*, shots of the eruption itself, with smoke and fire dominating the screen. Over a span of 20 years, the Archive acquired 31 films from Gaumont and other French film companies, many depicting landmark ceremonies and festivities.[97] The percentage of acquired films of natural disaster is therefore not irrelevant, and when viewed alongside the films shot in Paris, they evince Kahn's concern with archiving the intrusion of the natural world into the human world by way of the unexpected, calamitous event.

Taken together, these films are in some measure indicative of what Nigel Clark has recently theorised as the 'inhuman nature' of our 'dynamic planet', or what he calls 'the earth's eventfulness', that is, the idea that 'instability and upheaval, rhythmical movement and dramatic changes of state are ordinary aspects of the earth's own history'.[98] Covering an array of meteorological, climactic and geological instabilities in a number of geographical regions across the globe, these films ask us to reconsider the Kahn Archive's construction of planetarity beyond the sociopolitical and the everyday and in terms of the Earth's physical agency and disruption. That is, if the war images attested to the human destruction of the world, these filmic records of floods, glaciers, volcanic eruptions, tornados and earthquakes show that the planet of the Archives de la Planète also acted of its own volition and bit back.

Seen in this light, Dutertre's stereographs of the San Francisco earthquake ruins are significant not only because they are the very first visual documents to enter the Kahn Archive, but also because of the importance which that natural event acquired by means of the novelty associated with its recording and subsequent dissemination worldwide. As Stephen Morgan notes: 'San Francisco 1906 was undoubtedly the first major disaster to be widely photographed by amateurs, but it was also the first to gain widespread coverage through the moving image'.[99] Whereas far deadlier natural catastrophes had shaken the world before, destruction was often relayed by words, not the mediums of photography and film, the indexical

97 Jacques Siracusa, 'Les Archives de la Planète et Gaumont actualités: Les Écarts par rapport à la norme commerciale', *Les Archives de la Planète*, 324–325.

98 Nigel Clark, *Inhuman Nature: Sociable Life on a Dynamic Planet* (London: Sage, 2011), xi.

99 Stephen Morgan, 'A Tale of Two Cities: San Francisco 1906 and *Earthquake in Adelaide*' in Alan Wright (ed.) *Film on the Faultline* (Bristol and Chicago: Intellect, 2015), 21–45; 23.

dimension of which promised a previously unthinkable sense of 'being there'. Unsurprisingly, as Kahn's acquired actualities attest, natural catastrophes became a favourite theme in early cinema as they forcefully testified to the medium's ability to capture the ushering forth of the unexpected and its aftermath. In other words, these early films of disaster evince an obsession with recording the intrusion of difference into the fabric of daily life, the rupturing moment that seismically disorders a world, leaving ruins in its wake. Kahn's produced and acquired films of natural disturbance, together with the Archive's sustained focus on human-driven destruction, participate in this fascination to register the moment in which the ordinary is suddenly, sometimes literally, inundated with extraordinariness.

A Nonhuman Planet

To consider the Archives de la Planète and human geography in the anachronistic light of the Anthropocene helps us see the way in which both projects – one a multimedia archive supported by a finance trader, the other a newly established academic discipline – conceived of the indeterminate ways of the world as it presented itself in the first decades of the twentieth century. As I hope to have shown, change, transformation, disorder and disruption are essential categories for an understanding of Kahn's Archive, the resulting planetarity of which, like that of human geography, postulated a dynamic and sometimes unpredictable interaction between the human and the nonhuman. This is not to say that Kahn or Brunhes articulated an 'Anthropocenic' vision of the world, since, as noted earlier, the Anthropocene presupposes the idea of a human-induced geological intervention whose environmental disturbances are unprecedented and relatively new according to a number of scientific measures and criteria. But it is to say that, as a project, the Archives de la Planète both mirrored and reflected on the precarity of social and natural worlds in the context of a global modernity whose effects were also the cause of fear and anxiety. While invoking concepts of variability, mutability and evanescence, the ruins in Kahn's Archive seem also to embody his fear that the war could wipe humanity off the face of the planet: 'the storm that shaking the Earth touches every living being'.[100]

However, as much as Kahn conveyed in words the fear of such a fate, and as much as the numerous images of ruins in his Archive allow us a momentary glimpse into a world bereft of humans, the Archives de la

100 Kahn, *Des droits*, 11.

Planète remained human-centric when not ethnocentric. After the end of the war, Sauvageot and Le Saint travelled through France to record its ruins on film and, shortly after, the reconstruction. Variously catalogued as *régions dévastées* (destroyed regions), *régions libérées* (freed regions) and *villes reconstruites* (reconstructed cities), these films and autochromes unequivocally subscribe to patriotism in their documentation of national restoration, including the resumption of human life in extraordinary times.

Thus, in the remarkable three-part series *En dirigeable sur les champs de bataille* (Airship over the Battlefield, 1919), the ever-moving aerial perspective over the Western front surveys the staggering scale of the destruction while capturing from high above diminutive human figures going about their business amid the wreckage. In *La ville en ruines* (City in Ruins, AI107664), also shot in 1919, ground-level panoramas scanning the dilacerated commune of Albert allow the unprogrammed intrusion, in its penultimate shot, of a single man pushing a cart traversing the rubble. Though autochromes were also utilised in the immediate aftermath of the war, the properties of film heightened the reawakening of human life in cities as people went about through the wreckage, their movement reinscribing vitality amid dead matter.

Above all, the Kahn Archive is human-centric in that it presupposed a human spectator in the future. Today, Kahn's autochromes are now available in digitised form on our computer interfaces, a turning public of the Archive that comes to fulfill Brunhes's prophecy that it is 'for those who will come after us'.[101] Yet to say that Kahn's images were manufactured with humans in mind may sound redundant, even pointless: aren't mechanically recorded images always produced for and by humans?

Some contemporary projects – including the film *Homo Sapiens* (2016) and the photographic project *The Last Pictures* (2012), discussed below – have attempted to complicate a straightforward answer to this question, no doubt as a response to a widespread sense of precarity fuelled by unrelenting sociopolitical, economic and ecological turbulences. As in the Archives de la Planète, the driving forces behind these projects are fear of disappearance and anxieties over the passing of time. Unlike Kahn's Archive, they envision the worldhood of the world after humanity; the life they depict and/or imagine is of a nonhuman, sometimes extraterrestrial nature, even if they simultaneously make recourse to contemporary ruination and anticipate a human spectator in the present. By 'imaging and imagining a certain future "after the human" from the viewpoint of the here and now', to cite

101 Brunhes, 'Ethnographie et Géographie humaine', 38.

Joanna Zylinska, some of these projects must be therefore viewed as more than so-called 'ruin porn', since their visual power is located not in mere aestheticisation but in a temporal forking that attempts to 'reimagine our relation to the world just before the ruin'.[102]

Consider briefly Tong Lam's photobook *Abandoned Futures: A Journey to the Posthuman World* (2013), which defines itself as a 'global overview of [...] decay and abandonment'. Compiling images recorded in Japan, China and the US, among other countries, Lam justifies his focus on ruination as a march against time. Unlike ancient ruins, which 'deteriorate solemnly over centuries', their industrial and postindustrial counterparts have been abandoned or discarded in decades if not years. The alarming rate at which ruination proliferates makes it plausible to infer that 'there will be a time when ruins and wastelands begin to outnumber functioning spaces'. Lam concludes: 'globally, there are probably already more ruins created in the first decade of the Twenty First century alone than in the entire Twentieth century. Time, in short, is speeding up exponentially'.[103] Yet if *Abandoned Futures* testifies to an urgency to record disappearance, it equally proposes, as per its oxymoronic title, to use indexical records of the now as a means of envisioning and reflecting on what a world without humanity looks like.

Something similar happens in *Homo Sapiens*, by Nikolaus Geyrhalter. Inspired by Alain Weisman's bestseller *The World without Us* (2008), which starts from the premise of a 'world from which we all suddenly vanished', *Homo Sapiens* compiles durational shots of ruinous sites recorded in many corners of the world, many of which are featured in *Abandoned Futures* and possibly recognisable to the viewer.[104] These include the Buzludzha monument in Bulgaria, which bookends the film; the abandoned building blocks in the Hashima Island and the streets of Fukushima in Japan; and the postindustrial cityscape of Detroit in the US. For all intents and purposes, then, *Homo Sapiens* is a documentation of worldwide ruination authenticated through indexical imagery, which situates the film in the present and provokes thought on the world as ravaged by unchecked progress. Wars, consumerism, natural disasters, communism and capitalism – all are haunting absences 'living' in the film's decaying structures.

102 Joanna Zylinska, *Nonhuman Photography* (Cambridge, MA, and London: MIT Press, 2016), 85, 89. On ruin porn, see Dora Apel, *Beautiful Terrible Ruins: Detroit and the Anxiety of Decline* (New Brunswick, NJ, and London: Rutgers University Press, 2015).
103 All citations in this paragraph are from Tong Lam, *Abandoned Futures: A Journey to the Posthuman World* (Carpet Bombing Culture, 2013). The book is not paginated.
104 Weisman, *The World without Us*, 4.

Further, the film's protracted inspection of ruinous sites, with their unexpected arrangements of living and nonliving matter, capitalises on the excessive, uncanny materiality of such places, with their disorderly combination of the strange and the familiar. At one point in the film, for example, we see the remnants of a bar as though going through a process of mutation, with the stools lined along the counter covered with green mould. Another shot shows an oversized plant in a vase inside an abandoned office, the gigantic branches growing on both sides reminding the viewer of the plant's will to live. As Tim Ederson argues, to encounter a world in ruins is to attend to the heightened and hybrid physicality of a new ontology of things:

> Things in ruins [...] are charged with a more radical alterity. Lacking any obvious meaning, feeling different, unclear in their function, aesthetically indecipherable and out of place, these artefacts pose an alternative way of relating to objects that goes beyond buying and possessing them, domestically displaying and enfolding them, and using them as common-sense fixtures around which everyday life is organised.[105]

In *Homo Sapiens* this is reinforced in shots that show objects in unexpected places or juxtapositions, such as a boat in a field (figure 6.2), a giant teddy bear collapsed on the floor in an abandoned theme park (figure 6.3), a car inside a cave illuminated by a beam of sunlight, and a rollercoaster on the seashore partly submerged in water (figure 6.4). In their surrealist but indexically certified figurations, these images recall some of Kahn's autochromes, including the striking image of a train standing diagonally on railway tracks in Turkey after the Greco-Turkish war discussed earlier. Although the implication that the director may have tampered with reality to achieve some of these juxtapositions cannot be discounted, they call attention to the evidentiary properties of the image by highlighting the unexpected as resulting from natural calamities and the Earth's eventfulness: indeed, the rollercoaster was in reality pushed into the sea by Hurricane Sandy, in Seaside Heights, New Jersey.

Yet because there is no recourse here to conventional staples of the documentary mode – written titles, interviews, talking heads or voiceovers – *Homo Sapiens*, like *The World without Us*, must also be viewed as a post-apocalyptic fable. This reading is substantiated when we consider the temporal ordering of shots, which imparts a sense of narrative progression whereby the Earth, as a planet endowed with agency and volition, seems

105 Tim Edensor, *Industrial Ruins: Space, Aesthetics and Materiality* (Oxford: Berg, 2005), 123.

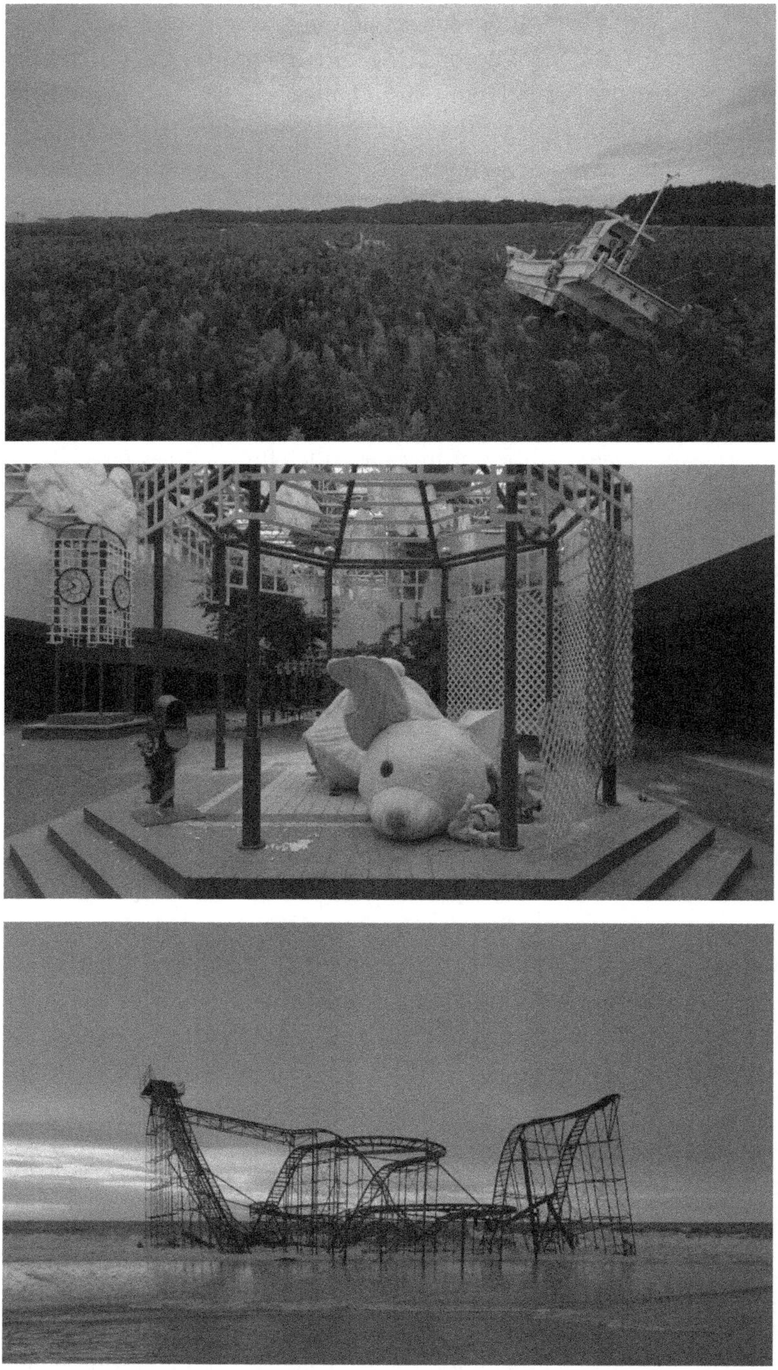

Figures 6.2, 6.3 & 6.4 Surrealist but all-too-real juxtapositions in *Homo Sapiens*, some of which, as in the image of a rollercoaster pushed into the sea by Hurricane Sandy, are the result of natural disasters. Courtesy of NGF.

to gradually reclaim its formerly occupied surface over an unspecified period of time. Whereas *Homo Sapiens* starts with short-distance images of human figures inscribed in the tiled walls of the Buzludzha monument and then proceeds in its first hour to show the inside of buildings – a cinema, an office, a hospital, a church, a dance club and so on – the film's last 30 minutes gesture towards a more complete planetary takeover. This is conveyed through outdoor shots in which the signs of 'civilisation' become less prominent, with some images displaying houses and buildings almost entirely submerged in sand and snow. The film's closing shot is telling in this context, for it goes back to the Buzludzha monument, which is now recorded from outside, engulfed in winter fog. Gradually the screen turns white, implying a sense of closure to this process of repossession – as if humanity was nothing but an interlude in the planet's temporal itinerary, with the Earth, now freed of humans and human signs, carrying on its business as usual.

Such a futurocentric outlook may raise suspicions in its seeming postulation of an inevitable fate against which human action is futile. Donna Haraway warns us: 'There is a fine line between acknowledging the extent and seriousness of the troubles and succumbing to abstract futurism and [...] its politics of sublime indifference'.[106] Yet to 'learn to be truly present', to follow Haraway's call, also requires adjustments to the differing space-time frameworks that the global environmental crisis has dramatically jumbled up.[107] This includes a reckoning with the intrusion of geohistory into human history, which Dipesh Chakrabarty has described (not coincidentally in relation to Weisman's *The World without Us*) as 'a universal that arises from a shared sense of catastrophe'.[108] In any case, *Homo Sapiens* is no 'abstract futurism' or even a thought experiment, as the book by which it was inspired defines itself. It is instead a record of the concrete world that simultaneously visualises its possible future. In other words, the film documents an imaginary situation that is embedded in present-day reality; its world-without-us is one that sends us back to *this* world.

That Geyrhalter is concerned with registering a changing world comes as no surprise: his films are often keenly interested in documenting the effects of globalisation in terms of its socio-cultural, political, technological and environmental effects across the planet. Of particular relevance here is

106 Donna Haraway, *Staying with the Trouble: Making Kin in the Chthulucene* (Durham, NC, and London: Duke University Press, 2016), 4, 2.

107 Ibid., 1.

108 Chakrabarty, 'The Climate of History', 222.

one of his most ambitious projects, the four-hour globe-trotting *Elsewhere* (2001), conceived as a world catalogue of remote cultures living at the dawn of the new millennium. It is worth quoting in full Geyrhalter's rationale for this project:

> We thought this special year would be a good opportunity to travel around and see how the world was changing. We knew that many traditions around the world were in what seemed to be their final stages, so we decided to search for instances of the changes that were taking place. I assumed that the film would be shown at festivals and in cinemas, but most of all I had the feeling that *I was doing this film for the human archive*. In a hundred years I think *Elsewhere* will be valuable as a mapping of the state of the world during this specific year.[109]

It would be no exaggeration to say that Geyrhalter's words reproduce almost ipsis litteris the justification for the Archives de la Planète over a hundred years ago. Like Kahn and Brunhes, Geyrhalter attributes his urgency to document to global modernity's wiping out of traditional localisms. He conceives of film as a suitable medium to preserve the state of the world as inhabited by humans at the beginning of a new century (and millennium). And he validates the significance of his project in terms of its storage of the world for 'the human archive' in the future.

In turn, Geyrhalter's projecting of *Elsewhere* into the far ahead makes us reassess *Homo Sapiens*, released 15 years later. While doubtless also made to be 'shown at festivals and in cinemas', and therefore for a present-day human viewership, we would be perhaps forgiven for speculating whether *Homo Sapiens* was produced with 'the human archive' in mind, in which case it would function, like *Elsewhere*, as a global catalogue of ruinous sites recorded in the first decades of the new millennium; or whether it instead presupposes, as conveyed by its visual content, that in a hundred years' time, such an archive may prove pointless in view of the non-existence of humanity, in which case it begs the question to whom exactly these images would be of service. Significantly, this oscillation between the human and the nonhuman is mapped on to the film's formal carapace and visual design.

On the one hand, despite its title, there are no humans in *Homo Sapiens*. If this is a film that speaks to 'the possibility of life in capitalist ruins', to cite the title of Tsing's book, these are not the people emerging from the

109 Geyrhalter, cited in Scott MacDonald, *The Sublimity of Document: Cinema as Diorama* (New York: Oxford University Press, 2019), 135–136, my emphasis.

skeleton towns of Kahn's films. The focus here, rather, is on nonhuman animals, forces, elements and entities: mould, vegetation, birds, insects, frogs, rain, snow, sand, and especially the wind. In fact, Geyrhalter even used leaf blowers in some scenes so as to inject a sense of motion, with the wind moving things as it moves about. One could go so far as to say that *Homo Sapiens* grants a new lease of life to 'the wind in the trees' that presumably enthralled early film spectators – but now there are no humans in the image feeding a baby and the wind blows discarded, humanmade things: bird cages, plastic bags, book pages, strips of paper.[110]

On the other hand, the human *is* a negative principle around which *Homo Sapiens* worlds its world. The majority of the film is devoted to recording dilapidated institutions, buildings and artefacts where humans are conspicuous by their absence. At the same time, the human is a spectre that haunts the film's poised machinic gaze, with stationary images featuring Geyrhalter's signature geometric shots, achieved through symmetric compositions and wide-angle lenses (figures 6.5 and 6.6). This produces a perspectival regime of vision that continually underlines the centrality of the human eye to which these images are directed, recalling Jean-Louis Baudry's assertion that the camera internalises and 'specifies the position of the "subject", the very spot it must necessarily occupy'.[111]

The film's vision, however, remains that of a machine: its assertion of the camera's mechanical eye constructs an ocular regime that implies in the structure of the image a human subject but never fully morphs into a human perspective. Could we therefore assume that this is a camera miraculously recording the world after we are gone? If so, *Homo Sapiens* invites some reflections on cinematic automatism as espoused by André Bazin and Stanley Cavell.

Bazin famously writes in his 'Ontology' essay, originally published in 1945: 'For the first time, between the originating object and its reproduction there intervenes only the instrumentality of a nonliving agent. *For the first time an image of the world is formed automatically,* without the

110 As the story goes, when confronted with one of Auguste and Louis Lumière's first films, *Le repas de bébé* (1895), audiences were more interested in the blowing leaves behind the eponymous baby and its feeding parents. Decades later, in 1944, D. W. Griffith claimed on his deathbed that films lacked the 'beauty of the moving wind in the trees'. See Nico Baumbach, 'Nature Caught in the Act: On the Transformation of an Idea of Art in Early Cinema' in Jeffrey Geiger and Karin Littau (eds.) *Cinematicity in Media History* (Edinburgh: Edinburgh University Press, 2013); 107–116; 107.

111 Jean-Louis Baudry, 'The Ideological Effects of the Basic Cinematographic Apparatus', *Film Quarterly* 28:2 (1974): 39–47; 41.

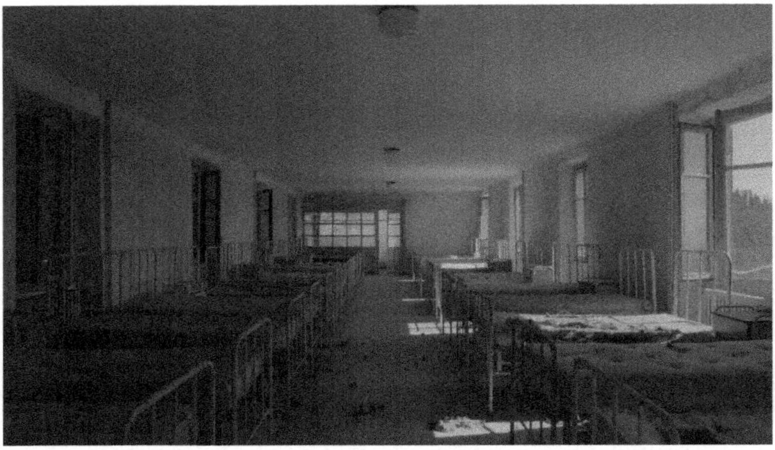

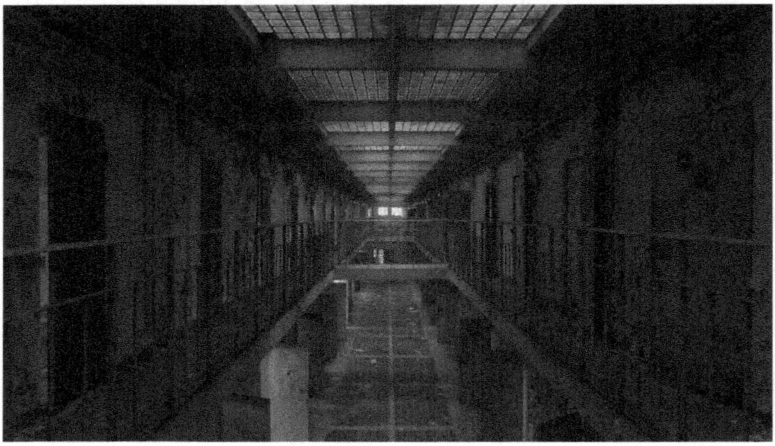

Figures 6.5 & 6.6 Ultra-perspectival images in *Homo Sapiens* underline the human eye to which they are directed. Courtesy of NGF.

creative intervention of man'.[112] Inspired by Bazin, Cavell similarly marvels at the fact that photography 'does not present us with "likeness" of things; it presents us, we want to say, with the things themselves'.[113] Ultimately the only certainty we have about photographs is that they 'are not hand-made; they are manufactured. And what is manufactured is *an image of the world*'.[114] By eliminating the human agent, Cavell concludes, photography 'overcame subjectivity', so much so that it 'maintains the presentness of the world by

112 Bazin, *What is Cinema?*, 13, my emphasis.

113 Stanley Cavell, *The World Viewed: Reflections on the Ontology of Film* (Cambridge, MA, and London: Harvard University Press, 1979), 17.

114 Ibid., 20, my emphasis.

accepting our absence from it'.[115] For both thinkers it is therefore imperative to consider the ontology of photography – and consequently cinema – as resulting from an automatic process whereby it is the world that presents itself to the camera and gives itself to be viewed. In this sense, not only is their thought still relevant to the digital capture of reality (as in *Homo Sapiens*), it also endows the world with a sense of agency and intentionality in the forming of its own image.

Now, this is certainly not an all-seeing vision of the world. Indeed, if Bazin defines the 'myth of total cinema' in terms of a desire to achieve a complete reproduction of reality – an 'integral realism, a recreation of the world in its own image, an image unburdened by the freedom of interpretation of the artist and the irreversibility of time' – such an idea, in its dream of totality, remains a myth given the technical limitations of cinema and the fact that there are always subjective elements in the image.[116] That said, for both Bazin and Cavell, it is the world itself that shapes its own image for the camera, and in this sense, the image does imply a totality of the world that is left out of the picture. As Cavell notes, unlike painting, 'a photograph is *of* the world', since 'the implied presence of the rest of the world, and its explicit rejection, are as essential in the experience of a photograph as what it explicitly presents. A camera [...] holding on an object is holding the rest of the world away'.[117]

No doubt both thinkers championed the cinema in relation to its values and benefits for a human subject and therefore as a phenomenology: in Bazin a subject in fear of the inexorability of time and who finds in cinema's reproduction of duration – its capacity to 'mummify' *change*, as we saw earlier – a way of deferring anxiety about death; in Cavell a subject whose consciousness has been unhinged from the world and who finds solace in the cinema precisely because cinema mechanically seals this unhinging 'through no fault of my subjectivity'.[118] But if Bazin and Cavell assessed the cinema in relation to a human viewer, it was because cinematic images of the world are not subjectively formed; cinema is valuable not so much because it allows us to see the world in different ways but because it allows the world to reveal itself in its own terms for a nonhuman entity: the camera.

Homo Sapiens resonates with these ideas and reflections. It does so, first, through its fictional universe, which manufactures the world as a world in

115 Ibid., 23.

116 Bazin, *What is Cinema?*, 21.

117 Cavell, *The World Viewed*, 24.

118 On Bazin's phenomenological subject, see Rosen, *Change Mummified*, 3–41; Cavell, *The World Viewed*, 23.

which only the world, without humans, exists. And, second, because this is a world that makes itself available in visual form through the lens aperture of film cameras whose perspectival vision, in its turn, reminds us of the role cinema once played by and for humans in an unspecified past. In other words, *Homo Sapiens* reduces a Bazinian-Cavellian ontology of cinema to its most fundamental principle by imagining a scenario that strips down the automatic process of cinema to the world-in-itself (the producer of images) and nonhuman film cameras (the recipient of images), while at the same time highlighting in the composition of the image the significance of cinema for humanity.

The uncanny mode of spectatorship the film promotes is therefore tied to the epistemologically restless recognition that, as viewers, we are not simply witness to a world without us but, rather, to a world that produces its own images, on its own, without us. In so doing, the film raises questions concerning the use and archiving of indexical images after humanity is no more. To put it bluntly: would other nonhuman entities, terrestrial or extraterrestrial, be able to discern, understand and perhaps be moved by images of our planet after we are gone?

A Non-Planet

Outlandish as it may seem, this is one of the driving questions of Trevor Paglen's *The Last Pictures* project, which he developed in partnership with the public arts organisation Creative Time and the satellite company EchoStar Corporation. With a creative team, Paglen selected 100 illustrations – most of them photographs – that are meant to act as a visual memento of a global humanity before its perishing. The images were etched onto a small silicon disk attached to the side of the satellite spacecraft EchoStar, which was launched in 2012 and will orbit around the Earth for at least billions of years or even indefinitely. Like *Homo Sapiens*, *The Last Pictures* is interested in the future in order to make people think about our planet now: whereas all efforts were made to ensure that the materials utilised in the archival disc would indeed last for billions of years, Paglen recognises that 'the idea someone in the future might actually find the Artifact was close to nil [...] but the probability of people on Earth thinking about it here and now was guaranteed'.[119] Thus, concomitant to the sending of the disk into

119 Trevor Paglen, *The Last Pictures* (New York, Berkeley, Los Angeles and London: Creative Time Books and University of California Press, 2012), 11–12.

orbital space, *The Last Pictures* was turned into a publication containing Paglen's ideas, inspirations and the selected pictures, most accompanied by statements explaining the reason behind their inclusion in the project.

Like all works discussed in this chapter, *The Last Pictures* is the fruit of anxieties over what Paglen defines as modernity's 'continual speeding up of time', which has now begun to interfere with geological temporalities. He goes on: 'The Anthropocene is a period of temporal contradictions, a period in which Marx's space-time annihilation chafes against the deep time of the earth'.[120] Tellingly, the opening image in the collection is a photograph of the back of Paul Klee's *Angelus Novus* (1920), with labels from the Israel Museum (Jerusalem) and the Museum of Contemporary Art (Chicago). As explained in the book's contextual note, the picture is an allusion to Benjamin's famous reflection on the monoprint as the 'angel of history', one that sees history as 'one single catastrophe, which keeps piling wreckage upon wreckage'.[121]

Many of the subsequent, sometimes iconic, images subscribe to this idea, offering glimpses into stories of human conquest, transformation but especially destruction of the world. They depict wars, the space age, extractivism, deforestation, surveillance, animal cruelty, epidemics, consumerism, nuclear power and biogenetics. At the same time, the trope of the ruin reappears, though it is no longer connected with terrestrial wastelands but set in relation to orbiting satellites, which are defined by Paglen as 'the cultural and material ruins of the late twentieth and early twenty-first century'; 'destined to remain in Earth orbit until Earth is no more' and thus 'far outlast anything else humans have created'.[122] Space junk becomes the final, ceaselessly floating wreckage piled upon a Benjaminian history.

As Paglen is quick to point out, the collection was never designed as 'a grand representation of humanity' or 'a portrait of life on earth', but rather, as the proof of how and why things went wrong down on terrestrial soil.[123] Three different images allude to the failure of the universal language project: a page of a dictionary of Volapük, hailed at the end of the nineteenth century as a 'world speak'; a detail of Pieter Bruegel's *Tower of Babel* painting (1563); and a page of John Wilkins's 1668 *An Essay Towards a Real Character and a Philosophical Language*, considered one of the first attempts to develop a universal language. A pair of successive pictures, the only ones depicting

120 Ibid., xii.
121 Benjamin, *Illuminations*, 12.
122 Paglen, *The Last Pictures*, 11.
123 Ibid., 11.

exactly the same place from the same angle, tells a story of disappearance in the age of the Anthropocene, showing the before and after of Grinnell Glacier in Glacier National Park, Montana, in 1940 and 2006 respectively, with the glacier in the latter picture entirely melted.

But disappearance appears in the project in another, perhaps even more significant guise: a disappearing world of images. Crucial for the execution of *The Last Pictures* was finding the right home, materials and shape for its archiving gold disk so that it would survive the fullness of time. Whereas the disk was housed in a satellite that will cease functioning 15 years after its launch, the spacecraft will remain in space, among hundreds of satellites, for potentially billions of years thanks to the timeless stability of the orbit in which it finds itself: the geostationary orbit (GSO), or Clarke Belt. Though we now produce trillions and trillions of images, most of which are transmitted across the surface of the globe by these very satellites, such images are ultimately evanescent and fleeting as material objects when examined from the perspective of cosmic time. As Sean Cubitt notes in his *Finite Media*:

> Media are finite, in the sense both that, as matter, they are inevitably tied to physics, especially the dimension of time; and that their constituent elements – matter and energy, information and entropy, time and space, but especially the first pair – are finite resources in the closed system of the planet Earth. Because they are finite, media not only cannot persist forever; they cannot proliferate without bounds.[124]

The Last Pictures must be viewed in the context of this problematic. To cite Paglen himself: 'One of the technical problems I was trying to solve was that of archivability: what materials and techniques exist, if any, to produce an "ultra-archival" medium for storing images'?[125]

When placed next to that other archive we have explored in this chapter, the Archives de la Planète, *The Last Pictures* invites some reflections on questions relating to the finitude and materiality of images both in present and future times and terrestrial and cosmic worlds. As Cubitt goes on to note: 'When we speak of film as a "living medium", we should take the term literally', with its archival life defined by a struggle 'between the order of the archive and the entropy of what the archivist understands as decay, but which can also be understood as the evolution of a new artifact from the

124 Sean Cubitt, *Finite Media: Environmental Implications of Digital Technologies* (Durham, NC, and London: Duke University Press, 2017), 7.
125 Paglen, *The Last Pictures*, 17.

old', with the artifact 'moving on through time, now and forever pinned to migration from format to format'.[126]

A hundred years after its genesis, the Kahn Archive has already undergone momentous technological mutations in relation to how its images are *archived*, that is: handled, stored, formatted, catalogued, juxtaposed and visualised. Indeed, the majority of researchers visiting the Musée Albert-Kahn's facilities today will most likely have access not to the original autochrome plates and reels of celluloid film, but to the digitised version of these materials housed in the FAKIR database, with all the navigation tools and tags that comprise, structure and constrain any digital catalogue, as we explored in chapter 3. In a way, as Lundermo argues, the return of these images in digital form doubles down on the disappearance that provided the impetus for the Kahn Archive by alluding to the disappearance of its originary formats and supports:

> The communities and people at the center of Kahn's archival project were on the brink of disappearance. The fleeting moments inscribed in the archive's films and autochromes were inhabited by ghosts and specters of a lost time [...] When this haunted archive returns in digital mode, it is a return of ghosts of ghosts.[127]

Although there is much debate concerning the unreliability of the digital as a vertiginously mutating format, the current digitisation of archives, as in the Kahn Archive, is meant to ensure that the images can be preserved for a longer time by sacrificing their original material bases. In other words, the digital appears as a precautionary measure to maximise the survival of original artefacts stored away from human contact, thus ensuring the continuance of their visual content in other formats – even if, of course, the digital also has a physical basis and will, too, be subject to decay.

For Bjorli and Jakobsen, the Archives de la Planète 'is a time capsule, designed to communicate with an unknown future'.[128] Today this capsule finds itself materialised as digitised photographs on a computer interface. In some ways, *The Last Pictures* also conceives of itself as a time capsule, but where it differs from Kahn's Archive is that its future lies inconceivably ahead. As Janine Marchessault observes: 'In Paglen's project, the planet is

126 Cubitt, *Finite Media*, 2–3

127 Lundermo, '"The Archives of the Planet" and Montage', 222.

128 Bjorli and Jakobsen, 'Introduction' in *Cosmopolitics of the Camera*, 3. Amad makes the same point in *Counter-Archive*, 154.

no more. The time capsule communicates the end of the end – an Oblivion which is, at least in part, countered by the gesture of sending the satellite up into space to tell the diversity of stories that make up the end'.[129] This desire to tell the end of the end dictated not only the fabrication of a specific disk but also its location: a satellite that may outlive the Earth because of the space orbit's time-free properties when compared with planetary time. In this context, though infinitely more modest in scope and literally miniscule in its physical constitution, *The Last Pictures* nonetheless appears surprisingly more ambitious than Kahn's Archive in its quest to 'transcend the Anthropocene and [...] deep time itself' and 'theoretically last for an eternity'.[130]

Although we might be tempted to scoff at the loftiness of these statements, perhaps we should see them simply as the most hyperbolic expression of profound anxieties over the passing of time, anxieties which recording media have been recruited to assuage since their inception. But whereas the promise of the indexical image for the archive is the promise of a preservation of life for the future, neither the archive nor the image may ultimately sidestep the incessant force of continuance and variability that is time, one whose inalienable presage is the degradation of materials and realities of which archives, images and worlds are constituted. From this perspective, *The Last Pictures* merely radicalises the preservationism of the Archives de la Planète by locating photographs literally outside the Earth and beyond terrestrial time. Or, to put it differently, it sends its collection of images up into the dark abyss of interstellar space so as to archive the planet into infinity.

129 Janine Marchessault, *Ecstatic Worlds: Media, Utopia, Ecologies* (Cambridge, MA, and London: MIT Press, 2017), 246.
130 Paglen, *The Last Pictures*, xiii; 18.

Bibliography

Abel, Richard (ed.). *French Film Theory and Criticism: A History/Anthology, 1907–1939*, Volume 2. Princeton, NJ: Princeton University Press, 1988.

Acland, Charles R. 'Imax Technology and the Tourist Gaze', *Cultural Studies* 12:3 (1998): 429–445.

Agathocleous, Tanya. *Urban Realism and the Cosmopolitan Imagination in the Nineteenth Century: Visible City, Invisible World*. Cambridge: Cambridge University Press, 2013.

Altick, Richard D. *The Shows of London*. Cambridge, MA/London: Belknap Press, 1978.

Amad, Paula. *Counter-Archive: Film, the Everyday, and Albert Kahn's Archives de la Planète*. New York: Columbia University Press, 2010.

Amad, Paula. 'Visual Riposte: Looking Back at the Return of the Gaze as Postcolonial Theory's Gift to Film Studies', *Cinema Journal* 52:3 (2013): 49–74.

Amad, Paula. 'Experimental Cosmopolitanism: The Limits of Autour du Monde-ism in the Kahn Archive' in Trond Erik Bjorli and Kjetil Ansgar Jakobsen (eds.) *Cosmopolitics of the Camera: Albert Kahn's Archives of the Planet*. Bristol/Chicago: Intellect, 2020, 133–154.

Appadurai, Arjun. *Modernity at Large: Cultural Dimensions of Globalization*. Minneapolis: University of Minnesota Press, 1996.

Arendt, Hannah. *The Origins of Totalitarianism*. Cleveland/New York: Meridian Books, 1958, 302.

Arendt, Hannah. 'Man's Conquest of Space', *The American Scholar* 32:4 (Autumn 1963): 527–540.

Arendt, Hannah. *The Human Condition*. Chicago/London: University of Chicago Press, 1998.

Arendt, Hannah. *On Revolution*. New York: Penguin Books, 2006.

Arvatov, Boris. 'Everyday Life and the Culture of the Thing (Toward the Formulation of the Question'), trans. Christina Kiaer, *October* 81 (1997): 119–128.

Augé, Marc. *Non-Places: An Introduction to Supermodernity*. London: Verso, 2008.

Azcona, Maria del Mar. *The Multi-Protagonist Film*. Oxford: Wiley-Blackwell, 2010.

Badiou, Alain. *Saint Paul: The Foundation of Universalism*, trans. Ray Brassier. Stanford, CA: Stanford University Press, 2003.

Badiou, Alain. *Ethics: An Essay on the Understanding of Evil*. London/New York: Verso, 2012.

Badiou, Alain. *Cinema*, texts selected and introduced by Antoine de Baecque; trans. Susan Spitzer. Cambridge/Malden, MA: Polity Press, 2013.

Baggesgaard, Mads Anders. 'Picturing the World – Cinematic Globalization in the Deserts of *Babel*', *Journal of Aesthetics & Culture*, 5:1 (2013): 1–10.

Balázs, Béla. *Béla Balázs: Early Film Theory; Visible Man* and *The Spirit of Film*, trans. Rodney Livingstone and edited by Erica Carter. New York/Oxford: Berghahn Books, 2010.

Bann, Stephen. *The Clothing of Clio: A Study of the Representation of History in Nineteenth-Century Britain and France*. New York: Cambridge University Press, 1984.

Bard, Perry. 'When Film and Database Collide' in Geert Lovink and Rachel Somer Miles (eds.) *Video Vortex Reader II: Moving Images Beyond YouTube*. Amsterdam: Institute of Network Cultures, 2011, 322–329.

Barthes Roland. *Camera Lucida: Reflections on Photography*, trans. Richard Howard. New York: Farrar, Straus and Giroux, 1981.

Barthes Roland. *Image, Music, Text*, trans. Stephen Heath. London: Flamingo, 1984.

Bates, Marston. *The Nature of Natural History*. Princeton, NJ: Princeton University Press, 1990.

Baudry, Jean-Louis. 'The Ideological Effects of the Basic Cinematographic Apparatus', *Film Quarterly* 28:2 (1974): 39–47.

Bazin, André. *What Is Cinema?* Volume 1, trans. Hugh Gray. Berkeley: University of California Press, 2005.

Beck, Richard. 'Costing *Planet Earth*', *Film Quarterly* 63:3 (2010): 63–66.

Belisle, Brooke. 'Picturing Networks: Railroads and Photographs', *Amodern* 2 (October 2013), https://amodern.net/article/picturing-nineteenth-century-networks/.

Belisle, Brooke. 'Total Archive: Picturing History from the Stereographic Library to the Digital Database', *Mediascape* (2013), http://www.tft.ucla.edu/mediascape/pdfs/Winter2013/PicturingHistory.pdf.

Belisle, Brooke. 'Nature at a Glance: Immersive Maps from Panoramic to Digital', *Early Popular Visual Culture* 13:4 (2015): 313–335.

Belisle, Brooke. 'Whole World Within Reach: Google VR', *Journal of Visual Culture* 19:1 (2020): 112–136.

Benjamin, Walter. *Illuminations*, trans. Harry Zohn and edited by Hannah Arendt. New York: Schocken Books, 2007.

Bennett, Jane. *Vibrant Matter: A Political Ecology of Things*. Durham, NC: Duke University Press, 2010.

Berg, Charles Ramírez. 'A Taxonomy of Alternative Plots in Recent Films: Classifying the "Tarantino Effect"', *Film Criticism* 31:1/2 (2006): 5–61.

Berger, John. *About Looking*. London: Bloomsbury, 2009.

Bergson, Henri. *Creative Evolution*, trans. Arthur Mitchell. New York: Random House, 1944.

Besse, Jean-Marc. *Face au monde: Atlas, jardins, géoramas*. Paris: Desclée de Brouwer, 2003.

Bevis, John. *The Keartons: Inventing Nature Photography*. Devon: Uniformbooks, 2016.

Bjorli, Trond Erik, and Kjetil Ansgar Jacobsen (eds.). *Cosmopolitics of the Camera: Albert Kahn's Archives of the Planet.* London: Intellect, 2020.

Bogost, Ian. *Alien Phenomenology, or What It's Like to Be a Thing.* Minneapolis/London: University of Minnesota Press, 2012.

Bonneuil, Christophe, and Jean-Baptiste Fressoz. *The Shock of the Anthropocene.* London: Verso, 2017.

Boon, Timothy. *Films of Fact: A History of Science and Documentary Films and Television.* London/New York: Wallflower Press, 2008.

Bordwell, David. *Poetics of Cinema.* New York: Routledge, 2008.

Bousé, Derek. *Wildlife Films.* Philadelphia: University of Pennsylvania Press, 2000.

Brady, Emily. *The Sublime in Modern Philosophy: Aesthetics, Ethics, and Nature.* New York: Cambridge University Press, 2013.

Braidotti, Rosi. *The Posthuman.* Cambridge: Polity Press, 2013.

Brantlinger, Patrick. *Dark Vanishings: Discourse on the Extinction of Primitive Races, 1800–1930.* Ithaca, NY/London: Cornell University Press, 2003.

Brown, Bill. *Other Things.* Chicago/London: University of Chicago Press, 2015.

Brown, Karl. *Adventures with D.W. Griffith.* London: Faber and Faber, 1973.

Brown, Robert. *The Peoples of the World; Being a Popular Description of the Characteristics, Condition, and Customs of the Human Family.* London/Paris/New York: Cassell, Petter, Galpin & Co., 1882.

Brown, Robert. *The Races of Mankind: Being a Popular Description of the Characteristics, Manners and Customs of the Principal Varieties of The Human Family.* Volume III. London/Paris/New York: Cassell, Petter & Galpin, 1887.

Brunhes, Jean. 'Ethnographie et Géographie humaine', *L'Ethnographie: Bulletin de la société d'ethnographie de Paris* (1913): 29–40.

Brunhes, Jean. *Human Geography*, abridged edition, trans. Ernest F. Row. London: George G. Harrap, 1952.

Burke, Edmund. *A Philosophical Enquiry into the Origin of Our Ideas of the Sublime and the Beautiful.* Mineola, NY: Dover Publications, 2008.

Buzan, Barry, and George Lawson. *The Global Transformation: History, Modernity and the Making of International Relations.* Cambridge: Cambridge University Press, 2015.

Byerly, Alison. *Are We There Yet? Virtual Travel and Victorian Realism.* Ann Arbor: University of Michigan Press, 2012.

Cahill, James Leo. 'A YouTube Bestiary: Twenty-six Theses on a Post-Cinema of Animal Attractions' in Paul Flaig and Katherine Groo (eds.) *New Silent Cinema.* New York: Routledge, 2016, 263–293.

Casetti, Francesco. *Eye of the Century: Film, Experience, Modernity*, trans. Erin Larkin with Jennifer Pranolo. New York: Columbia University Press, 2008.

Castells, Manuel. *The Rise of the Network Society.* Malden, MA: Wiley-Blackwell, 2010.

Castro, Teresa. *La pensée cartographique des images: Cinéma et culture visuelle.* Lyon: Aléas Éditeur, 2011.

Castro, Teresa. 'Le Quotidian est politique: Le cas des films 1914–1918' in Valérie Perlès and Anne Singaud (eds.) *Réalités (In)Visibles*, 188–192.

Castro, Teresa. 'Vues panoramiques et travellings: Les Paysages en mouvement des Archives de la Planète' in Valérie Perlès, with Manon Demurger (eds.) *Les Archives de la Planète*, 390–399.

Cavell, Stanley. *The World Viewed: Reflections on the Ontology of Film.* Cambridge, MA/London: Harvard University Press, 1979.

Chakrabarty, Dipesh. 'The Climate of History', *Critical Inquiry* 35:2 (2009): 197–222.

Chaplin, Joyce E. *Round About the Earth: Circumnavigation from Magellan to Orbit.* New York: Simon and Schuster, 2012.

Chris, Cynthia. *Watching Wildlife.* Minneapolis: University of Minnesota Press, 2006.

Ciafone, Amanda. 'The Magical Neoliberalism of Network Films', *International Journal of Communication* 8 (2014), http://ijoc.org/index.php/ijoc/article/view/2452/1213.

Clark, Nigel. *Inhuman Nature: Sociable Life on a Dynamic Planet.* London: Sage, 2011.

Coeuré, Sophie, and Frédéric Worms (eds.). *Henri Bergson et Albert Kahn, Correspondances.* Strasbourg/Bolougne: Desmaret/Musée départemental Albert-Kahn, 2003.

Comment, Bernard. *The Panorama.* London: Reaktion Books, 1999.

Comolli, Jean-Louis. 'Machines of the Visible' in Stephen Heath and Teresa de Lauretis (eds.) *The Cinematic Apparatus.* London: Palgrave, 1980, 121–142.

Connor, Ralph O. *The Earth on Show: Fossils and the Poetics of Popular Science, 1802–1856.* Chicago/London: University of Chicago Press, 2007.

Cosgrove, Denis. *Apollo's Eye: A Cartographic Genealogy of the Earth in the Western Imagination.* Baltimore/London: Johns Hopkins University Press, 2001.

Cowan, Michael. 'Cutting Through the Archive: *Querschnitt* Montage and Images of the World in Weimar Visual Culture', *New German Critique* 120, 40:3 (Fall 2013), 1–40.

Cowan, Michael. *Walter Ruttmann and the Cinema of Multiplicity: Avant-Garde–Advertising–Modernity.* Amsterdam: Amsterdam University Press, 2015.

Cubitt, Sean. *Finite Media: Environmental Implications of Digital Technologies.* Durham/London: Duke University Press, 2017.

Dammann, C., and F. W. Dammann. *Ethnological Photographic Gallery of the Various Races of Man.* London: Trübner & Co. Ludgate Hill, 1876.

Danowski, Déborah, and Eduardo Viveiros de Castro. *The Ends of the World*, trans. Rodrigo Nunes. Cambridge/Malden, MA: Polity Press, 2017.

Darrah, William C. *Cartes de Visite in Nineteenth Century Photography.* Gettysburg: W. C. Darrah Publisher, 1981.

Darwin, Charles. *The Expression of the Emotions in Man and Animals.* London: John Murray, 1892.

Davis, Therese. *The Face on the Screen: Death, Recognition and Spectatorship.* Bristol/ Portland: Intellect, 2004.

de Luca, Tiago. 'Earth Networks: *The Human Surge* and Cognitive Mapping', *NECSUS* 7:2 (2018): 121–140, https://necsus-ejms.org/earth-networks-the-human-surge-and-cognitive-mapping/.

de Luca, Tiago, and Nuno Barradas Jorge (eds.). *Slow Cinema.* Edinburgh: Edinburgh University Press, 2016.

Demos, T. J. *Against the Anthropocene: Visual Culture and Environment Today.* Berlin: Stenberg Press, 2017.

Dennison, Stephanie, and Song Hwee Lim (eds.). *Remapping World Cinema: Identity, Culture and Politics in Film.* London/New York: Wallflower Press, 2006.

Doane, Mary Ann. *The Emergence of Cinematic Time: Modernity, Contingency, the Archive.* Cambridge, MA/London: University of Harvard Press, 2002.

Doane, Mary Ann. 'Facing a Universal Language', *New German Critique* 41:2 (2014): 111–124.

Dovey, Jon. 'Documentary Ecosystems: Collaboration and Exploitation' in Kate Nash, Craig Hight and Catherine Summerhayes (eds.) *New Documentary Ecologies: Emerging Platforms, Practices and Discourses.* Hampshire: Palgrave Macmillan, 2014.

Drew, William M. *D.W. Griffith's Intolerance: Its Genesis and Its Vision.* Jefferson, NC: McFarland, 1986.

Drouin, Jean-Marc, and Bernadette Bensaude-Vincent. 'Nature for the People' in N. Jardine, J.A. Second and E. C. Spary (eds.) *Cultures of Natural History.* Cambridge: Cambridge University Press, 1996, 408–425.

Druick, Zoë. '"A Wide-Angle View of the Fragile Earth": Capitalist Aesthetics in the Work of Yann Arthus-Bertrand', *Open Cultural Studies*, 2 (2018): 396–405.

Druick, Zoë. 'On (Not) Falling from the Sky: Fly-Over Global Documentary as Capitalist Body Genre' in Steve Anderson and Christie Milliken (eds.) *Reclaiming Popular Documentary.* Bloomington: University of Indiana Press, 2021.

Durovicová, Nataša, and Kathleen E. Newman (eds.). *World Cinemas, Transnational Perspectives.* London: Taylor & Francis, 2009.

Edensor, Tim. *Industrial Ruins: Space, Aesthetics and Materiality.* Oxford: Berg, 2005.

Edwards, Elizabeth. 'Evolving Images: Photography, Race, and Popular Darwinism' in D. Donald and J. Munro (eds.) *Endless Forms: Darwin, Natural Sciences and the Visual Arts.* New Haven/London: Yale University Press, 2009, 166–193.

Eisenstein, Sergei. *Film Form: Essays in Film Theory*, trans. and edited by Jay Leyda. San Diego/New York/London: Harcourt, 2011.

Elsaesser, Thomas. *Film History as Media Archaeology: Tracking Digital Cinema.* Amsterdam: Amsterdam University Press, 2016.

Ewen, Elizabeth, and Stuart Ewen. *Typecasting: On the Arts and Sciences of Human Inequality*. New York/Oakland/ London: Seven Stories Press, 2008.

Ezra, Elizabeth. *The Cinema of Things: Globalization and the Posthuman Object*. London: Bloomsbury, 2017.

Ferreira da Silva, Denise. *Toward a Global Idea of Race*. Minneapolis/London: University of Minnesota Press, 2007.

Fisher, Philip. *Wonder, The Rainbow and the Aesthetics of Rare Experiences*. Cambridge/London: Harvard University Press, 1998.

Fitzpatrick, Sheila. *Everyday Stalinism: Ordinary Life in Extraordinary Times: Soviet Russia in the 1930s*. New York: Oxford University Press, 2000.

Foucault, Michel. *The Archaeology of Knowledge and the Discourse on Language*, trans. A. M. Sheridan Smith. New York: Vintage Books, 2010.

Friedman, Ryan Jay. *The Movies as a World Force: American Silent Cinema and the Utopian Imagination*. New Brunswick, NJ: Rutgers University Press, 2019.

Galperin, William. *The Return of the Visible in British Romanticism*. Baltimore, MD: Johns Hopkins University Press, 1993.

Galt, Rosalind, and Karl Schoonover (eds.). *Global Art Cinema: New Theories and Histories*. New York: Oxford University Press, 2010.

Gates, Kelly A. *Our Biometric Future: Facial Recognition Technology and the Culture of Surveillance*. New York/London: New York University Press, 2011.

Gaycken, Oliver. *Devices of Curiosity: Early Cinema and Popular Science*. New York: Oxford University Press, 2015.

Ghosh, Amitav. *The Great Derangement: Climate Change and the Unthinkable*. Chicago: University of Chicago Press, 2016.

Gilroy, Paul. *Against Race: Imagining Political Culture Beyond the Color Line*. Cambridge, MA: Belknap Press, 2000.

Gish, Lillian, with Ann Pinchot. *The Movies, Mr. Griffith, and Me*. Englewood Cliffs, NJ: Prentice Hall, 1973.

Gorfinkel, Elena, and Tami Williams (eds.). *Global Cinema Networks*. New Brunswick, NJ: Rutgers University Press, 2018.

Gouyon, Jean-Baptiste. 'The BBC Natural History Unit: Instituting Natural History Film-Making in Britain', *History of Science*, 49:4 (2011): 425–451.

Gouyon, Jean-Baptiste. 'From Kearton to Attenborough: Fashioning the Telenaturalist's Identity', *History of Science* 49:1 (2011): 25–60.

Gouyon, Jean-Baptiste. *BBC Wildlife Documentaries in the Age of Attenborough*. Cham: Palgrave, 2019.

Grau, Oliver. *Virtual Art: From Illusion to Immersion*, trans. Gloria Custance. Cambridge/London: MIT Press, 2003.

Greene, Doyle. *The American Worker on Film: A Critical History, 1909–1999*. Jefferson/ London: McFarland and Company, 2010.

Griffiths, Alison. *Wondrous Difference: Cinema, Anthropology, and Turn-of-the-Century Visual Culture*. New York: Columbia University Press, 2002.

Griffiths, Alison. *Shivers Down Your Spine: Cinema, Museums, and the Immersive View*. New York: Columbia University Press, 2013.

Groo, Katherine. *Bad Film Histories: Ethnography and the Early Archive*. Minneapolis/London: University of Minnesota Press, 2019.

Guido, Laurent. 'Toward an Archaeology of Global Rhythms: *Melodie der Welt* and Its Reception in France' in Elena Gorfinkel and Tami Williams (eds.) *Global Cinema Networks*. New Brunswick, NJ: Rutgers University Press, 2018, 97–116.

Gunning, Tom. 'The World as Object Lesson: Cinema Audiences, Visual Culture and the St. Louis World's Fair', *Film History* 6:4 (1994): 422–444.

Gunning, Tom. 'An Aesthetic of Astonishment: Early Film and the (In)Credulous Spectator' in Linda Williams (ed.) *Viewing Positions: Ways of Seeing Film*. New Brunswick, NJ: Rutgers University Press, 1995, 114–133.

Gunning, Tom. 'Re-Newing Old Technologies: Astonishment, Second Nature, and the Uncanny in Technology from the Previous Turn-of-the-Century' in David Thorburn and Henry Jenkins (eds.) *Rethinking Media Change: The Aesthetics of Transition*. Cambridge/London: MIT Press, 2003, 39–60.

Gunning, Tom. '"The Whole World Within Reach": Travel Images without Borders' in Jeffrey Ruoff (ed.) *Virtual Voyages: Cinema and Travel*. Durham/London: Duke University Press, 2006, 25–41.

Gunning, Tom. 'Early Cinema as Global Cinema: the Encyclopedic Ambition' in Richard Abel, Giorgio Bertellini and Rob King (eds.) *Early Cinema and the 'National'*. Bloomington: Indiana University Press, 2008, 11–16.

Gunning, Tom. 'Before Documentary: Early Non-fiction Films and the "View" Aesthetic' in Jonathan Kahana (ed.) *The Documentary Film Reader: History, Theory, Criticism*. New York: Oxford University Press, 2016, 52–61.

Haffner, Jeanne. *The View from Above: The Science of Social Space*. Cambridge/London: MIT Press, 2013.

Hagener, Malte. *Moving Forward, Looking Back: The European Avant-Garde and the Invention of Film Culture, 1919–1939*. Amsterdam: Amsterdam University Press, 2007.

Hamilton, Clive. *Defiant Earth: The Fate of Humans in the Anthropocene*. Sydney: Allen & Unwin, 2017.

Hanley, Keith, and Greg Kucich (eds.). *Nineteenth-Century Worlds: Global Formations Past and Present*. London/New York: Routledge, 2008.

Hansen, Miriam. 'Early Cinema, Late Cinema: Permutations of the Public Sphere', *Screen* 34:3 (1993): 197–210.

Hansen, Miriam. *Babel and Babylon: Spectatorship in American Silent Film*. Cambridge/London: Harvard University Press, 1991.

Hansen, Miriam. *Cinema and Experience: Siegfried Kracauer, Walter Benjamin, and Theodor W. Adorno*. Berkeley/Los Angeles/London: University of California Press, 2012.

Haraway, Donna. *Staying with the Trouble: Making Kin in the Chthulucene*. Durham/London: Duke University Press, 2016.

Harman, Graham. *Object-Oriented Ontology: A New Theory of Everything*. London: Penguin, 2018.

Heidegger, Martin. *The Question Concerning Technology and Other Essays*, trans. William Lovitt. New York/London: Garland, 1977.

Heise, Ursula K. *Sense of Place and Sense of Planet: The Environmental Imagination of the Global*. New York: Oxford University Press, 2008.

Heise, Ursula K. *Imagining Extinction: the Cultural Meanings of Endangered Species*. Chicago/London: Chicago University Press, 2016.

Hell, Julia, and Andreas Schönle, 'Introduction' in Julia Hell and Andreas Schönle (eds.) *Ruins of Modernity*. Durham/London: Duke University Press, 2010.

Hicks, Jeremy. *Dziga Vertov: Defining Documentary Film*. London/New York: I.B. Tauris, 2007.

Homes, Oliver Wendell. *Soundings from the Atlantic*. Boston: James R. Osgood and Company, 1872.

Holmes, Richard. *The Age of Wonder: How the Romantic Generation Discovered the Beauty and Terror of Science*. London: HarperPress, 2008.

Huhtamo, Erkki. *Illusions in Motion: Media Archaeology of the Moving Panorama and Related Spectacles*. Cambridge/London: MIT Press, 2013.

Huhtamo Erkki, and Jussi Parikka (eds.). *Media Archaeology: Approaches, Applications, and Implications*. Berkeley/Los Angeles/London: University of California Press, 2011.

Humboldt, Alexander von. *Cosmos: Sketch of a Physical Description of the Universe*, Volumes 1–2, edited by Edward Sabine. New York: Cambridge University Press, 2010 (1848).

Ivakhiv, Adrian J. *Ecologies of the Moving Image: Cinema, Affect, Nature*. Ontario: Wilfrid Laurier University Press, 2013.

Jacobs, Steven, Eva Hielscher and Anthony Kinik (eds.). *The City Symphony Phenomenon: Cinema, Art, and Urban Modernity Between the Wars*. New York: Routledge, 2018.

Jay, Martin. *Marxism and Totality: The Adventures of a Concept from Lukács to Habermas*. Berkeley/Los Angeles: University of California Press, 1984.

Jean Brunhes, autour du monde: regards d'un géographe/regards de la géographie. Boulogne-Billancourt: Musée Albert Kahn, 1992.

Jeffries, Michael. 'BBC Natural History Versus Science Paradigms', *Science as Culture*, 12:4 (2003): 527–545.

Kaganovsky, Lilya. '"The Threshold of the Visible World:" Dziga Vertov's *A Sixth Part of the World* (1926)' in Lilya Kaganovsky, Scott MacKenzie and Anna Westerstahl Stenport (eds.) *Arctic Cinemas and the Documentary Ethos.* Bloomington: Indiana University Press, 2019, 46–67.

Kant, Immanuel. *Critique of Judgement*, trans. J.H. Bernard. E-book. Digireads.com, 2012.

Kearton, Cherry. *Photographing Wild Life Across the World.* London: J.W. Arrowsmith, 1923.

Kenaan, Hagi. 'The Selfie and the Face' in Julia Eckel, Jens Ruchatz and Sabine Wirth (eds.) *Exploring the Selfie: Historical, Theoretical, and Analytical Approaches to Digital Self-Photography.* Cham, Switzerland: Palgrave Macmillan, 2018.

Kendrick, James. *Hollywood Bloodshed: Violence in 1980s American Cinema.* Carbondale: Southern Illinois University Press, 2009.

Kerr, Paul. '*Babel*'s Network Narrative: Packaging a Globalized Art Cinema', *Transnational Cinemas*, 1:1 (2014): 37–51.

Kahn, Albert. *Des Droits et des devoirs des gouvernements.* Paris: Imprimerie de Vaugirard, 1918.

Kier, Christina. *Imagine No Possessions: The Socialist Objects of Russian Constructivism.* Cambridge, MA: MIT Press, 2005.

Kracauer, Siegfried. *From Caligari to Hitler: A Psychological History of the German Film.* Princeton, NJ: Princeton University Press, 1966.

Kracauer, Siegfried. *The Mass Ornament: Weimar Essays*, trans. and edited by Thomas Y. Levin. Cambridge/London: Harvard University Press, 1995.

Kracauer, Siegfried. *Theory of Film: The Redemption of Physical Reality.* Princeton, NJ: Princeton University Press, 1997.

Lam, Tong. *Abandoned Futures: A Journey to the Posthuman World.* London: Carpet Bombing Culture, 2013.

Latour, Bruno. *Down to Earth: Politics in the New Climactic Regime*, trans. Catherine Porter. Cambridge: Polity Press, 2018.

Latour, Bruno. *Reassembling the Social: An Introduction to Actor-Network-Theory.* Oxford: Oxford University Press, 2005.

Levin, Miriam R., et al. *Urban Modernity: Cultural Innovation in the Second Industrial Revolution.* Cambridge/London: MIT Press, 2010.

Levinas, Emmanuel. *Totality and Infinity: An Essay on Exteriority*, trans. Alphonso Lingis. The Hague/Boston/London: Martinus Nijhoff Publishers, 1979.

Lightman, Bernard. 'Spectacle in Leicester Square: James Wyld's Great Globe' in Joe Kember, John Plunkett and Jill A. Sullivan (eds.). *Popular Exhibitions, Science and Showmanship, 1840–1910.* London: Pickering & Chatto, 19–39.

Lim, Song Hwee. 'Temporal Aesthetics of Drifting: Tsai Ming-Liang and a Cinema of Slowness' in Tiago de Luca and Nuno Barradas Jorge (eds.) *Slow Cinema*, 87–98.

Lindsay, Vachel. *The Art of the Moving Picture*. New York/Toronto: Random House, 2000.

Lorimer, Douglas A. *Science, Race Relations and Resistance: Britain, 1870–1914*. Manchester/New York: Manchester University Press, 2013.

Louson, Eleanor. 'Taking Spectacle Seriously: Wildlife Film and the Legacy of Natural History Display', *Science in Context* 31:1 (2018): 15–38.

Lovelock, James. *Gaia: A New Look at Life on Earth*. Oxford: Oxford University Press, 2000.

Luckett, Moya. 'Space, Gender, Oversight, and Social Change: Progressivism and the Films of D.W. Griffith, 1909–1916' in Charlie Keil (ed.) *A Companion to D.W. Griffith*. Oxford: Wiley Blackwell, 2018, 309–329.

Lundermo, Trond. '"The Archives of the Planet" and Montage: The Movement of the Crowd and "the Rhythm of Life"' in Eivind Røssaak (ed.) *Between Stillness and Motion: Film, Photography, Algorithms*. Amsterdam: Amsterdam University Press, 2011, 207–224.

Lutz, Catherine A., and Jane L. Collins. *Reading National Geographic*. Chicago/London: University of Chicago Press, 1993.

Lydon, Jane. *Photography, Humanitarianism, Empire*. London/New York: Bloomsbury, 2016.

Lynn-Jagoe, Eva. 'Jumping the Break: Wildfire and the Logic of Separation', *Discourse* 40:2 (2018): 231–251.

MacDonald, Scott. *The Sublimity of Document: Cinema as Diorama*. New York: Oxford University Press, 2019

Manovich, Lev. *The Language of New Media*. Cambridge, MA: MIT Press, 2001.

Marchessault, Janine. Ecstatic *Worlds: Media, Utopia, Ecologies*. Cambridge/London: MIT Press, 2017.

Marinone, Isabelle (ed.). *Un monde et son double: Regards sur l'entreprise visuelle des Archives de la Planète (1919–1931)*. Perpignan: Presses Universitaires de Perpignan and Institut Jean Vigo, 2019.

Margulies, Ivone. *Nothing Happens: Chantal Akerman's Hyperrealist Everyday*. Durham/London: Duke University Press, 1996.

Marx, Karl, and Friedrich Engels. *The Communist Manifesto*. New York: International Publishers, 1948.

Marx, Karl, and Friedrich Engels. *The German Ideology: Part One*, edited by C. J. Arthur. London: Lawrence & Wishart, 1970.

Mattelart, Armand. *Networking the World, 1794–2000*, trans. Liz Carey-Libbrecht and James A. Cohen. Minneapolis/London: University of Minnesota Press, 2000.

Maxwell, Anne. *Colonial Photography and Exhibitions: Representations of the 'Native' and the Making of European Identities*. London/New York: Leicester University Press, 1999.

McGowan, Todd. 'The Contingency of Connection: The Path to Politicization in *Babel*', *Discourse* 30:3 (2008): 401–418.

McKernan, Luke. *Charles Urban: Pioneering the Non-Fiction Film in Britain and American, 1897–1925*. Exeter: University of Exeter Press, 2013.

McKibben, Bill. *Eaarth: Making a Life on a Tough New Planet*. New York: Henry Holt and Company, 2010.

Mendibil, Didier. 'À la recherche de l'effet Brunhes' in Valérie Perlès, with Manon Demurger (eds.) *Les Archives de la Planète*, 130–139.

Metz, Christian. *Language and Cinema*. The Hague: De Gruyter, 1974.

Miller, Angela. 'The Panorama, the Cinema, and the Emergence of the Spectacular', *Wide Angle* 18:2 (1996): 34–69.

Mirzoeff, Nicholas. *The Right to Look: A Counterhistory of Visuality*. Durham/London: Duke University Press, 2011.

Mirzoeff, Nicholas. 'It's Not the Anthropocene, It's the White Supremacy Scene; or, The Geological Color Line' in Richard Grusin (ed.) *After Extinction*. Minneapolis/London: University of Minnesota Press, 2018, 123–149.

Mirzoeff, Nicholas. *How to See the World: An Introduction to Images, from Self-Portraits to Selfies, Maps to Movies, and More*. New York: BasicBooks, 2016.

Mitchell, Timothy. 'The World as Exhibition', *Comparative Studies in Society and History* 31: 2 (1989): 217–236.

Mitchell, W. J. T. 'World Pictures: Globalization and Visual Culture', *Neohelicon* 34:2 (2007): 49–59.

Mitman, Gregg. *Reel Nature: America's Romance with Wildlife on Film*. Seattle/London: University of Washington Press, 2009.

Moholy-Nagy, Lazlo. *Painting, Photography, Film*, trans. Janet Seligman. London: Lund Humphries, 1969.

Moltke, Johannes von. 'Ruin Cinema' in Julia Hell and Andreas Schönle (eds.) *Ruins of Modernity*. Durham/London: Duke University Press, 2010, 395–417.

Morgan, Stephen. 'A Tale of Two Cities: San Francisco 1906 and *Earthquake in Adelaide*' in Alan Wright (ed.) *Film on the Faultline*. Bristol/Chicago: Intellect, 2015, 21–45.

Mulvey, Laura. *Death 24x a Second: Stillness and the Moving Image*. London: Reaktion Books, 2006.

Nagib, Lúcia, Chris Perriam and Rajinder Dudrah (eds.). *Theorizing World Cinema*. London/New York: I.B. Tauris, 2012.

Nancy, Jean-Luc, and Aurélien Barrau, *What's These Worlds Coming to?*, trans. Travis Holloway and Flor Méchain. New York: Fordham University Press, 2015.

Nichols, Bill. 'Documentary Film and the Modernist Avant-Garde', *Critical Enquiry* 27:4 (Summer 2001): 580–610.

Nott, J. C., and G.R. Gliddon. *Types of Mankind, or Ethnological Researchers*. Philadelphia: Lippincott, Grambo & Co; London: Trübner & Co, 1854.

Nye, David E. *American Technological Sublime*. Cambridge, MA: MIT Press, 1994.
Oettermann, Stephan. *The Panorama: History of a Mass Medium*, trans. Deborah Lucas Schneider. New York: Zone Books, 1997.
Oleksijczuk, Denise Blake. *The First Panoramas: Visions of British Imperialism*. Minneapolis: University of Minnesota Press, 2011.
Oliver, Kelly. *Earth and World: Philosophy after the Apollo Missions*. New York: Columbia University Press, 2015.
Osterhammel, Jürgen. *The Transformation of the World: A Global History of the Nineteenth Century*. Princeton/Oxford: Princeton University Press, 2009.
Paglen, Trevor. *The Last Pictures*. New York/Berkeley/Los Angeles/London: Creative Time Books and University of California Press, 2012.
Parikka, Jussi. *What is Media Archaeology?* Cambridge: Polity Press, 2012.
Parikka, Jussi. *A Geology of Media*. Minneapolis: University of Minnesota Press, 2015.
Parshall, Peter F. *Altman and After: Multiple Narratives in Film*. Plymouth: Scarecrow, 2012.
Peeples, Jennifer. 'Toxic Sublime: Imaging Contaminated Landscapes', *Environmental Communication: A Journal of Nature and Culture*, 5:4 (2011): 373–392.
Perlès, Valérie, and Anne Singaud (eds.). *Réalités (In)Visibles: Autour d'Albert Kahn, les archives de la Grande Guerre*. Paris: Musée départemental Albert-Kahn, Bernard Chaveau Édition, 2019.
Perlès, Valérie, with Manon Demurger (eds.). *Les Archives de la Planète*. Paris/Boulogne-Billancourt: Musée départemental Albert-Kahn and Lienart, 2019.
Peterson, Jennifer Lynn. *Education in the School of Dreams: Travelogues and Early Nonfiction Film*. Durham/London: Duke University Press, 2013.
Pisters, Patricia. 'The Mosaic Film: Nomadic Style and Politics in Transnational Media Culture' in Mieke Bal and Miguel Á. Hernández-Navarro (eds.). *Art and Visibility in Migratory Culture: Conflict, Resistance, and Agency*. New York: Rodopi, 2011, 175–190.
Podalsky, Laura. *The Politics of Affect and Emotion in Contemporary Latin American Cinema: Argentina, Brazil, Cuba, and Mexico*. New York: Palgrave, 2011.
Poignant, Roslyn. 'Surveying the Field of View: The Making of the RAI Photographic Collection' in Elizabeth Edwards (ed.) *Anthropology and Photography 1860–1920*. New Haven/London: Yale University Press and RAI, 1994, 42–73.
Poivert, Michel. 'Les yeux d'Albert Kahn: les photographies d'Albert Dutertre durant le voyage autour du monde (1908–1909)', in Valérie Perlès, with Manon Demurger (eds.) *Les Archives de la Planète*, 358–373.
Poole, Deborah. Vision*, Race, and Modernity: A Visual Economy of the Andean Image World*. Princeton, NJ: Princeton University Press, 1997.
Poole, Robert. *Earthrise: How Man First Saw the Earth*. New Haven/London: Yale University Press, 2008.

Pratt, Mary Louise. *Imperial Eyes: Travel Writing and Transculturation*. New York/London: Routledge, 2008.

Prichard, James Cowles. *The Natural History of Man; Comprising Inquiries into the Modifying Influence on Physical and Moral Agencies on the Different Tribes of the Human Family*. London: Hippolyte Baillere, 1845.

Rancière, Jacques. *The Future of the Image*, trans. Gregory Eliott. London/New York: Verso, 2009.

Reclus, Élisée. 'A Great Globe', *The Geographical Journal*, 12:4 (1898): 401–406.

Richards, Bruce. *Feeling Global: Internationalism in Distress*. New York/London: New York University Press, 1999.

Richards, Morgan. 'Greening Wildlife Documentary' in Libby Lester and Brett Hutchins (eds.). *Environmental Conflict and the Media*. New York: Peter Lang, 2013, 171–185.

Richards, Morgan. 'Global Nature, Global Brand: BBC Earth and David Attenborough's Landmark Wildlife Series', *Media International Australia* 146: 1 (2013): 143–154.

Roberts, Martin. '*Baraka*: World Cinema and the Global Culture Industry', *Cinema Journal* 37:3 (1998): 62–82.

Robertson, Roland. *Globalization: Social Theory and Global Culture*, London: Thousand Oaks and New Delhi: SAGE Publications, 1992.

Robic, Marie-Claire. 'Administrer la preuve par l'image: géographie physique et géographie humaine', in *Jean Brunhes*, 221–233.

Robic, Marie-Claire. 'La géographie dans le mouvement scientifique', in *Jean Brunhes*, 52–65.

Robic, Marie-Claire. 'Un laboratoire: l'Egypte', in *Jean Brunhes*, 164–180.

Rorty, Richard. *Contingency, Irony, and Solidarity*. New York: Cambridge University Press, 1989.

Rosen, Philip. *Change Mummified: Cinema, Historicity, Theory*. Minneapolis/London: University of Minnesota Press, 2001.

Rust, Stephen, Salma Monani and Sean Cubitt (eds.). *Ecocinema Theory and Practice*. New York/London: Routledge, 2013.

Sagan, Carl. *Pale Blue Dot: A Vision of the Human Future in Space*. New York: Ballantine Books, 1994.

Sartre, Jean-Paul. *The Writings of Jean-Paul Sartre*, edited by Michel Contat and Michel Rybalka. Evanston, IL: Northwestern University Press, 1974.

Schoonover, Karl, and Rosalind Galt. *Queer Cinema in the World*. Durham/London: Duke University Press, 2016.

Schor, Naomi. *Bad Objects: Essays Popular and Unpopular*. Durham/London: Duke University Press, 1995, 3.

Schwartz, Vanessa R., and Jeannene Przyblyski (eds.). *The Nineteenth-Century Visual Culture Reader*. London/New York: Routledge, 2004.

Sconce, Jeffrey. 'Irony, Nihilism and the New American "Smart" Film', *Screen* 43:4 (2002): 349–369.

Sekula, Allan. *Photography against the Grain: Essays and Photo Works, 1973–1983*. Nova Scotia: The Press of the Nova Scotia, 1984.

Shaviro, Steven. *The Universe of Things: On Speculative Realism*. Minneapolis: University of Minnesota Press, 2014.

Shaw, Deborah. *The Three Amigos: The Transnational Filmmaking of Guillermo del Toro, Alejandro González Iñárritu, and Alfonso Cuáron*. Manchester: Manchester University Press, 2013.

Sigaud, Anne. 'The Archives of the Planet: Between Cosmopolitanism and Patriotism' in Trond Erik Bjorli and Kjetil Ansgar Jacobsen (eds.) *Cosmopolitics of the Camera*, 178–205.

Sigaud, Anne. 'Entre documentation et propagande: vocation et usage politiques des Archives de la Planète' in Valérie Perlès, with Manon Demurger (eds.) *Les Archives de la Planète*, 280–291.

Sigaud, Anne. 'Des Archives de la Planète aux archives de la guerre: "Réalités invisibles, notre raison doit les discerner"', in Valérie Perlès and Anne Singaud (eds.) *Réalités (In)Visibles*, 30–54.

Simmel, Georg. 'Two Essays', *The Hudson Review* 11:3 (1958): 371–385.

Simmon, Scott. *The Films of D.W. Griffith*. Cambridge/New York/Melbourne: Cambridge University Press, 1993.

Steimatsky, Noa. *The Face on Film*. New York: Oxford University Press, 2017.

Stokes, Melvyn. *D.W. Griffith's The Birth of a Nation: A History of the Most Controversial Motion Picture of All Time*. New York: Oxford University Press, 2007.

Stone, Rob et al. (eds.). *The Routledge Companion to World Cinema*. London/New York: Routledge, 2018.

Strauven, Wanda. 'Media Archaeology: Where Film History, Media Art, and New Media (Can) Meet' in Julia Noordegraaf et al. (eds.) *Preserving and Exhibiting Media Art: Challenges and Perspectives*. Amsterdam: Amsterdam University Press, 2013, 59–79.

Tagg, John. *The Burden of Representation: Essays on Photographies and Histories*. Basingstoke/New York: Palgrave, 1988.

Tierney, Dolores. *New Transnationalisms in Contemporary Latin American Cinema*. Edinburgh: Edinburgh University Press, 2018.

Tode, Thomas, Barbara Wurm and Österreichischen Filmmuseum (eds.). *Dziga Vertov: The Vertov Collection at the Austrian Film Museum* (Vienna: SYNEMA, 2006).

Tsing, Anna Lowenhaupt. *The Mushroom at the End of the World: On the Possibility of Life under Capitalist Ruins*. Princeton/Oxford: Princeton University Press, 2015.

Turvey, Malcolm. *The Filming of Modern Life*. Cambridge, MA and London: MIT Press.

Uradike, Nwachukwu Frank. *Black African Cinema*. Berkeley/Los Angeles/London: University of California Press, 1994.

Vertov, Dziga. *Kino-Eye: The Writings of Dziga Vertov*, trans. Kevin O'Brien and edited by Annete Michelson. Berkeley/London: University of California Press, 1984.

Virilio, Paul. *The Information Bomb*, trans. Chris Turner. London/New York: Verso, 2005.

Wallace-Wells, David. *The Uninhabitable Earth: Life after Warming*. New York: The Duggan Books, 2019.

Walls, Laura Dassow. *The Passage to Cosmos: Alexander von Humboldt and the Shaping of America*. Chicago/London: University of Chicago Press, 2009.

Walls, Laura Dassow. 'Natural History in the Anthropocene' in John Parham and Louise Westling (eds.) *A Global History of Literature and the Environment*. Cambridge: Cambridge University Press, 2016, 187–200.

Weisman, Alan. *The World Without Us*. London: Virgin Books, 2008.

Wells, H. G. *A Modern Utopia*. London: W. Collins Sons & Co, 1905.

Wheatley, Helen. 'Beautiful Images in Spectacular Clarity: Spectacular Television, Landscape Programming and the Question of (Tele)Visual Pleasure', *Screen* 52:2 (Summer 2011): 233–248.

Wheatley, Helen. *Spectacular Television: Exploring Televisual Pleasure*. London/ New York: I.B. Tauris, 2016.

White, Frank. *The Overview Effect: Space Exploration and Human Evolution*. Boston: Houghton Mifflin Company, 1987.

Whitney, Allison Patricia. *The Eye of Daedalus: A History and Theory of IMAX Cinema*. PhD thesis, University of Chicago, 2005.

Willoquet-Maricondi, Paula (ed.). *Framing the World: Explorations in Ecocriticism and Film*. Charlottesville/London: University of Virginia Press, 2015.

Winter, Jay. *Dreams of Peace and Freedom: Utopian Moments in the 20th Century*. New Haven/London: Yale University Press, 2006.

Winter, Jay. 'Pacifist Photography: Seeing the Face of Humanity', in *Cosmopolitics of the Camera*, 116–132.

Wood, John G. *The Uncivilized Races of Men in All Countries of the World; Being a Comprehensive Account of Their Manners and Customs, and of their Physical, Social, Mental, Moral and Religious Characteristics*. San Francisco: J.A. Brainerd & Company, 1883.

Wood, John G. *The Natural History of Man; Being an Account of the Manners and Customs of the Uncivilised Races of Man*. London: George Routledge and Sons, 1868.

Wulf, Andrea. *The Invention of Nature: The Adventures of Alexander von Humboldt, the Lost Hero of Science*. London: John Murray, 2015.

Yacavone, Daniel. *Film Worlds: A Philosophical Aesthetics of Cinema*. New York: Columbia University Press, 2015.

Young, Paul. *Globalization and the Great Exhibition: The Victorian New World Order.* Basingstoke: Palgrave, 2009.

Yusoff, Kathryn. *A Billion Black Anthropocenes or None.* Minneapolis: University of Minnesota Press, 2018.

Zalasiewicz, Jan, with contributions from Kim Freedman. *The Earth after Us: What Legacy Will Humans Leave in the Rocks?* New York: Oxford University Press, 2008.

Žižek, Slavoj. 'Against an Ideology of Human Rights' in Kate E. Tunstall (ed.) *Displacement, Asylum, Migration: The Oxford Amnesty Lectures 2004*. Oxford: Oxford University Press, 2006, 56–85.

Zylinska, Joanna. *Nonhuman Photography.* Cambridge, MA/London: MIT Press, 2017.

Index

Page numbers in *italics* refer to illustrations.